Yunzhi Zou
Single Variable Calculus

Also of Interest

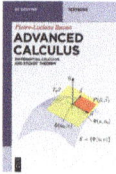

Advanced Calculus.
Differential Calculus and Stokes' Theorem
Pietro-Luciano Buono, 2016
ISBN 978-3-11-043821-5, e-ISBN (PDF) 978-3-11-043822-2,
e-ISBN (EPUB) 978-3-11-042911-4

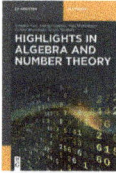

Algebra and Number Theory. A Selection of Highlights
Benjamin Fine, Anthony Gaglione, Anja Moldenhauer,
Gerhard Rosenberger, Dennis Spellman, 2017
ISBN 978-3-11-051584-8, e-ISBN (PDF) 978-3-11-051614-2,
e-ISBN (EPUB) 978-3-11-051626-5

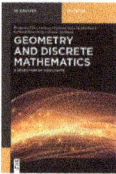

Geometry and Discrete Mathematics. A Selection of Highlights
Benjamin Fine, Anthony Gaglione, Anja Moldenhauer,
Gerhard Rosenberger, Dennis Spellman, 2019
ISBN 978-3-11-052145-0, e-ISBN (PDF) 978-3-11-052150-4,
e-ISBN (EPUB) 978-3-11-052153-5

Probability Theory.
A First Course in Probability Theory and Statistics
Werner Linde, 2016
ISBN 978-3-11-046617-1, e-ISBN (PDF) 978-3-11-046619-5,
e-ISBN (EPUB) 978-3-11-046625-6

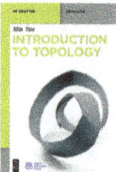

Introduction to Topology
Min Yan, 2016
ISBN 978-3-11-037815-3, e-ISBN (PDF) 978-3-11-037816-0,
e-ISBN (EPUB) 978-3-11-041302-1

Yunzhi Zou

Single Variable Calculus

A First Step

DE GRUYTER

Mathematics Subject Classification 2010
26A06

Author
Yunzhi Zou
Department of Mathematics
Sichuan University
Chengdu, Sichuan
China
610065
609181697@qq.com, zouyz@scu.edu.cn

ISBN 978-3-11-052462-8
e-ISBN (PDF) 978-3-11-052778-0
e-ISBN (EPUB) 978-3-11-052785-8

Library of Congress Control Number: 2018935086

Bibliographic information published by the Deutsche Nationalbibliothek
The Deutsche Nationalbibliothek lists this publication in the Deutsche Nationalbibliografie;
detailed bibliographic data are available on the Internet at http://dnb.dnb.de.

© 2018 Walter de Gruyter GmbH, Berlin/Boston and Beijing World Publishing Corporation, P.R. China
Typesetting: VTeX UAB, Lithuania
Printing and binding: CPI books GmbH, Leck
Cover image: Agsandrew/iStock/Getty Images
♾ Printed on acid-free paper
Printed in Germany

www.degruyter.com

Preface

In recent years, more and more Chinese students are going overseas, either as exchange students or to pursue degrees. At the same time, more and more students are coming from other countries to Chinese universities to further their studies. We have noticed that, in terms of both knowledge transfer and intercultural communication, the English language has played an indispensable role. Furthermore, it is in the interest of these students to have a smooth transition from one system to another, for their credits to transfer, and for them to immerse themselves in the new environment as quickly as possible. To this end, more and more Chinese schools are offering courses delivered bilingually or in English to enhance students' international outlook. One of the greatest challenges in offering Chinese students a course in English is finding a suitable textbook; the textbook must seriously consider what students have done in high school, it must meet the national and local official course standards, and it should resonate with significant international flavors so as to benefit the students. *Single Variable Calculus* is a textbook that meets all of these challenges. This textbook offers a rigorous approach to single variable calculus and incorporates graphical and numerical approaches and historical notes for many key concepts. All of the important topics of a traditional single variable calculus course are thoroughly covered, but the size of the text is minimized in order to reduce the cost for potential users. It is also the first ever calculus textbook in China printed in color – a tremendous benefit for understanding the applications in differential and integral calculus.

In writing the first edition of *Single Variable Calculus*, we benefited from the contributions of many people, and here I would like to recognize and thank them. We received useful assistance and valuable suggestions from:
Stanley D Bristol, editor, Rio Salado College, USA
Julie M Clark, primary editor, Hollins University, USA
John Jensen, Rio Salado College, USA
Liao Wenyuan, University of Calgary, Canada
Min Xinchang, Sichuan University, China
Xia Fuquan, Sichuan Normal University, China
Xu Xiaozhan, Sichuan University, China
Xu Youcai, Sichuan University, China
Xiao Yibin, University of Electronic Science and Technology of China, China
Yang Liang, Sichuan University, China
Zhang Liangcai, Chongqing University, China
and students of the years of 2013 to 2017 at Wu Yuzhang College. In particular, I want to thank Mr. Zhang Shenhang and Wu Zengbao for collecting materials and student feedbacks for me.

During the past decade, I have used many calculus books either as textbooks or reference books. These include *Calculus* by Gilbert Strang; *Calculus: Ideas and Appli-*

https://doi.org/10.1515/9783110527780-201

cation by Alex Himonas and Alna Howard; *Calculus* (3rd edition) by Michael Spivak; *Calculus* (7th edition) by Robert Adams and Christopher Essex; *Calculus* (6th edition) by James Stewart; *Calculus for Engineers* (4th edition) by Donald Trim; *Calculus of a Single Variable: Early Transcendental Functions* (9th edition) by Ron Larson, Bruce Edwards and Robert Hostetler; *Calculus: Graphical, Numerical, Algebraic* (4th edition) by Ross Finney, Franklin Demana, Bert Waits and Daniel Kennedy; *Calculus* (10th edition) by Ron Larson and Bruce Edwards; *Calculus: for Business, Economics, and the Social and Life Sciences* (10th edition) by Laurence E. Hoffmann and Gerald L. Brandley; *Calculus I and II* by Jigang Ma, Yunzhi Zou and Peter Aitchision; *Calculus textbook* (6th and 7th edition) authored collectively by the department of mathematics at Tongji University, and *Calculus textbook* authored collectively by the College of Mathematics at Sichuan University. I thank those wonderful calculus textbooks for giving me inspirations and insights in developing this calculus text. Without borrowing ideas and adapting materials, this work would have been impossible.

I am very glad that this work is published in Germany. I thank people working hard on this. These include editor Liu Hui at the World Publishing Corporation, and editor Ina Talandienė at VTeX for De Gruyter without whose work, this book would not have achieved this level.

I also want to thank the Department of Mathematics and the Academic Affairs Office at Sichuan University for their financial support.

The responsibility for any errors in this text lies entirely with me. Corrections and feedback are very welcome and can be sent to zouyz@scu.edu.cn.

Zou Yunzhi
Department of Mathematics
Sichuan University
Wang Jiang Road 29
Chengdu, China
610065
zouyz@scu.edu.cn; 609181697@qq.com;

Contents

Preface —— V

1	**Prerequisites for calculus** —— 1	
1.1	Overview of calculus —— 1	
1.2	Sets and numbers —— 6	
1.2.1	Sets —— 6	
1.2.2	Numbers —— 8	
1.2.3	The least upper bound property —— 9	
1.2.4	The extended real number system —— 11	
1.2.5	Intervals —— 12	
1.3	Functions —— 14	
1.3.1	Definition of a function —— 14	
1.3.2	Graph of a function —— 17	
1.3.3	Some basic functions and their graphs —— 18	
1.3.4	Building new functions —— 20	
1.3.5	Fundamental elementary functions —— 31	
1.3.6	Properties of functions —— 32	
1.4	Exercises —— 37	

2	**Limits and continuity** —— 41	
2.1	Rates of change and derivatives —— 41	
2.2	Limits of a function —— 42	
2.2.1	Definition of a limit —— 42	
2.2.2	Properties of limits of functions —— 49	
2.2.3	Limit laws —— 50	
2.2.4	One-sided limits —— 55	
2.2.5	Limits involving infinity and asymptotes —— 59	
2.3	Limits of sequences —— 68	
2.3.1	Definitions and properties —— 68	
2.3.2	Subsequences —— 77	
2.4	Squeeze theorem and Cauchy's theorem —— 78	
2.5	Infinitesimal functions and asymptotic functions —— 86	
2.6	Continuous and discontinuous functions —— 91	
2.6.1	Continuity and discontinuity —— 91	
2.6.2	Continuous functions —— 94	
2.6.3	Theorems on continuous functions —— 99	
2.6.4	Uniform continuity —— 107	
2.7	Some proofs in Chapter 2 —— 108	
2.8	Exercises —— 114	

3 **The derivative —— 121**
3.1   Derivative of a function at a point —— **121**
3.1.1   Instantaneous rates of change and derivatives revisited —— **121**
3.1.2   One-sided derivatives —— **128**
3.1.3   A function may fail to have a derivative at a point —— **129**
3.2   Derivative as a function —— **133**
3.2.1   Graphing the derivative of a function —— **134**
3.2.2   Derivatives of some basic functions —— **135**
3.3   Derivative laws —— **139**
3.4   Derivative of an inverse function —— **143**
3.5   Differentiating a composite function – the chain rule —— **147**
3.6   Derivatives of higher orders —— **152**
3.7   Implicit differentiation —— **154**
3.8   Functions defined by parametric and polar equations —— **159**
3.8.1   Functions defined by parametric equations —— **159**
3.8.2   Polar curves —— **163**
3.9   Related rates of change —— **166**
3.10   The tangent line approximation and the differential —— **167**
3.10.1 Linearization —— **167**
3.10.2 Differentials —— **170**
3.11   Derivative rules – summary —— **174**
3.12   Exercises —— **175**

4 **Applications of the derivative —— 181**
4.1   Extreme values and the candidate theorem —— **181**
4.2   The mean value theorem —— **188**
4.3   Monotonic functions and the first derivative test —— **196**
4.3.1   Monotonic functions —— **196**
4.3.2   The first derivative test —— **199**
4.4   Extended mean value theorem and the L'Hôpital rules —— **201**
4.4.1   Extended mean value theorem —— **201**
4.4.2   The indeterminate forms $\frac{0}{0}$, $\infty - \infty$, $\frac{\infty}{\infty}$, and $0 \times \infty$ —— **203**
4.5   Taylor's theorem —— **209**
4.5.1   The error analysis for the linear approximation —— **209**
4.5.2   The quadratic approximation —— **210**
4.5.3   Taylor's theorem —— **214**
4.6   Concave functions and the second derivative test —— **219**
4.6.1   Concave functions —— **219**
4.6.2   The second derivative test —— **225**
4.7   Extreme values of functions revisited —— **227**
4.8   Curve sketching —— **231**
4.9   Solving equations numerically —— **234**

4.9.1	Decimal search ——	234
4.9.2	Newton's method ——	236
4.10	Curvatures and the differential of the arc length ——	238
4.11	Exercises ——	243

5	**The definite integral —— 249**	
5.1	Definite integrals and properties ——	249
5.1.1	Introduction ——	249
5.1.2	Properties of the definite integral ——	259
5.1.3	Interpreting $\int_a^b f(x)\,dx$ in terms of area ——	265
5.1.4	Interpreting $\int_a^b v(t)\,dt$ as a distance or displacement ——	267
5.2	The fundamental theorem of calculus ——	267
5.3	Numerical integration ——	275
5.3.1	Trapezoidal rule ——	276
5.3.2	Simpson's rule ——	277
5.4	Exercises ——	279

6	**Techniques for integration and improper integrals —— 285**	
6.1	Indefinite integrals ——	285
6.1.1	Definition of indefinite integrals and basic antiderivatives ——	285
6.1.2	Differential equations ——	289
6.1.3	Substitution in indefinite integrals ——	293
6.1.4	Further results using integration by substitution ——	297
6.1.5	Integration by parts ——	300
6.1.6	Partial fractions in integration ——	304
6.1.7	Rationalizing substitutions ——	312
6.2	Substitution in definite integrals ——	313
6.3	Integration by parts in definite integrals ——	317
6.4	Improper integrals ——	318
6.4.1	Improper integrals of the first kind ——	318
6.4.2	Improper integrals of the second kind ——	322
6.5	Exercises ——	326

7	**Applications of the definite integral —— 333**	
7.1	Areas, volumes, and arc lengths ——	333
7.1.1	The area of the region between two curves ——	333
7.1.2	Volumes of solids ——	337
7.1.3	Arc length ——	339
7.2	Applications in other disciplines ——	344
7.2.1	Displacement and distance ——	344
7.2.2	Work done by a force ——	345

X —— Contents

7.2.3	Fluid pressure —— **346**	
7.2.4	Center of mass —— **347**	
7.2.5	Probability —— **349**	
7.3	Exercises —— **350**	

8 **Infinite series, sequences, and approximations —— 355**
8.1 Infinite sequences —— **355**
8.2 Infinite series —— **357**
8.2.1 Definition of infinite series —— **357**
8.2.2 Properties of convergent series —— **359**
8.3 Tests for convergence —— **363**
8.3.1 Series with nonnegative terms —— **363**
8.3.2 Series with negative and positive terms —— **371**
8.4 Power series and Taylor series —— **375**
8.4.1 Power series —— **375**
8.4.2 Working with power series —— **381**
8.4.3 Taylor series —— **383**
8.4.4 Applications of power series —— **391**
8.5 Fourier series —— **393**
8.5.1 Fourier series expansion with period 2π —— **394**
8.5.2 Fourier cosine and sine series with period 2π —— **399**
8.5.3 The Fourier series expansion with period $2l$ —— **400**
8.5.4 Fourier series with complex terms —— **403**
8.6 Exercises —— **404**

Index —— 411

1 Prerequisites for calculus

1.1 Overview of calculus

In this chapter, you will:
- *see a big picture of calculus;*
- *review the definition of a function;*
- *review properties of a function;*
- *review basic functions and form new functions using basic functions.*

When first introduced to calculus, one probably will ask questions such as "what is calculus?" and "what can calculus do?". This book has answers to these questions. To begin, we start with two problems, the area problem and the tangent problem.

The area problem

We start with an example. Many students start their school day by traveling in cars or buses from their homes to their schools. We will simplify our discussion by assuming such a trip is along a straight road of 100 km and the trip duration is 2 hours. If you kept looking at the speedometer (see Figure 1.1.1), then you would be able to sketch an approximate speed/velocity versus time graph.

The velocity graph in Figure 1.1.2 shows a period of acceleration, followed by a brief period of constant velocity and then a period of deceleration.

Figure 1.1.1: A speedometer.

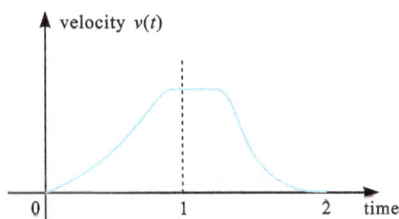

Figure 1.1.2: Velocity versus time graph.

https://doi.org/10.1515/9783110527780-001

How do we know the distance traveled up to time = 1 hour? We note that

$$\text{distance traveled} = \text{speed} \times \text{time}.$$

However, the problem here is that the speed/velocity on this trip is not constant, and in order to determine the distance traveled, we subdivide the time interval $[0,1]$ into several subintervals. For each subinterval $[t_{i-1}, t_i]$, since the elapsed time is very little, we can approximate the distance traveled by the car during that time interval Δt_i by considering the motion as one with constant speed. Multiplying Δt_i by $v(t_i^*)$, where t_i^* is a sample point in $[t_{i-1}, t_i]$, we obtain $v(t_i^*)\Delta t_i$, which approximates the real distance traveled by the car during that time interval. Notice that this is exactly the area of the shaded rectangle in Figure 1.1.3. Adding up the areas of all these rectangles (assume there are n such rectangles), we have an approximation of the area of the region R (see Figure 1.1.4) that is between the t-axis, the curve $v(t)$, and the two lines $t = 0$ and $t = 1$, which is also an approximation of the total distance traveled by the car during this time interval:

$$\text{area of region R} = \text{distance traveled by the car}$$
$$\approx v(t_1^*) \cdot (t_1 - t_0) + v(t_2^*) \cdot (t_2 - t_1) + \cdots + v(t_n^*) \cdot (t_n - t_{n-1}).$$

As seen in Figure 1.1.5, it is easy to see that the more rectangular the region, the better the approximation. As the width of each rectangle tends to 0, the sum of the areas of all rectangles will approximate the area that we want to calculate closer and closer, so we need to investigate the behavior of the limiting procedure when the partition gets smaller and smaller.

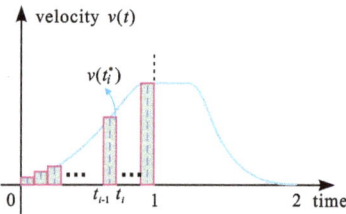

Figure 1.1.3: Area of a shaded rectangle approximates the real distance traveled by the car in that time interval.

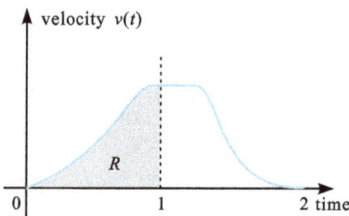

Figure 1.1.4: The area of region R is the distance travelled by the car.

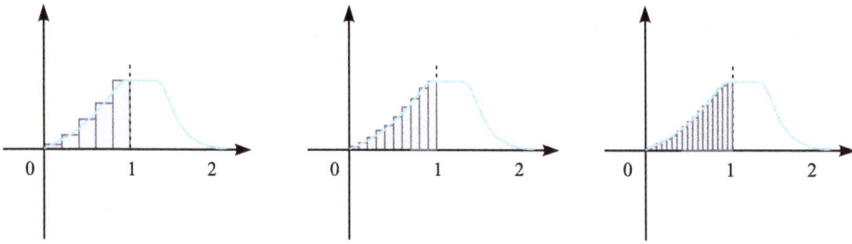

Figure 1.1.5: More rectangles, better approximation.

As seen above, finding the distance traveled is equivalent to finding the area of some region which may not have a familiar shape such as a triangle, rectangle, circle, etc. The formulas studied in high school do not apply. Calculus can do the job, as we will see later.

This idea has a long history and was used by the ancient Greeks and Chinese thousands of years ago. For example, the astronomer Eudoxus used the method of exhaustion, while Liu Hui used a similar method in order to find the area of a circle, as shown in Figure 1.1.6.

Figure 1.1.6: Regular polygons approximate a circle.

The tangent problem

Now suppose the distance versus time graph for your trip is shown in Figure 1.1.7.

How can you find the velocity of the car at the instant when $t = 1$? If we use the equation

$$\frac{\Delta s}{\Delta t} = \frac{100 - 0}{2 - 0} = 50 \text{ km/h},$$

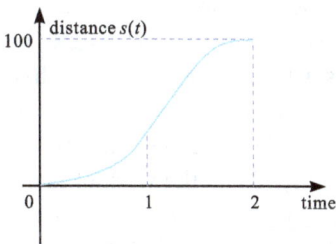

Figure 1.1.7: Distance versus time graph.

we obtain the average velocity of the car during the trip, but the question is to obtain the velocity at the instant $t = 1$. We can still use the equation, but it now becomes

$$\frac{\Delta s}{\Delta t} = \frac{s(t) - s(1)}{t - 1}.$$

This still gives the average velocity of the car, but during the time interval $[1, t]$.

Starting at $t = 1$ and imagining the additional distance traveled from $t = 1$ to $t = 1.1$ seconds gives $\frac{\Delta s}{\Delta t}$ as a very close approximation of the real instantaneous velocity at $t = 1$. As shown in Figure 1.1.8, extrapolating, we can make Δt smaller and smaller, each time improving our approximation of the true value. That is, as $t \to 1$ and $\Delta t \to 0$, $\frac{\Delta s}{\Delta t}$ gets closer and closer to the instantaneous velocity that we want to calculate.

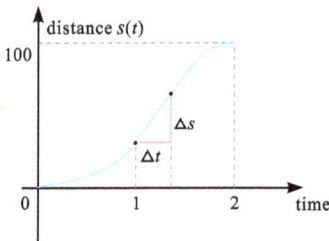

Figure 1.1.8: The slope of the tangent is the instantaneous velocity at the point.

Also, we notice that $\frac{\Delta s}{\Delta t}$ is the slope of the secant line to the velocity curve and when $\Delta t \to 0$, the secant line approaches the tangent line of the curve at $t = 1$. This problem is the same as finding the tangent line to the graph at $t = 1$.

NOTE. The word *tangent* is derived from the Latin word *tangens*, which means "touching".

Equivalently, we need to determine the limiting process of some function when the change in one variable tends to 0.

Many of the applications in the practical world fall into two categories of mathematical problems, the tangent problem and the area problem. The area problem can be solved by *integral calculus*, whose origins can be traced back to 2 000 years ago. The tangent problem can be solved by *differential calculus*, which came much later than integral calculus. Modern calculus is generally considered to have been developed by the English physicist and mathematician *Isaac Newton* and the German mathematician *Gottfried Wilhelm Leibniz* in the seventeenth century. Using the work of many mathematicians, such as Fermat, Descartes, Cavalieri, and Barrow, Newton and Leibniz discovered the *fundamental theorem of calculus*, which is the connection between integral and differential calculus. It allows one to go from nonconstant rates of change to the total net change or vice versa. In many problems, we usually know one and try to find the other, as illustrated previously.

Sir Isaac Newton
(1642–1726) was an English physicist and mathematician. He is widely recognized as one of the most influential scientists of all time. He was also a key figure in the scientific revolution.

Gottfried Wilhelm von Leibniz
(1646–1716) was a German polymath and philosopher. He occupies a prominent place in the history of mathematics and the history of philosophy. Most scholars believe Leibniz developed calculus independently of Isaac Newton, and Leibniz's notation has been widely used ever since it was published. He was the first to describe a pinwheel calculator in 1685 and invented the Leibniz wheel, used in the arithmometer, the first mass-produced mechanical calculator. He also refined the binary number system, which is the foundation of virtually all digital computers. http://en.wikipedia.org/wiki/Gottfried_Wilhelm_Leibniz

Gottfried Wilhelm Leibniz.

Both types of calculus have enormous applications in many scientific fields, for example in areas such as the physical sciences, actuarial science, computer science, statistics, engineering, economics, business, medicine, demography, and in other fields wherever a problem can be mathematically modeled and an optimal solution is desired. In physics, classical mechanics, and electromagnetism, Newton's second law, Maxwell's theory of electromagnetism, and Einstein's theory of general relativity all need calculus. Economists use calculus for the determination of maximal profit. In the realm of medicine, calculus can be used to find the optimal branching angle of a blood vessel to maximize blood flow. Calculus is also used to derive dosing concentrations from the decay laws for the elimination of a particular drug from the body. In nuclear medicine, calculus is used to build models of radiation transport in targeted tumor therapies. Chemists also use calculus in determining reaction rates and radioactive decay.

As already seen, calculus deals with changes in functions. A *function* is used to describe, in a mathematical way, the relationship between two or more changing quantities or variables. In this book, all the functions determine the value of one variable (the *dependent variable*) from the value of another variable (the *independent variable*), using a specific rule or formula relating the variables. Such a function is called a function of one variable because there is only one independent variable. All the functions in this book are defined on sets of real numbers, so we will first review some basic mathematical notations and concepts that provide the building blocks for the development of calculus. These include sets, numbers, intervals, and functions.

1.2 Sets and numbers

1.2.1 Sets

A *set* is a collection of *elements*, or *members*, that are often numbers but may be other mathematical or nonmathematical objects. Sets can have a finite or infinite number of members. Sets are denoted by capital letters such as A, B, S, or T and can be defined simply by listing all elements, such as

$$S = \{1, 2, 3, 4, 5, 6, 7, 8, 9, 10\}.$$

Sets can also be defined by giving the characterizing properties of the elements. For example, the set S defined above can also be defined by any of the following characterizing properties:

$$S = \{\text{natural numbers from 1 to 10 inclusive}\} \quad \text{or}$$
$$S = \{x \mid |1 \leqslant x \leqslant 10, \text{ where } x \text{ is a natural number}\}.$$

Another example of a set defined by a property is

$$T = \{\text{real numbers that are zeros of } \sin(x)\}$$
$$= \{x \mid x = n\pi, \ n \text{ is an integer}\}.$$

We write

$$x \in S \text{ if } x \text{ is an element of } S \text{ (or shorter: ``}x \text{ is in } S\text{''),}$$
$$x \notin S \text{ if } x \text{ is not an element of } S \text{ (or shorter: ``}x \text{ is not in } S\text{'').}$$

For example, if $\mathbf{N} = \{\text{all natural numbers}\}$, then $1 \in \mathbf{N}$ means that 1 is a member of the set of natural numbers (or shorter, "1 is a natural number"). Also, $\pi \notin \mathbf{N}$ means that $\pi = 3.1415926535\ldots$ is not a member of the set of natural numbers (or shorter, "π is not a natural number").

If A, B are two sets, then they can be compared in various ways:

$$A \subseteq B \text{ or in words: ``}A \text{ is a subset of } B\text{''}$$
means: every element in A is also an element of B.
$$A = B \text{ or in words: ``}A \text{ equals } B\text{''}$$
means: A and B have precisely the same elements.

$A \subset B$ or in words: "A is a *proper subset* of B"

means: A is a subset of B, but A is not equal to B.

A' (sometimes $\bar{A}$) or in words: "the complement of A"

means: the set of all elements that are not in A.

If A, B are sets, then new sets can be created in various ways:

$A \cup B$ or in words: "A *union* B"

means: the set of all elements that are either in A or in B or both.

$A \cap B$ or in words: "A *intersect* B"

meaning: the set of all elements that are both in A and in B.

$A \setminus B$ or in words: "A *minus* B"

means: the set of all elements that are in A but not in B.

$A \times B$ or in words "the *product set* of A and B"

(also called the *direct product* or the *Cartesian product*)

means: the set of all *pairs* (a, b) where $a \in A$ and $b \in B$.

In particular, $\mathbf{R} \times \mathbf{R}$ is the set of all real number pairs $(x, y) \in \mathbf{R}^2$ (also referred to as *points* of the *real plane* with respect to a two-dimensional Cartesian coordinate system).

It is useful in some circumstances to be able to refer to a special set that has no members at all. Hence, $\varnothing$ denotes the *empty* set, the only set that contains no elements.

To illustrate relationships between sets, we often use a *Venn diagram*, conceived around 1880 by John Venn. An example of a Venn diagram is shown in Figure 1.2.1.

John Venn (1834–1923) was an English logician and philosopher noted for introducing the Venn diagram, used in the fields of set theory, probability, logic, statistics, and computer science. http://en.wikipedia.org/wiki/John_Venn

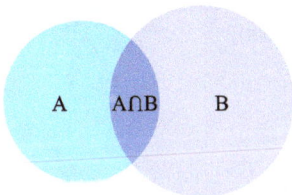

Figure 1.2.1: A Venn diagram.

1.2.2 Numbers

In the following, the use of the three dots "…" means that the pattern established by the preceding numbers is followed forever (an infinite number of times). For example,

N denotes the set of all *natural numbers*, $\{1, 2, 3, 4, \ldots\}$.

Z denotes the set of all *integers*: $\{\ldots, -3, -2, -1, 0, 1, 2, 3, \ldots\}$.

Q denotes the set of all *rational numbers*. A rational number is of the form $\frac{p}{q}$, where p and q are integers. There are also *irrational numbers*. The number $\sqrt{2}$ cannot be written as a quotient of two integers; $\sqrt{2}$ is an irrational number.

R denotes the set of all *real numbers*, which includes the rational and irrational numbers.

C denotes the set of all *complex numbers*. A complex number has the form $a + bi$, where a, b are two real numbers and i is a special number (an imaginary number, not in **R**) satisfying $i^2 = -1$. If $b = 0$, then $a + 0i$ is also a real number.

Obviously, we have

$$\mathbf{N} \subset \mathbf{Z} \subset \mathbf{Q} \subset \mathbf{R} \subset \mathbf{C}, \quad \text{as shown in Figure 1.2.2.}$$

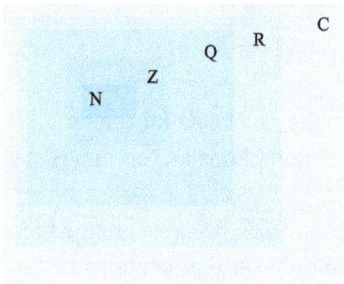

Figure 1.2.2: Sets of numbers.

NOTE. If not specified, all the numbers in this calculus course are real numbers.

We now review some basic knowledge of the real number system **R**. If a, b, c are any three real numbers, then the following properties hold (where the notation ab means multiplication: $a \times b$):

1. $a + b$ and ab are both in the set **R** (closed under addition and multiplication);
2. $a + b = b + a$ and $ab = ba$ (commutative laws);
3. $(a + b) + c = a + (b + c)$ and $(ab)c = a(bc)$ (associative laws);
4. $a + 0 = a$ (0 is the additive identity);
5. $a \times 0 = 0$;
6. $1 \times a = a$ (1 is the multiplicative identity);
7. $a(b + c) = ab + ac$ (distributive law);
8. for any number a, there is another number written as $-a$ (the negative of a) such that $a + (-a) = 0$ (additive inverses);

9. for any number $b \neq 0$, there is another number written $\frac{1}{b}$ (the reciprocal) such that $b \times \frac{1}{b} = 1$ (multiplicative inverses).

NOTE. *Subtraction* is a special kind of addition in which "*a* minus *b*" is written as $a - b$ and defined by $a - b = a + (-b)$. *Division* is a special kind of multiplication in which "*a* divided by *b*" is written as $\frac{a}{b}$ and defined by $\frac{a}{b} = a \times (\frac{1}{b})$. However, the operations of subtraction and division do not obey the commutative and associative laws. In other words, $a - b \neq b - a$ and $a/b \neq b/a$.

Also, there is an order relation "$<$" for real numbers which satisfies the following properties:

1. For any two real numbers a and b, exactly one of the following is true

$$a = b, \quad a < b, \quad \text{or} \quad b < a.$$

2. If $a < b$ and $b < c$, then $a < c$ (transitive property).
3. If $a < b$, then $a + c < b + c$. If $c > 0$, then $ac < bc$.

The notation "$\geq$" means greater than or equal to. For example $a \geq b$ means a is greater than or equal to b, and $a \leq b$ mean a is less than or equal to b.

The real numbers, and their subsets, share the unifying geometric property that any straight line that is infinitely long in both directions can be made into a *number line*, as seen in Figure 1.2.3.

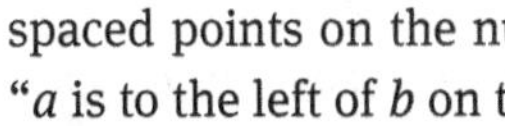

Figure 1.2.3: Number line/axis.

This means that we can create a one-to-one correspondence between the real numbers and points of the number line. A number line preserves our intuitive ideas of ordering and size of numbers. In particular, the integers $\ldots, -3, -2, -1, 0, 1, 2, 3, \ldots$ are at equally spaced points on the number line and "$a < b$" is exactly identical to the statement "a is to the left of b on the number line". Number lines are used as axes in Cartesian coordinate systems and in other important applications.

1.2.3 The least upper bound property

The term "bounded" is the same as in daily language.

Definition 1.2.1. A set S of real numbers is said to be:

1. *bounded above* if there is a real number M such that $x \leq M$ for all $x \in S$; M is called an *upper bound* of S;

2. *bounded below* if there is a real number m such that $x \geqslant m$ for all $x \in S$; m is called a *lower bound* of S;
3. bounded of it is bounded above and bounded below.

If a set S of real numbers is bounded above, then S has infinitely many upper bounds. For example, for the set $\{x \mid x < 1\}$, 1, 2, 3 are all upper bounds of S.

The *least upper bound*, β, of S, if it exists, is unique. The least upper bound of S is called the *supremum* of S. We write

$$\beta = \sup S.$$

In mathematical language, the formal definition of supremum is given as follows.

Definition 1.2.2. β is the *supremum* of a nonempty set S of real numbers if:
1. $x \leqslant \beta$ for all $x \in S$;
2. if $\varepsilon > 0$, then there is a number x_0 in S such that $x_0 > \beta - \varepsilon$.

Similarly, if S is bounded below, the *greatest lower bound* of S (the *infimum* of S), if it exists, is denoted by $\inf S$. The supremum or infimum of a set S may or may not be in the set.

Example 1.2.1. The supremum of the set $\{x \mid x^2 < 4\}$ is 2. The infimum of the set $\{x \mid 0 \leqslant x < 2\}$ is 0.

An interesting question is whether or not a set of real numbers has an infimum or supremum. For this purpose, we have to adopt an important axiom that is necessary to make the real number system a complete one. This axiom is also known as the *least upper bound property*, which assumes the existence of a supremum for any nonempty set of real numbers that is bounded above.

Axiom 1.2.1. *If a nonempty set of real numbers is bounded above, then it has a supremum.*

Similarly, any nonempty set of real numbers that is bounded below has an infimum. In fact, for any set of real numbers that is bounded below, we take the negative of each member of the set to form a new set which is bounded above. Then the negative of the supremum of the new set is the infimum of the original set, so the Axiom could also be stated as follows.

Axiom 1.2.2. *If a nonempty set of real numbers is bounded above/below, then it has a supremum/infimum.*

1.2.4 The extended real number system

For an unbounded set of real numbers, we have the following definition.

Definition 1.2.3. A nonempty set S of real numbers is *unbounded above* if it has no upper bound and *unbounded below* if it has no lower bound. It is *unbounded* if it is either unbounded above or unbounded below.

Example 1.2.2. Sets N, Q, Z, R are all unbounded. The set of whole numbers $\{0, 1, 2, \ldots\}$ is unbounded above, and the set of negative even integers $\{\ldots, -6, -4, -2\}$ is unbounded below.

To better describe the term unbounded, it is convenient to introduce two fictitious points, $+\infty$ (which we usually write more simply as ∞) and $-\infty$. They are called *points at infinity*. We define the order relationships between them and any real number x by $-\infty < x < \infty$, as seen in Figure 1.2.4.

Figure 1.2.4: $-\infty$ and ∞.

In addition, we define

$$\infty + \infty = \infty, \quad (-\infty) \times (-\infty) = \infty,$$
$$(-\infty) + (-\infty) = -\infty, \quad \infty \times \infty = \infty,$$
$$\infty \times (-\infty) = -\infty, \quad \infty^\infty = \infty, \quad |-\infty| = \infty.$$

If a is a finite real number, then it is natural to define the relations

$$\infty + a = \infty, \quad a - \infty = -\infty, \quad \frac{a}{\infty} = 0, \quad \text{and} \quad \frac{a}{-\infty} = 0.$$

Furthermore, if $a > 0$, then we define

$$a \times \infty = \infty, \quad a \times (-\infty) = -\infty.$$

However,

$$\infty - \infty, \quad \infty^0, \quad 0^\infty, \quad 1^\infty, \quad \text{and} \quad \frac{\infty}{\infty}$$

are not defined; they are *indeterminate forms*.

If S is a nonempty set of real numbers without an upper bound, we write

$$\sup S = \infty.$$

If S is a nonempty set of real numbers without a lower bound, we write

$$\inf S = -\infty.$$

The real number system with the two points at infinity is called the *extended real number system*.

1.2.5 Intervals

Intervals are special and important subsets of real numbers. They often appear as solution sets to inequalities and are important in the definition of many functions. Intervals come in various forms as summarized below. In the following, it is assumed that a and b are real numbers satisfying $a < b$, except in the definition of the closed interval $[a, b]$, where we allow the possibility that $a = b$.

The *open* interval (a, b), as seen in Figure 1.2.5, is the set of real numbers

$$\{x \in \mathbf{R} \mid a < x < b\}.$$

Figure 1.2.5: Open interval (a, b).

The *closed* interval $[a, b]$ is the set of real numbers

$$\{x \in \mathbf{R} \mid a \leqslant x \leqslant b\}.$$

The *half-open* interval $(a, b]$, as seen in Figure 1.2.6, is the set of real numbers

$$\{x \in \mathbf{R} \mid a < x \leqslant b\}.$$

Figure 1.2.6: Half-open interval $(a, b]$.

The *half-open* interval $[a, b)$ is the set of real numbers

$$\{x \in \mathbf{R} \mid a \leqslant x < b\}.$$

The above intervals are intervals with finite length $b - a$. The following intervals have infinite length:
the *infinite* interval $[a, \infty)$, as seen in Figure 1.2.7, is the set of real numbers

$$\{x \in \mathbf{R} \mid a \leqslant x < \infty\};$$

Figure 1.2.7: Infinite interval $[a, \infty)$.

the *(open) infinite* interval (a, ∞) is the set of real numbers

$$\{x \in \mathbf{R} \mid a < x < \infty\};$$

the *infinite* interval $(-\infty, b]$ is the set of real numbers

$$\{x \in \mathbf{R} \mid -\infty < x \leqslant b\};$$

the *infinite* interval $(-\infty, \infty)$ is the same as the set $\mathbf{R}$ of all real numbers.

NOTES. 1.   The closed interval $[a, a]$ is just the single value a.

2.   The symbol (a, b), where a and b are real numbers, has many uses in mathematics and it is used in particular to denote a point in the plane defined by a two-dimensional Cartesian coordinate system. Thus, sometimes it is necessary to make this clear using words such as "the interval (a, b)" or "the point (a, b)".

3.   The notation $\{a, b\}$ indicates the set with members a and b, **not** an interval or a point.

A *neighborhood* of $x = a$ means an open interval (b, c) containing a, so that $b < a < c$. A δ-neighborhood $(\delta > 0)$ of $x = a$, denoted by $U(a, \delta)$, is the open interval $(a - \delta, a + \delta)$, or equivalently, all x satisfying $|x - a| < \delta$. That is,

$$U(a, \delta) = \{x \mid |x - a| < \delta\}.$$

Since $|x - a|$ stands for the distance between x and a, $U(a, \delta)$ represents all the x whose distance from a is less than δ. Consequently a is called the *center of the neighborhood*, while δ is sometimes called the *radius of the neighborhood*.

We use the notation $\mathring{U}(a, \delta)$ to represent the set $\{x \mid 0 < |x - a| < \delta\}$. This means that $\mathring{U}(a, \delta)$ is the δ-neighborhood without its center, often called the *deleted δ-neighborhood* of $x = a$.

Usually, we say that $(a - \delta, a)$ is the *left δ-neighborhood* of $x = a$, while $(a, a + \delta)$ is the *right δ-neighborhood* of $x = a$. Figure 1.2.8 shows some neighborhoods of a.

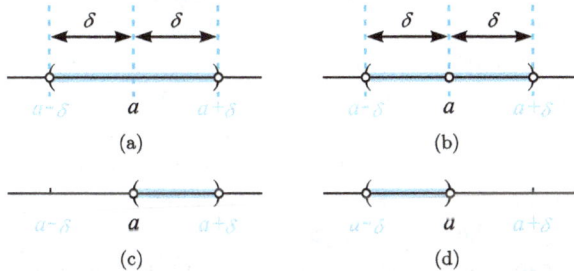

Figure 1.2.8: Some neighborhoods of a.

1.3 Functions

1.3.1 Definition of a function

In previous courses, you have encountered many functions defined by equations. For example, $y = kx + b$ is a linear function, $y = ax^2 + bx + c$ is a quadratic function, and $f(x) = \sin x$ is a trigonometric function. In the practical world, the volume V of a sphere is also a function of the radius r of the sphere. The rule is given by

$$V = V(r) = \frac{4}{3}\pi r^3.$$

The temperature of a particular day at a particular place is a function of the time t, although the explicit rule linking temperature and time may be difficult to find. The amount of money that you save at a bank is also a function of time, since you receive interest from the bank.

So what is a function? We now give the formal definition of a *function*.

Definition 1.3.1 (Function, domain, and range). Let D and R be two subsets of the set of real numbers **R**. A *function f* from D to R is a *rule/mapping/correspondence* that assigns to each element x in D exactly one element y in R.

The set D is called the *domain* of the function. The set of elements in R that are assigned to one or more elements of D is called the *range* of the function, as shown in Figure 1.3.1.

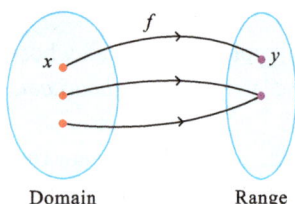

Figure 1.3.1: A function is a mapping between two sets.

There are some conventional notations and terms for describing a function. We sometimes write a function as $f : D \rightarrow R$ or $y = f(x)$ and we say that y is a function of x. In this case, x is a variable, called the *independent variable*, representing all values from the domain D. Then y is a variable, called the *dependent variable*, representing all values in the range. The essential property of a function is that, for each $x \in D$, there is just one value of $f(x)$, and if $x \notin D$, then the function has no value. The domain D of a function f is sometimes denoted as D_f. For $x \in D$, the corresponding value in R is denoted as $f(x)$ and we refer to $f(x)$ as the *value of the function* at x.

The definition of a function can be extended, in an obvious way, to cases where the domain and/or range are not sets of real numbers. In multivariable calculus, we will meet functions whose domains are pairs of real numbers. However, for most of the functions that we will work with in this course, the domain and range are subsets of real numbers.

Most often, functions are described by equations, although functions can be described in other ways. Three other common ways to represent a function are verbally (by a description in words), numerically (by a table of values), and visually (by a graph). Thus, graphs and tables that appear in newspapers or magazines may also represent functions, though we usually do not regard them as "functions" when we see such things.

NOTE. It may help to think of a function as a machine. When you put in an x-value from the domain of f, the machine gives a unique output value y, depending on this x-value and the rule f that links x and y, as seen in Figure 1.3.2.

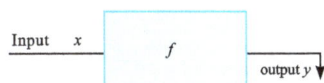

Figure 1.3.2: A function behaves like a machine.

In many situations, the domain of a function is not given. Then we generally assume that the domain is the largest possible set for which the function is defined. We can then try to determine this domain by using knowledge of algebra and trigonometry.

Example 1.3.1. Let f be the function defined by

$$f(x) = \sqrt{2+x}.$$

1. Find the domain and range of f.
2. Find the values of $f(1), f(a), f(\frac{1}{a})$, and $\frac{f(x_0+h)-f(x_0)}{h}$.

Solution. To find the domain, we notice that the real square root function cannot take a negative input, which means

$$2+x \geqslant 0,$$

so the domain of f is $\{x \mid x \geqslant -2\}$, or in interval notation $[-2, \infty)$.

The range of f is the set of numbers that are assigned to each value of x in $[-2, \infty)$. The minimum value of f is 0 when $x = -2$ and the range of f is $\{y \mid y \geqslant 0\}$ or $[0, \infty)$.

To find the value of $f(1)$, we replace x by 1 in the formula for $f(x)$ and obtain

$$f(1) = \sqrt{2+1} = \sqrt{3}.$$

Similarly,

$$f(a) = \sqrt{2+a}, \quad f\left(\frac{1}{a}\right) = \sqrt{2+\frac{1}{a}}$$

and

$$\frac{f(x_0 + h) - f(x_0)}{h} = \frac{\sqrt{2+(x_0+h)} - \sqrt{2+x_0}}{h}.$$

The graph of $f(x) = \sqrt{2+x}$ is shown in Figure 1.3.3.

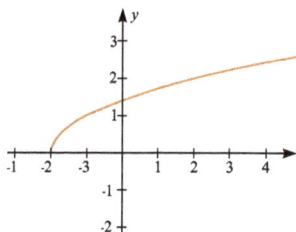

Figure 1.3.3: Graph of $y = \sqrt{2+x}$.

Example 1.3.2. A rectangular storage container has a volume of 16 cm³. The length of its base is twice its width. The materials for making the container have a cost of $12 per square centimeter. Express the cost of material as a function of the width of the base.

Solution. First we draw a diagram, as in Figure 1.3.4. The volume V of the container is $V = 2w \times w \times h$, so

$$2w^2 h = 16 \text{ cm}^3.$$

The surface area S of the container is

$$S = 2 \times (2w \times w + h \times 2w + h \times w).$$

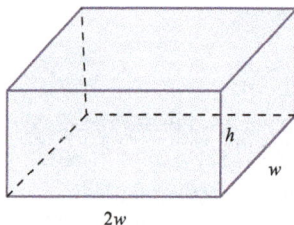

Figure 1.3.4: A container with dimension $2w \times w \times h$.

Therefore, the cost C is

$$C = 12 \times S = 12 \times 2 \times (2w \times w + h \times 2w + h \times w)$$
$$= 24 \times (2w \times w + 3wh).$$

Substituting $h = \frac{16}{2w^2}$, we have

$$C = 24 \times \left(2w^2 + 3w\frac{16}{2w^2}\right)$$
$$= 24 \times \left(2w^2 + \frac{24}{w}\right)$$
$$= 48w^2 + \frac{24^2}{w}, \quad \text{the domain is } w > 0.$$

1.3.2 Graph of a function

The graph of a function $y = f(x)$ consists of the points (x, y) in the xy-plane whose coordinates x and y satisfy the equation $y = f(x)$. Graphs of functions are curves in the xy-plane, but not all curves in the plane are graphs of functions. For example, the curve in Figure 1.3.5 is a graph of a function and the curve in Figure 1.3.6 is not a graph of any function. This is because, for each x-value, there must be one and exactly one y-value assigned to this x-value. The method of determining whether or not a given curve in the plane is a graph of some function is called the *vertical line test*. It says that, if a vertical line intersects a curve more than once, the curve is not a graph of any function.

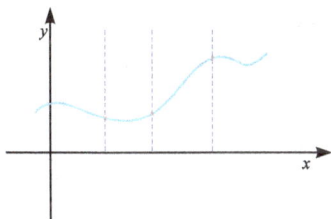

Figure 1.3.5: Vertical line test: it is the graph of some function.

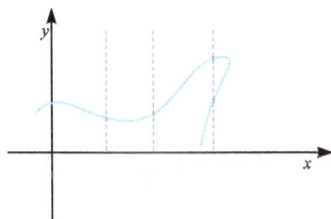

Figure 1.3.6: Vertical line test: it is not a graph of any function.

1.3.3 Some basic functions and their graphs

We now review the following basic functions, which you have already seen in previous studies.

Constant functions

The constant functions are given by $f(x) = C$, where C is a real number. The graph of a constant function is a horizontal line in the xy-plane, as seen in Figure 1.3.7. The domain of a constant function is **R** and the range of a constant function is the set with only one element, $\{C\}$.

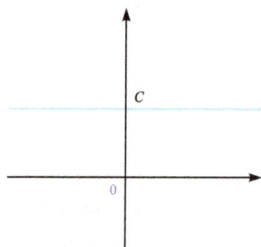

Figure 1.3.7: Graph of $f(x) = C$.

Power functions

The power functions are $f(x) = x^k$. For example, $f(x) = x$, $f(x) = x^2$, and $f(x) = x^{-3}$ are power functions. There are two special power functions: the reciprocal function $f(x) = \frac{1}{x}$ and the root functions, such as $f(x) = x^{\frac{1}{2}}$, $f(x) = x^{\frac{1}{3}}$, $f(x) = x^{\frac{4}{3}}$,.... We can obtain the domain of a power function by using our knowledge of algebra. For example, we cannot take the square root of a negative number so the domain of $f(x) = \sqrt{x}$ is $\{x \mid x \geq 0\}$. The domain of $f(x) = x^{-\frac{1}{4}}$ is $\{x \mid x > 0\}$ because the denominator cannot be zero. Graphs of some power/root functions are shown in Figure 1.3.8 and Figure 1.3.9.

The sine and cosine functions

The sine function is $y = \sin x$ and the cosine function is $y = \cos x$. Their domains are both all real numbers **R** and their ranges are also the same, $[-1, 1]$. Figure 1.3.10 and Figure 1.3.11 show graphs of $\sin x$ and $\cos x$ respectively.

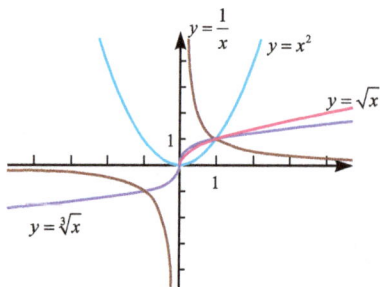

Figure 1.3.8: Graphs of some power functions.

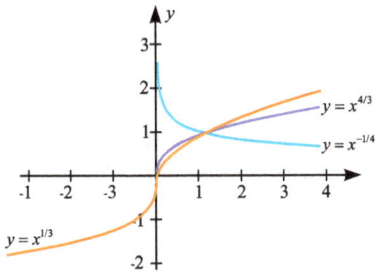

Figure 1.3.9: Graphs of some root functions.

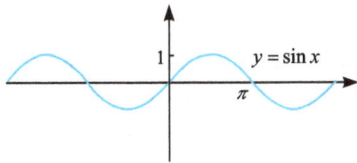

Figure 1.3.10: Graph of $y = \sin x$.

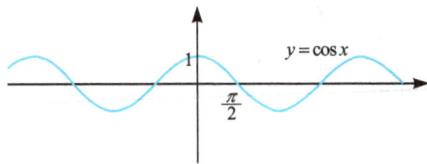

Figure 1.3.11: Graph of $y = \cos x$.

Exponential functions

The exponential functions are $f(x) = a^x$ where $a > 0$ and $a \neq 1$. For example $f(x) = e^x$, $y = 2^x$, and $y = (\frac{1}{2})^x$ are all exponential functions. Figure 1.3.12 shows graphs of some exponential functions.

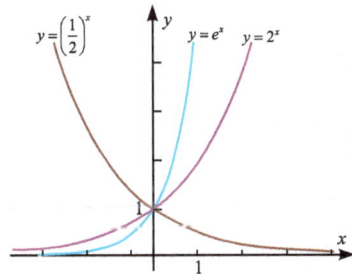

Figure 1.3.12: Graphs of some exponential functions.

1.3.4 Building new functions

Combining functions
New functions can be formed from simpler functions by addition, subtraction, multiplication, and division. Given two functions f and g, with domains D_f, D_g, respectively, we can add, multiply, and divide these functions to create new functions in a very natural way.

$$(f + g)(x) = f(x) + g(x), \quad (fg)(x) = f(x)g(x), \quad \text{and}$$
$$\left(\frac{f}{g}\right)(x) = \frac{f(x)}{g(x)}, \quad \text{if } g(x) \neq 0.$$

The domains of these new functions will be $D_f \cap D_g$ (with the additional restriction for $\frac{f}{g}$ that $g(x) \neq 0$). This set must be nonempty for the new functions to exist.

For example, the other four basic trigonometric functions are combinations of the sine and cosine functions in accordance with above rules.

1. The tangent function is $\tan x = \frac{\sin x}{\cos x}$, with domain

$$D = \left\{ x \,\middle|\, x \neq k\pi + \frac{\pi}{2}, k \in \mathbf{Z} \right\}$$

and range $R = \mathbf{R}$. Figure 1.3.13 shows the graph of $y = \tan x$.

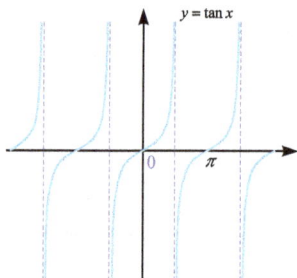

Figure 1.3.13: Graph of $y = \tan x$.

2. The cotangent function is $\cot x = \frac{\cos x}{\sin x}$ with domain

$$D = \{x \mid x \neq k\pi, k \in \mathbf{Z}\}$$

and range $R = \mathbf{R}$. Figure 1.3.14 shows the graph of $y = \cot x$.

3. The secant function is $\sec x = \frac{1}{\cos x}$ with domain

$$D = \left\{ x \,\middle|\, x \neq k\pi + \frac{\pi}{2}, k \in \mathbf{Z} \right\}$$

and range $R = (-\infty, -1] \cup [1, \infty)$. Figure 1.3.15 shows the graph of $y = \sec x$.

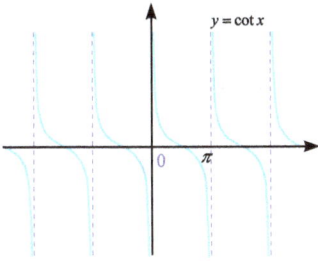

Figure 1.3.14: Graph of $y = \cot x$.

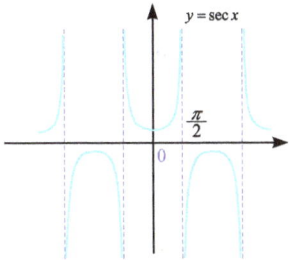

Figure 1.3.15: Graph of $y = \sec x$.

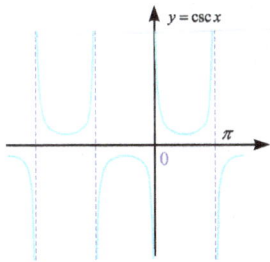

Figure 1.3.16: Graph of $y = \csc x$.

4. The cosecant function is $\csc x = \frac{1}{\sin x}$ with domain

$$D = \{x \mid x \neq k\pi + \pi, k \in \mathbf{Z}\}$$

and range $R = (-\infty, -1] \cup [1, \infty)$. Figure 1.3.16 shows the graph of $y = \csc x$.

A *linear function* has the form $f(x) = mx + b$, or $y = mx + b$, and its graph is a straight line. The number m is its slope and b is the y-intercept. The equation $y = mx + b$ is called the *slope-intercept form* of the line. If we know a line has slope m and passes through a point (x_0, y_0), then we can obtain its equation in its point-slope form as follows:

$$y - y_0 = m(x - x_0).$$

A *polynomial function* of degree n can be written

$$f(x) = a_0 + a_1 x + \cdots + a_n x^n, \quad \text{where } a_0, a_1, \ldots, a_n \text{ are constants and } a_n \neq 0,$$

and is a combination of power functions and the constant function. Some examples of polynomial functions include $f(x) = 1 + 2x$, $f(x) = 2x^2 - 3x - 5$, and $f(x) = x^3$. The domain of a polynomial function is the set of all real numbers, **R**.

A *rational function* is

$$P(x) = \frac{a_0 + a_1 x + \cdots + a_n x^n}{b_0 + b_1 x + \cdots + b_m x^m},$$

where $a_0, a_1, \ldots, a_n$ and $b_0, b_1, \ldots, b_m$ are constants, $a_n \neq 0$, and $b_m \neq 0$. A rational function is simply the quotient of two polynomial functions. Its domain is all real numbers except those that make the denominator zero.

Composite functions

Another way of constructing new functions from simpler functions is by substituting the equation for one function into the equation for a second function (a process called *composition of functions*) to form a *composite function*. The formal definition is as follows.

Definition 1.3.2. Let f and g be two functions with domains D_f and D_g, respectively. The composition $f \circ g$ of f with g is defined by

$$(f \circ g)(x) = f(g(x)),$$

provided that x is in the domain of g and $g(x)$ is in the domain of f. Therefore, the domain of $f \circ g$ is

$$D_{f \circ g} = \{x \mid x \in D_g \text{ and } g(x) \in D_f\}.$$

The composition of two functions $f(x)$ and $g(x)$ is illustrated in Figure 1.3.17.

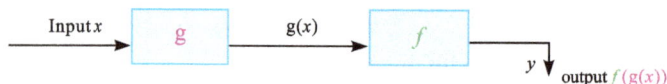

Figure 1.3.17: Composition of two functions.

Example 1.3.3. The function $f(x) = \sin x^2$ is the composition of the function $u(x) = \sin x$ and the function $v(x) = x^2$, so $f(x) = u(v(x))$, or, using another notation, $f(x) = u \circ v$.

Example 1.3.4. The function f is given by

$$f(x) = 2^{\cos(x^2 + 1)}.$$

Describe f as a composition of three functions.

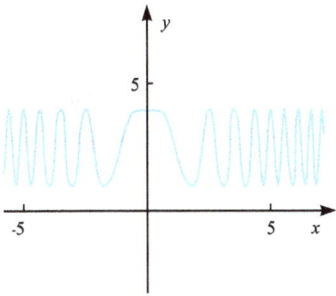

Figure 1.3.18: Graph of $y = 2^{\cos(x^2+1)}$.

Solution. Let $g(x) = 2^x$, $u(x) = \cos x$, and $v(x) = x^2 + 1$. Then $f(x) = g(u(v(x)))$, or, in another notation, $f = g \circ u \circ v$. Figure 1.3.18 shows the graph of $f(x)$.

Example 1.3.5. Find $f \circ g$ and $g \circ f$ for the functions defined by the equations

$$f(x) = (x+1)^2 \quad \text{and} \quad g(x) = \sqrt{x-9}.$$

Give the domain of each composition.

Solution. The domain of f is $(-\infty, +\infty)$, the set of all real numbers. The domain of g is $[9, \infty)$.

First consider $f \circ g$, which is given by

$$(f \circ g)(x) = f(g(x)) = f(\sqrt{x-9}) = (\sqrt{x-9}+1)^2$$

and is defined whenever $g(x) = \sqrt{x-9}$ is defined. Hence $f \circ g$ has domain $[9, \infty)$.
Similarly, the function $g \circ f$ is given by

$$(g \circ f)(x) = g(f(x)) = g((x+1)^2) = \sqrt{(x+1)^2 - 9}$$

and is defined only when $f(x) = (x+1)^2$ is in the domain of g, which means

$$(x+1)^2 \geqslant 9 \implies$$
$$x+1 \geqslant 3 \quad \text{or} \quad x+1 \leqslant -3.$$

This implies $x \leqslant -4$ or $x \geqslant 2$. Hence the domain of $g \circ f$ is $(-\infty, -4] \cup [2, \infty)$.

Inverse functions
Definition 1.3.3. A function f is called a *one-to-one function* if, for each number y in the range of f, there is only one number x in the domain of f such that $y = f(x)$.

A function is one-to-one if and only if every horizontal line intersects its graph in at most one place. This method of determining whether or not a function is one-to-one is called the *horizontal line test*. If a horizontal line intersects the graph of $y = f(x)$ at x-values x_1 and x_2, then $f(x_1) = f(x_2)$ and this means that f is not one-to-one. The horizontal line test is illustrated in Figure 1.3.19 and Figure 1.3.20.

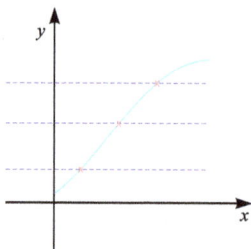

Figure 1.3.19: Horizontal line test: one-to-one.

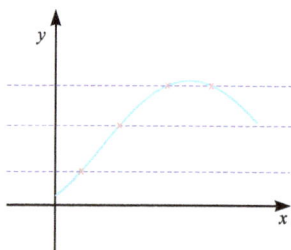

Figure 1.3.20: Horizontal line test: not one-to-one.

The horizontal and vertical line tests suggest that a one-to-one function, $y = f(x)$, also has a function of the form $x = g(y)$ that goes the other way, and this is why one-to-one functions are important. That is, one-to-one functions are precisely those functions that possess *inverse functions* in accordance with the following definition.

Definition 1.3.4. Let f be a one-to-one function with domain D and range R. Then it has an inverse function, denoted f^{-1}, with domain R and range D, defined as

$$x = f^{-1}(y), \quad \text{for any } y \in R,$$

where $x \in D$ is the unique value linked to y by the original function

$$f(x) = y.$$

Another way of putting this is that f^{-1} reverses the action of f. We have

$$f : x \longrightarrow y : f \text{ maps } x \text{ to } y = f(x),$$
$$f^{-1} : y \longrightarrow x : f^{-1} \text{ maps } y = f(x) \text{ to } x.$$

This is illustrated in Figure 1.3.21.

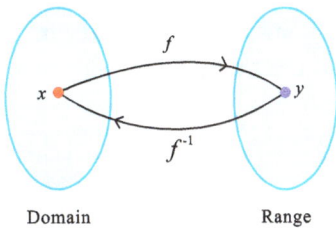

Domain Range

Figure 1.3.21: f and f^{-1} are inverse to each other.

NOTES. 1. The "reversing" property shows that

$$f^{-1}(f(x)) = x \quad \text{for all } x \in D,$$
$$f(f^{-1}(y)) = y \quad \text{for all } y \in R.$$

2. If f were not one-to-one, then x would not be determined uniquely by $y = f(x)$, so it would not be possible to define f^{-1}, because the choice for $x = f^{-1}(y)$ would not be uniquely defined.

3. The functions f and f^{-1} exchange their domains and ranges. We have

$$\text{domain of } f^{-1} = \text{range of } f,$$
$$\text{range of } f^{-1} = \text{domain of } f.$$

4. Though we defined the inverse function as $x = f^{-1}(y)$, we usually switch the roles of x and y and write the inverse function, with x, y in their usual roles, as $y = f^{-1}(x)$. In this case, the graphs of f and f^{-1} are symmetrical about the line $y = x$. That is, if (x,y) is a point on the graph of $y = f(x)$, then (y,x) is a point on the graph of $y = f^{-1}(x)$ and vice versa.

Definition 1.3.5. The logarithm function $y = \log_a x$ is the inverse of the exponential function $y = a^x$, where the constant $a > 0$ and $a \neq 1$.

That is, $y = \log_a x$ is the same relationship as $x = a^y$. The function $y = a^x$ has domain $\mathbf{R}$ and range $\{y \mid y > 0, y \in \mathbf{R}\}$, whereas $y = \log_a x$ has domain $\{x \mid x > 0, x \in \mathbf{R}\}$ and range $\mathbf{R}$.

We list some **properties of logarithms.**
For any real number $a > 0$, $a \neq 1$, $b > 0$, $b \neq 1$, $x > 0$, and $y > 0$, we have:
1. $\log_a(xy) = \log_a x + \log_a y$ (product rule);
2. $\log_a \frac{y}{x} = \log_a y - \log_a x$ (quotient rule);
3. $\log_a x^y = y \log_a x$ (power rule);
4. $a^{\log_a x} = x$ and $\log_a a^x = x$ (undo each other);
5. $\log_a x = \frac{\log_b x}{\log_b a}$ (change of base formula);
6. $a^0 = 1$ and $\log_a 1 = 0$ (definitions).

Example 1.3.6. A famous inverse function pair is $y = e^x$ and $y = \ln x$, where the transcendental number $e \approx 2.718$ is the base of the natural logarithm function, $y = \ln x$. Their graphs are shown in Figure 1.3.22, along with the line of symmetry, $y = x$.

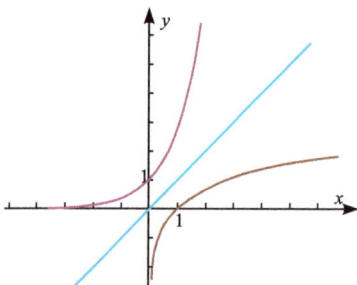

Figure 1.3.22: Graphs of $y = e^x$, $y = \ln x$ and $y = x$.

Example 1.3.7. Find the inverse function of $f(x) = x^3 + 5$ and draw a graph showing both functions and the line of symmetry $y = x$.

Solution. We write the function as $y = x^3 + 5$ and we solve this equation for x. We have

$$x^3 = y - 5,$$
$$x = \sqrt[3]{y - 5}.$$

Hence the inverse function is given by the equation $x = \sqrt[3]{y - 5}$. In order to change this to the usual notation for functions, $y = f^{-1}(x)$, we interchange x and y. We have

$$y = \sqrt[3]{x - 5}.$$

Therefore, the inverse function is $f^{-1}(x) = \sqrt[3]{x - 5}$. Figure 1.3.23 shows the graphs of $y = x^3 + 5$ and $y = \sqrt[3]{x - 5}$ as two solid curves and $y = x$ (line of symmetry for the two graphs) as a dashed line.

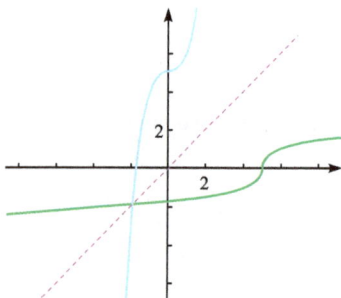

Figure 1.3.23: Graphs of $y = x^3 + 5$, $y = \sqrt[3]{x - 5}$ and $y = x$.

The six trigonometric functions $\sin x$, $\cos x$, $\tan x$, $\cot x$, $\sec x$, and $\csc x$ do not have inverse functions because none of them is one-to-one on their domain. Furthermore, you can see that a horizontal line intersects the graph of any of these functions in infinitely many points. However, if we just take a part of each function for which a horizontal line intersects the graph in a single point, then we can create an inverse function for that part of the trigonometric function. These give the well-known so-called inverse trigonometric functions.

Definition 1.3.6. The definition of the six inverse trigonometric functions are as follows:

1. The *inverse sine* function

$$y = \sin^{-1} x \text{ has domain } [-1,1] \text{ and range } \left[-\frac{\pi}{2}, \frac{\pi}{2}\right].$$

   It is equivalent to $\sin y = x$, $-\frac{\pi}{2} \leqslant y \leqslant \frac{\pi}{2}$.
2. The *inverse cosine* function

$$y = \cos^{-1} x \text{ has domain } [-1,1] \text{ and range } [0, \pi].$$

   It is equivalent to $\cos y = x$, $0 \leqslant y \leqslant \pi$.
3. The *inverse tangent* function

$$y = \tan^{-1} x \text{ has domain } (-\infty, \infty) \text{ and range } \left(-\frac{\pi}{2}, \frac{\pi}{2}\right).$$

   It is equivalent to $\tan y = x$, $-\frac{\pi}{2} < y < \frac{\pi}{2}$.
4. The *inverse cotangent* function

$$y = \cot^{-1} x \text{ has domain } (-\infty, \infty) \text{ and range } (0, \pi).$$

   It is equivalent to $\cot y = x$, $0 < y < \pi$.
5. The *inverse secant* function

$$y = \sec^{-1} x \text{ has domain } \{x \mid x \leqslant -1 \text{ or } x \geqslant 1\}$$
$$\text{and range } \left[0, \frac{\pi}{2}\right) \cup \left(\frac{\pi}{2}, \pi\right].$$

   It is equivalent to $\sec y = x$, $0 \leqslant y \leqslant \pi$, $y \neq \frac{\pi}{2}$.
6. The *inverse cotangent* function

$$y = \csc^{-1} x \text{ has domain } \{x \mid x \leqslant -1 \text{ or } x \geqslant 1\}$$
$$\text{and range } \left[-\frac{\pi}{2}, 0\right) \cup \left(0, \frac{\pi}{2}\right].$$

   It is equivalent to $\csc y = x$, $-\frac{\pi}{2} < y \leqslant \frac{\pi}{2}$, $y \neq 0$.

NOTE. The notations $\arcsin x$, $\arccos x$, $\arctan x$, $\text{arccot}\, x$, $\text{arcsec}\, x$, and $\text{arccsc}\, x$ are also used for the corresponding inverse trigonometric functions.

Figures 1.3.24–1.3.29 show the graphs of these functions.

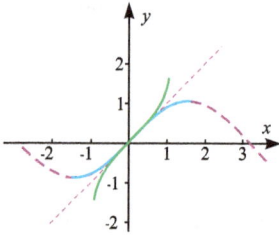

Figure 1.3.24: Graphs of $y = \sin x$ and $y = \sin^{-1} x$.

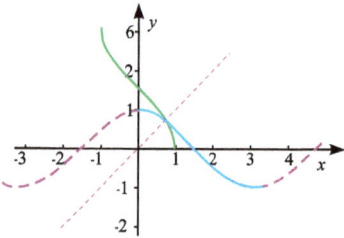

Figure 1.3.25: Graphs of $y = \cos x$ and $y = \cos^{-1} x$.

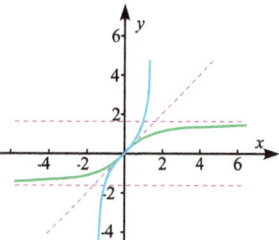

Figure 1.3.26: Graphs of $y = \tan x$ and $y = \tan^{-1} x$.

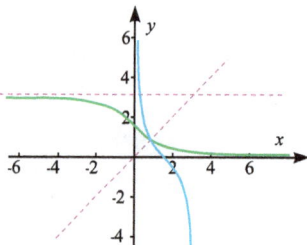

Figure 1.3.27: Graphs of $y = \cot x$ and $y = \cot^{-1} x$.

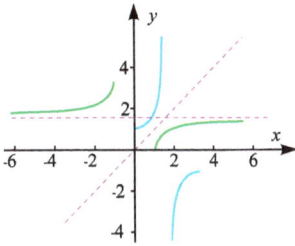

Figure 1.3.28: Graphs of $y = \sec x$ and $y = \sec^{-1} x$.

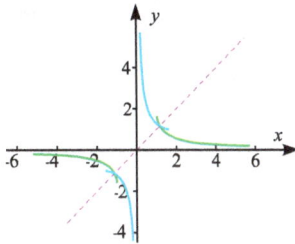

Figure 1.3.29: Graphs of $y = \csc x$ and $y = \csc^{-1} x$.

Piecewise defined functions

A function can be defined by different equations for different parts of its domain. For instance, the *absolute value function* $y = |x|$ is defined in two parts, as shown in Figure 1.3.30, as follows:

$$y = |x| = \begin{cases} -x, & \text{if } x < 0 \\ x, & \text{if } x \geqslant 0. \end{cases}$$

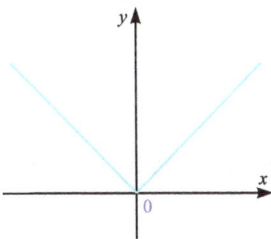

Figure 1.3.30: Graph of $y = |x|$.

Example 1.3.8. Let the function f be defined by

$$f(x) = \begin{cases} x + 1 & \text{if } x \leqslant 1 \\ 2 - (x - 1)^2 & \text{if } x > 1. \end{cases}$$

Evaluate $f(0), f(1)$, and $f(2)$ and sketch the graph of f.

Solution. Recall that a function is a rule. For this particular function, the rule is the following. First look at the value of the input x. If it happens that $x \leqslant 1$, then the value of $f(x)$ is $x + 1$. On the other hand, if $x > 1$, then the value of $f(x)$ is $2 - (x - 1)^2$.

Since $0 \leqslant 1$, we have $f(0) = 0 + 1 = 1$.

Since $1 \leqslant 1$, we have $f(1) = 1 + 1 = 2$.

Since $2 > 1$, we have $f(2) = 2 - (2 - 1)^2 = 1$.

The graph of f also consists of two parts. When $x \leqslant 1$, then $f(x) = y = x + 1$ is the part of the graph of f that lies to the left of the vertical line $x = 1$, and it is a straight line with slope 1 and y-intercept 1. When $x > 1$, then $f(x) = y = 2 - (x - 1)^2$ is the part of the graph of f that lies to the right of the line $x = 1$, and it is a parabola. This is shown in Figure 1.3.31.

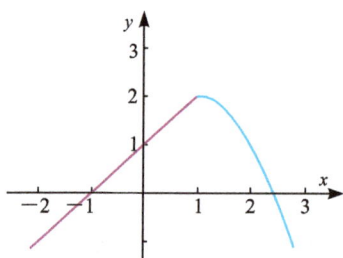

Figure 1.3.31: Graph of $f(x)$ in Example 1.3.8.

Example 1.3.9. The *sign function*

$$y = \text{sgn}(x) = \begin{cases} -1, & \text{if } x < 0 \\ 0, & \text{if } x = 0 \\ 1, & \text{if } x > 0, \end{cases}$$

which takes the sign of x, is also a piecewise defined function with domain $D = (-\infty, +\infty)$ and range $R = \{-1, 0, 1\}$. Its graph is shown in Figure 1.3.32. In the figure,

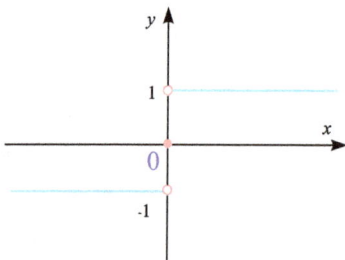

Figure 1.3.32: Graph of $y = \text{sgn}(x)$.

the solid dot indicates that the point $(0,0)$ is included on the graph; the open dots (circles) indicate that the points $(0,1)$ and $(0,-1)$ are excluded from the graph. Often we will show only the open dot.

Example 1.3.10. The *greatest integer function*, $y = [x]$, is the function that gives as output the largest possible integer which is less than or equal to x. For example, $[\frac{4}{5}] = 0$, $[\pi] = 3$, $[-1] = -1$, and $[-2.5] = -3$. This is therefore a piecewise defined function with range $\mathbf{Z}$ (set of all integers). A part of its graph is shown in Figure 1.3.33.

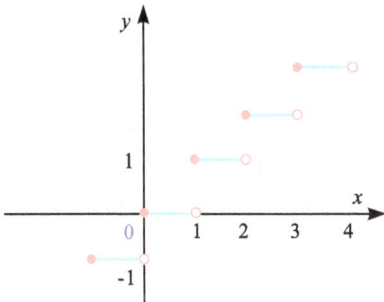

Figure 1.3.33: Graph of $y = [x]$.

1.3.5 Fundamental elementary functions

So far, we have reviewed all the basic functions. These simple functions are often referred to as the *fundamental elementary functions* and include:
-   the constant functions, given by $f(x) = C$, where C is a real number;
-   the absolute value function, $y = |x|$;
-   the power functions $y = x^n$, including the reciprocal function $y = \frac{1}{x}$;
-   the general power functions $y = x^k$ (where k is any real number), including the root functions $y = \sqrt[m]{x^n}$;
-   the six basic trigonometric functions $y = \sin x$, $y = \cos x$, $y = \tan x$, $y = \cot x$, $y = \sec x$, and $y = \csc x$;
-   the six inverse trigonometric functions $y = \sin^{-1} x$, $y = \cos^{-1} x$, $y = \tan^{-1} x$, $y = \cot^{-1} x$, $y = \sec^{-1} x$, and $y = \csc^{-1} x$;
-   the logarithmic functions $y = \log_a x$, $a > 0$, $a \neq 1$;
-   the exponential functions $y = a^x$, $a > 0$, $a \neq 1$.

An *elementary function* referred to in this book is a function built from a finite number of exponential, logarithm, constant, power, and trigonometric functions and their inverses through composition and combinations using the four elementary operations $(+, -, \times, \div)$. For example, the function $y = 2\sin e^{x^2} + \tan^{-1}(\ln(|x| + 1))$ is an elementary function.

1.3.6 Properties of functions

Boundedness

Definition 1.3.7. $f(x)$ is said to be:

1. *bounded above on the interval* **I** if there is a number M, called an *upper bound*, such that $f(x) \leqslant M$ for all $x \in I$;
2. *bounded below on the interval* **I** if there is a number m, called a *lower bound*, such that $m \leqslant f(x)$ for all $x \in I$;
3. *bounded on the interval* **I** if it is bounded above and below;
4. *unbounded above on the interval* **I** if $f(x)$ has no upper bound;
5. *unbounded below on the interval* **I** if $f(x)$ has no lower bound.

If $f(x)$ is bounded on the interval **I**, then there must be a positive number, say, B, such that $|f(x)| \leqslant B$ for each $x \in I$. Upper and lower bounds of a function are not unique. If an upper bound M is found for a function, then any other number greater than M is also an upper bound. Similarly, if a lower bound m is found for a function, then any other number less than m is also a lower bound.

Example 1.3.11. Find an upper bound and a lower bound for the following functions:
(a) $y = x^2$; (b) $f(x) = 2\sin(\frac{x^2+1}{2})$.

Solution. (a) $y = x^2$ is defined for all x, so its domain is all real numbers. It is not bounded above since x^2 becomes arbitrarily large as x becomes large. It is bounded below by 0, since $x^2 \geqslant 0$ for all x. The graphs of the two functions are shown in Figure 1.3.34 and Figure 1.3.35.

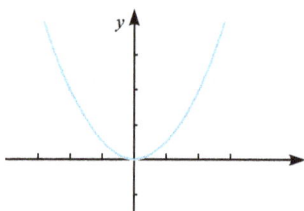

Figure 1.3.34: Graph of $y = x^2$.

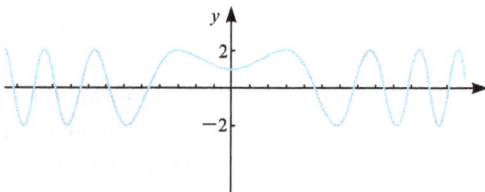

Figure 1.3.35: Graph of $y = 2\sin(\frac{x^2+1}{2})$.

(b) $f(x) = 2\sin(\frac{x^2+1}{2})$ is defined for all x, so its domain is all real numbers. The sine function satisfies $-1 \leqslant \sin x \leqslant 1$ for all values of x. Hence, $-2 \leqslant 2\sin(\frac{x^2+1}{2}) \leqslant 2$ for all x. Hence $f(x)$ has an upper bound 2 and a lower bound -2.

Monotone functions

A function f is called *monotone increasing* (or *increasing*) on an interval **I** if

$$f(x_1) \leqslant f(x_2) \quad \text{whenever } x_1 < x_2 \text{ and } x_1, x_2 \in I.$$

Figure 1.3.36 shows the graph of an increasing function.

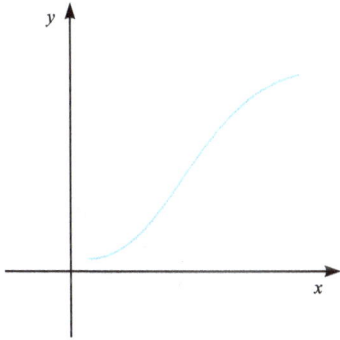

Figure 1.3.36: An increasing function.

A function f is called *monotone decreasing* (or *decreasing*) on an interval **I** if

$$f(x_1) \geqslant f(x_2) \quad \text{whenever } x_1 < x_2 \text{ and } x_1, x_2 \in I.$$

Figure 1.3.37 shows the graph of a decreasing function.

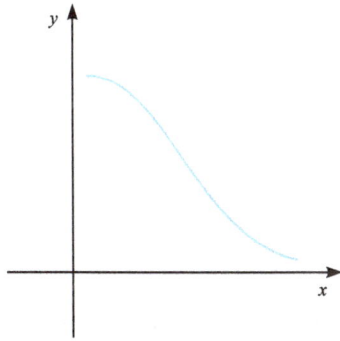

Figure 1.3.37: A decreasing function.

Following the definition of increasing functions, it is important to realize that the inequality $f(x_1) \leqslant f(x_2)$ must be satisfied for every pair of numbers x_1 and x_2 in **I** with $x_1 < x_2$.

A function f is called *strictly increasing* on an interval **I**, if

$$f(x_1) < f(x_2) \quad \text{whenever } x_1 < x_2 \text{ and } x_1, x_2 \in I.$$

A function f is called *strictly decreasing* on an interval **I**, if

$$f(x_1) > f(x_2) \quad \text{whenever } x_1 < x_2 \text{ and } x_1, x_2 \in I.$$

For example, the function $f(x) = x^2$ is strictly decreasing on the interval $(-\infty, 0]$ and strictly increasing on the interval $[0, +\infty)$. The exponential function $f(x) = a^x$ is strictly increasing on $(-\infty, \infty)$ when $a > 1$ and strictly decreasing when $0 < a < 1$.

Symmetry: even and odd functions

Definition 1.3.8. A function f is said to be an *even function* if, for each x in the domain, the number $-x$ is also in the domain and $f(-x) = f(x)$.

The functions $y = \cos x$, $y = |x|$, and $y = x^2$ are examples of even functions. The geometric significance of an even function is that its graph is symmetric about the y-axis, as seen in Figure 1.3.38 (a). That is, if (x, y) is a point on the graph, then $(-x, y)$ is also a point on the graph.

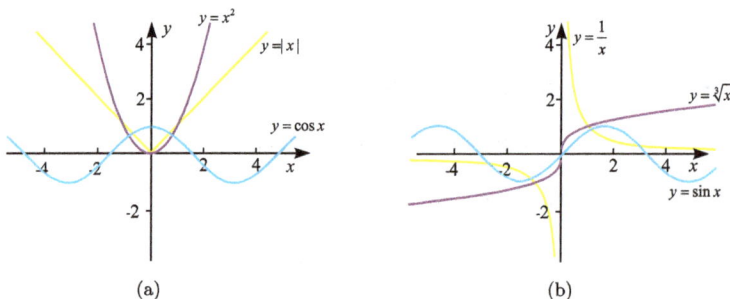

(a)   (b)

Figure 1.3.38: Graphs of some even/odd functions.

Definition 1.3.9. A function f is an *odd function* if, for each x in the domain, the number $-x$ is also in the domain and $f(-x) = -f(x)$.

The functions $y = \sin x$, $y = x$, $y = \frac{1}{x}$, and $y = \sqrt[3]{x}$ are examples of odd functions. The geometric significance of an odd function is that its graph is symmetric about the origin, as seen in Figure 1.3.38 (b). That is, if (x, y) is a point on the graph, then $(-x, -y)$ is also a point on the graph.

NOTE. Many functions are neither even nor odd. For example, $f(x) = x^2 - x$ is neither even nor odd.

Example 1.3.12. Determine whether each of the following functions is even, odd, or neither:

(a) $f(x) = x^3 - 2x$, (b) $g(x) = 1 - x^2$, (c) $t(x) = \sin x + \cos x$,

(d) $h(x) = \ln\left(\dfrac{1-x}{1+x}\right)$, for $-1 < x < 1$.

Solution. (a) f is an odd function, because

$$f(-x) = (-x)^3 - 2(-x) = -x^3 + 2x = -(x^3 - 2x) = -f(x).$$

(b) g is an even function, because

$$g(-x) = 1 - (-x)^2 = 1 - x^2 = g(x).$$

(c) t is neither even nor odd, since

$$t(-x) = \sin(-x) + \cos(-x) = -\sin x + \cos x$$
$$\text{and} \quad t(-x) \neq t(x), \ t(-x) \neq -t(x).$$

(d) h is odd, because

$$h(-x) = \ln\left(\frac{1+x}{1-x}\right) = \ln\left(\frac{1-x}{1+x}\right)^{-1} = -\ln\left(\frac{1-x}{1+x}\right) = -h(x).$$

Figure 1.3.39 shows the graphs of the functions. Notice that the graph of $t(x)$ is neither symmetric about the y-axis nor about the origin.

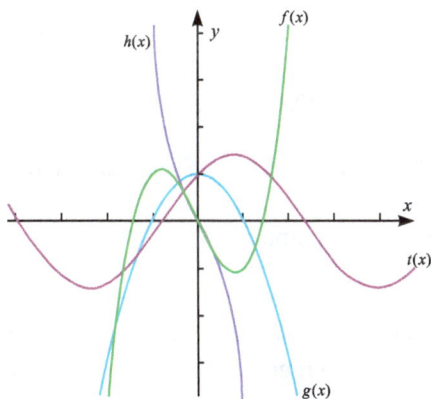

Figure 1.3.39: Graphs of functions in Example 1.3.12.

Periodicity

Definition 1.3.10. For a function $y = f(x)$, If there exists a positive constant T such that $f(x + T) = f(x)$ for all values of x in its domain D, then the function $f(x)$ is a *periodic function*. The least possible positive value of T is called the *period*.

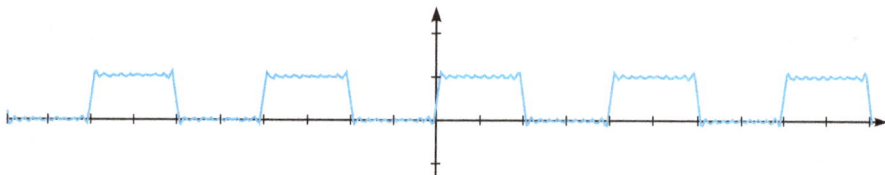

Figure 1.3.40: Graph of a periodic function with period 2.

For example, the four trigonometric functions $y = \sin x$, $y = \cos x$, $y = \sec x$, and $y = \csc x$ are periodic functions with period 2π. The two trigonometric functions $y = \tan x$ and $y = \cot x$ are periodic functions with period π. Figure 1.3.40 shows a periodic function with period 2.

Transformations of a function

Given a function $y = f(x)$ and its graph, we can obtain the graphs and equations of certain related functions. For example, we can shift the graph of $y = f(x)$ upward by k units, where $k > 0$. This is called a vertical translation. The new graph is still a graph of a function, but for each x-value, the corresponding y-value is k more than the original one, so we know the new function is $y = f(x) + k$. Similarly, we can obtain a new function by shifting the old one k units downward, or k units to the right or k units to the left. This is summarized as follows.

Vertical and horizontal shifts laws

If the constant k is positive ($k > 0$), then the graph of:
1. the function $y = f(x) + k$ is obtained by shifting the graph of $y = f(x)$ k units upward;
2. the function $y = f(x) - k$ is obtained by shifting the graph of $y = f(x)$ k units downward;
3. the function $y = f(x + k)$ is obtained by shifting the graph of $y = f(x)$ k units to the left;
4. the function $y = f(x - k)$ is obtained by shifting the graph of $y = f(x)$ k units to the right.

Also, we can stretch or shrink the graph of $f(x)$ by k units ($k > 1$) to obtain a graph of a new, related function. To find an equation of the new function, let us first consider the case where we stretch the graph of $f(x)$ vertically by k units. For the same x-value, the corresponding y-value is now k times the original one, so the new function must be $y = kf(x)$. Similarly, we can obtain the new function by stretching or shrinking the old one k units vertically or horizontally. This is summarized as follows.

Vertical and horizontal stretch/shrink law

If the constant $k > 1$, then the graph of the function:
1. $y = kf(x)$ is obtained by stretching the graph of $y = f(x)$ vertically by a factor of k;

2. $y = \frac{1}{k}f(x)$ is obtained by shrinking the graph of $y = f(x)$ vertically by a factor of k;
3. $y = f(kx)$ is obtained by shrinking the graph of $y = f(x)$ horizontally by a factor of k;
4. $y = f(\frac{x}{k})$ is obtained by stretching the graph of $y = f(x)$ horizontally by a factor of k.

If the graph of a function $y = f(x)$ is reflected over the x-axis or the y-axis, then a graph of a new function is obtained. We now consider the case that the reflection is over the x-axis. This means that, for each value of x, the corresponding y-value is the negative of the original one, so the new function must be $y = -f(x)$. Similarly, we can obtain the new function when reflecting the graph of the old one over the y-axis. This is summarized as follows.

Reflection laws

The graph of the function:
1. $y = -f(x)$ is obtained by reflecting the graph of $y = f(x)$ over the x-axis;
2. $y = f(-x)$ is obtained by reflecting the graph of $y = f(x)$ over the y-axis.

A combination of the above laws is illustrated in Figure 1.3.41.

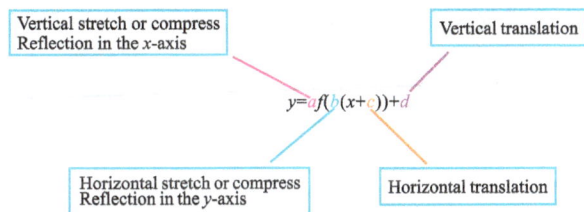

Vertical stretch or compress Reflection in the x-axis

Vertical translation

$$y = af(b(x+c))+d$$

Horizontal stretch or compress Reflection in the y-axis

Horizontal translation

Figure 1.3.41: Transformations of a function.

For example, to obtain the graph of the sinusoid

$$y = 3\sin\left(2\left(x + \frac{\pi}{4}\right)\right) + 2,$$

we first shift the graph of $\sin x$ to the left by $\frac{\pi}{4}$ units, then shrink the graph horizontally by a factor of 2, then stretch the graph vertically by a factor of 3, and, lastly, shift the graph 2 units vertically and we are done.

1.4 Exercises

1. Find the 0.02-neighborhood and the 0.001-deleted neighborhood of the point $a = 3.5$. Give your answer in terms of intervals.
2. Find an equation for the line which:
    (a) passes through the points $(1, 3)$ and $(-3, 5)$;

(b) passes through the point $(4,9)$ and has slope -2;

(c) passes through the point $(5,3)$ and is parallel to the line $y = 4x - 5$;

(d) passes the point $(7,-3)$ and is perpendicular to the line $y = -x + 7$.

3. Find $f(g(x)), g(f(x)), f(f(x)), g(g(x))$, and $f(f(f(x)))$ for each of the following functions:

   (a) $f(x) = \frac{1}{x}$, $g(x) = x$; (b) $f(x) = \sqrt{x}$, $g(x) = x^3 + 1$;

   (c) $f(x) = a^x$, $g(x) = \log_a x$, $a > 0$, $a \neq 1$; (d) $f(x) = 1 - x$, $g(x) = \sin x$.

4. Write the function $f(x) = \tan(e^{\sin x^2})$ as a composition of four functions.

5. Find the inverse function f^{-1} for each of the following functions, and then graph f and f^{-1} on the same diagram:

   (a) $f(x) = e^{2x}$; (b) $f(t) = t^2 - 1$, $t > 1$;

   (c) $f(\theta) = 3\theta - 1$; (d) $f(x) = \sec x$, $0 < x < \frac{\pi}{2}$.

6. Find the domain and range for each of the following functions, and graph each of them:

   (a) $y = x^2 + 2x - 1$; (b) $y = 1 - \ln(x - 2)$; (c) $y = \sqrt{25 - x^2}$;

   (d) $y = 3^{2-x} + 2$; (e) $y = 2\sin(2x + \frac{\pi}{3}) - 1$;

   (f) $y = \begin{cases} \tan(2x - \pi), & x < 0 \\ x^{2/3}, & 0 \leqslant x \leqslant 9 \\ \sqrt{x - 4}, & x > 9. \end{cases}$

7. Discuss the properties (boundedness, symmetry, periodicity) for each of the following functions:

   (a) $y = |x| - 1$; (b) $y = e^{x^2} + \cos x$; (c) $y = 1 - \sin x$; (d) $y = \sec x \tan x$;

   (e) $y = \sqrt{x^4 - 1}$; (f) $y = \frac{x^2 + 1}{x^3 + x}$; (g) $y = x^{2/3}$;

   (h) $y = \begin{cases} \cos x, & x > \pi \\ \sin x + 2, & x \leqslant \pi. \end{cases}$

8. If $f(x) = 2 - 5\cos(3x)$, then:

   (a) find the domain, range and period of f;

   (b) find all zeros of f in $[-\frac{\pi}{2}, \frac{\pi}{2}]$;

   (c) graph $f(x)$ for $-\pi \leqslant x \leqslant \pi$.

9. Describe the transformations required in order to obtain the graphs of the following functions from a basic trigonometric function:

   (a) $y = \frac{1}{2}\sin(3x)$; (b) $f(x) = -2\cos(\frac{2\pi x}{3}) - 1$; (c) $y = \tan\frac{x}{2} + 2$;

   (d) $y = -\sec(2x)$; (e) $f(x) = -2\sin(2x + \frac{\pi}{3}) + 1$.

10. Given the graph of $f(x)$, how would you graph (a) $f(|x|)$, (b) $|f(x)|$, and (c) $\frac{1}{f(x)}$?

11. The Dirichlet function $D(x)$ is defined by

$$D(x) = \begin{cases} 1, & \text{when } x \text{ is rational} \\ 0, & \text{when } x \text{ is irrational.} \end{cases}$$

Is $D(x)$ a periodic function? If so, what is its period? Explain.

12. Prove that:

   (a) the *sum or difference* of two odd functions is an odd function;

(b) the *sum or difference* of two even functions is an even function;
(c) the *product* of an even and an odd function is an odd function;
(d) the *product* of two even functions and the product of two odd functions are both even functions.

13. In mathematics, **hyperbolic functions** are analogs of the ordinary trigonometric, or circular functions. The hyperbolic sine, $\sinh x$, hyperbolic cosine, $\cosh x$, hyperbolic tangent, $\tanh x$, and hyperbolic cotangent, $\coth x$, are defined as

$$\sinh x = \frac{e^x - e^{-x}}{2}, \quad \cosh x = \frac{e^x + e^{-x}}{2},$$

$$\tanh x = \frac{\sinh x}{\cosh x}, \quad \text{and} \quad \coth x = \frac{\cosh x}{\sinh x}.$$

Show that:
(a) $\sinh x$ is odd and $\cosh x$ is even;
(b) $\cosh^2 x - \sinh^2 x = 1$;
(c) $\sinh 2x = 2\sinh x \cosh x$;
(d) $\cosh(x + y) = \cosh x \cosh y + \sinh x \sinh y$;
(e) $\sinh^{-1}(x) = \ln(x + \sqrt{x^2 + 1})$ ($\sinh^{-1}$ denotes the inverse of $\sinh$);
(f) Can you deduce more identities involving hyperbolic functions?

14. (**Telephone bills**) A telecom company sells two packages including the basic fees plus a variable fee charged per minute. Package A has basic fees of $5 per month and $0.05 per minute. Package B has no basic fees, but has a cost of $0.10 per minute. Alice estimates she will use approximately 120 minutes each month. Which package is better for her? Explain.

15. (**Earthquake intensity**) The magnitude R of an earthquake (measured by the Richter scale) is defined as

$$R = \log\left(\frac{A}{A_0(\delta)}\right),$$

where A is the amplitude in μm (micrometers) and the empirical function A_0 depends only on the epicentral distance of the station δ.
(a) Find the earthquake magnitude if $A = 1\,200$ and $A(\delta) = 2.5$.
(b) How many times more severe was the 2008 Wenchuan earthquake ($R = 8.0$, Sichuan) than the 2014 Ya'an earthquake ($R = 6.0$, Sichuan), if the $A_0(\delta)$ is the same value for both earthquakes?

16. (**Compounded interest**) If Carina invests $10\,000 into her saving account with a yearly interest rate of 5%, how long will it take until she has doubled her money?

17. The following are economic definitions related to the marketing of a particular commodity:
(a) The **supply function** $S(x)$ is a function that determines the unit price, p, of a product that must be charged if x units of the product are available (supplied).
(b) The **demand function** $D(x)$ for the commodity determines the unit price $p = D(x)$ that must be charged if x units are demanded by consumers.

(c) The **revenue function** $R(x)$ is defined as the number of units sold times the unit price: $R(x) = xp(x)$.

(d) The **cost function** $C(x)$ is the cost of producing x units of the commodity.

(e) The **profit function** is the profit obtained from selling x units of the goods and is defined to be $P(x) = R(x) - C(x)$. Production is profitable if $P(x) > 0$.

(f) The **utility function** $U(x, y)$ measures the total utility (or satisfaction) the consumer derives from having x units of the first commodity and y units of the second. This is *a function of two variables.*

When $S(x) > D(x)$ there is a surplus and when $D(x) > S(x)$ there is a shortage, as seen in the figure below. When the supply function and demand function intersect, the supply and demand are equal. At this point, the amount of goods being supplied is exactly the same as the amount of goods being demanded. The allocation of goods is therefore at its most efficient point, and the economy is said to be at *equilibrium.*

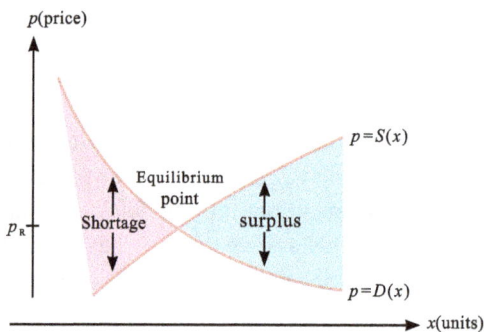

Market research indicates that the demand and supply functions for a particular coffeemaker are given by $D(x) = -3x + 27$ and $S(x) = x^2 + 2$.

(a) At what level of production x and unit price p is market equilibrium achieved?

(b) Sketch the supply and demand curves on the same graph and interpret.

2 Limits and continuity

In this chapter, you will learn about:
- *the definition of a limit;*
- *the properties of a limit;*
- *one-sided limits;*
- *how to evaluate limits;*
- *the squeeze theorem;*
- *limits of sequences;*
- *the monotonic and bounded sequence theorems;*
- *infinitesimal and asymptotic functions;*
- *continuity and discontinuities;*
- *the properties of continuous functions defined on closed intervals.*

2.1 Rates of change and derivatives

As introduced in Chapter 1, the *average rate of change* of a particle moving along a straight line with displacement s, expressed as a function of time t on a time interval $[t_1, t_2]$, is

$$\frac{\Delta s}{\Delta t} = \frac{s(t_2) - s(t_1)}{t_2 - t_1} = \frac{\text{change in } s}{\text{change in } t}.$$

In general, the average rate of change of a function $y = f(x)$ over an interval $[a, b]$ is defined by

$$\frac{\Delta y}{\Delta x} = \frac{f(b) - f(a)}{b - a} = \frac{\text{change in } y}{\text{change in } x}. \tag{2.1}$$

The geometric significance of the average rate of change of $f(x)$ over the interval $[a, b]$ is that it is the slope of the secant line PQ, as shown in Figure 2.1.1.

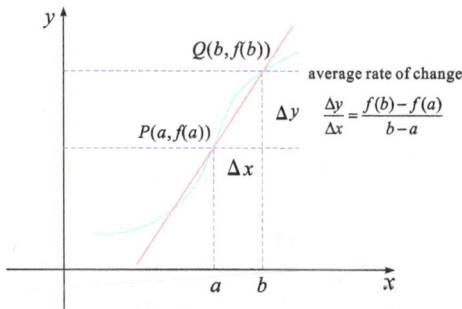

Figure 2.1.1: Average rate of change of $f(x)$ over the interval $[a, b]$.

https://doi.org/10.1515/9783110527780-002

Example 2.1.1. Find the average rate of change of the function $f(x) = x^3 + x$ over the interval $[2, 5]$.

Solution. Since $f(2) = 2^3 + 2 = 10$ and $f(5) = 5^3 + 5 = 130$, the average rate of change over the interval $[2, 5]$ is

$$\frac{f(5) - f(2)}{5 - 2} = \frac{130 - 10}{3} = 40.$$

When finding the *instantaneous rate of change* of $f(x)$ at the point P, one is determining how fast $f(x)$ is changing at that point. Now consider equation (2.1) over $[a, x]$ by varying b as a variable x. Then $\Delta x = x - a$ and $x = a + \Delta x$. Then we let $\Delta x \to 0$. This means $x \to a$ but x never actually equals a (otherwise Δx would be 0). Hence, we have derived the following limit of a quotient:

$$\lim_{\Delta x \to 0} \frac{\Delta y}{\Delta x} = \lim_{\Delta x \to 0} \frac{f(a + \Delta x) - f(a)}{\Delta x}. \tag{2.2}$$

We use this limit wherever we want to know the rate of change of a function at a specific point. For example, a chemist may want to know the rate of change of a chemical reaction at a particular point in time and a physicist may want to know the velocity of an object at a particular instant. This limit, if it exists, is given a special name: the *derivative* of $f(x)$ at the point $x = a$. It is denoted by $f'(a)$ or $y'(a)$.

From a geometric point of view, as $\Delta x \to 0$, the secant line PQ approaches the line tangent to the graph of $f(x)$ at the point P and the slope of the secant line PQ approaches the slope of the tangent line at P. Therefore, the slope of the tangent line at P can be defined as the derivative of $f(x)$ at P.

So far, we have used the term "limit" several times. It is not hard to understand the limit $\lim_{x \to a} f(x)$ from an intuitive point of view. If it exists, it is a number that $f(x)$ approaches in the limiting process. For example, $\lim_{x \to 1} x^2 = 1$, since $x^2 \to 1$ as $x \to 1$ and $\lim_{x \to \pi} \sin x = 0$ since $\sin x \to \sin \pi = 0$ as $x \to \pi$. We will discuss the limit in further details in the next section.

2.2 Limits of a function

2.2.1 Definition of a limit

To introduce the concept of a limit of a function, let us first investigate the behavior of the function f defined by $f(x) = x^2 + 2x - 3$ for values of x near 2. Table 2.2.1 gives values of $f(x)$ for some values of x close to 2, but not equal to 2.

In Figure 2.2.1, we see that, when x is close to 2 (on either side of 2), $f(x)$ is close to $5 = f(2)$. In fact, it appears that we can make the value of $f(x)$ as close as we please to 5 by taking x sufficiently close to 2. We express this by saying that "the limit of the

Table 2.2.1: Some values of $f(x)$ when x is near 2.

x	$f(x)$	x	$f(x)$
1.0	0.000 000	3.0	12.000 000
1.5	2.250 000	2.5	8.250 000
1.8	3.840 000	2.2	6.240 000
1.9	4.410 000	2.1	5.610 000
1.95	4.702 500	2.05	5.302 500
1.99	4.940 100	2.01	5.060 100
1.995	4.970 025	2.005	5.030 025
1.999	4.994 001	2.001	5.006 001

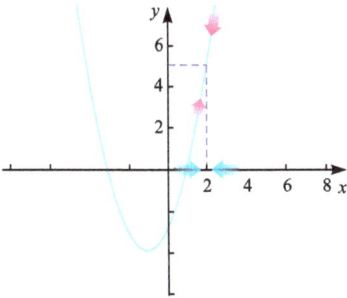

Figure 2.2.1: Behaviors of $f(x)$ when $x \to 2$.

function $f(x) = x^2 + 2x - 3$ as x approaches 2 is equal to 5". The notation for this is

$$\lim_{x \to 2}(x^2 + 2x - 3) = 5.$$

This example shows a result that is probably quite obvious to you, that is, as x approaches the value 2, the values of $f(x)$ approach the value $f(2) = 5$. However, in later applications of the limit, the results will not be so obvious.

In general, we use the following notations for a limit:

$$\lim_{x \to a} f(x) = L, \quad \text{or} \quad \lim_{x \to a} f(x) = L.$$

In words: "the limit of $f(x)$, as x approaches a, equals L". In plainer but not formal language, this means: when x gets sufficiently close to a, $f(x)$ gets arbitrarily close to L (as close to L as we please). The only values of $f(x)$ that matter in defining $\lim_{x \to a} f(x)$ are those values for x close to a, but with $x \neq a$. The limit $\lim_{x \to a} f(x)$ does not require $f(x)$ to exist at $x = a$ and even if $f(a)$ exists, then this value may or may not be equal to $\lim_{x \to a} f(x)$.

NOTE. The limit of a function refers to the value that the function approaches, but this may or may not be the actual function value if it exists. In Figure 2.2.2, $\lim_{x \to 2} f(x) = 2$, not 3.

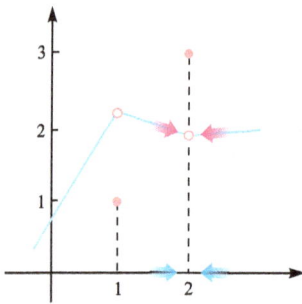

Figure 2.2.2: The limit of $f(x)$ at a may not necessarily equal $f(a)$.

An alternative way of writing $\lim_{x \to a} f(x) = L$ is

$$f(x) \to L \quad \text{as } x \to a,$$

or, in words, "$f(x)$ tends to L as x tends to a".

Example 2.2.1. Guess the value of $\lim_{x \to 1} \frac{x^2-1}{x-1}$.

Solution. We first notice that the function $f(x) = \frac{x^2-1}{x-1}$ is not defined when $x = 1$, but that doesn't matter because the limit $\lim_{x \to 1} f(x)$ only concerns the values of x that are close to 1, but not equal to 1. Table 2.2.2 gives values of $f(x)$ for values of x that approach 1.

Table 2.2.2: Some values of $f(x)$ when x is near 1.

$x < 1$	$f(x)$	$x > 1$	$f(x)$
0.5	1.5	1.5	2.5
0.9	1.9	1.1	2.1
0.99	1.99	1.01	2.01
0.999	1.999	1.001	2.001
0.9999	1.9999	1.0001	2.0001

It appears from the table that the limit might be 2. That is,

$$\lim_{x \to 1} \frac{x^2 - 1}{x - 1} = 2.$$

To emphasize this point, one needs to be aware that the limit $\lim_{x \to a} f(x)$ is the value that $f(x)$ approaches when $x \to a$ (but does not equal a). It is not a surprise that $\lim_{x \to 1} \frac{x^2-1}{x-1}$ exists, although the function $\frac{x^2-1}{x-1}$ is undefined at $x = 1$. Because x is never

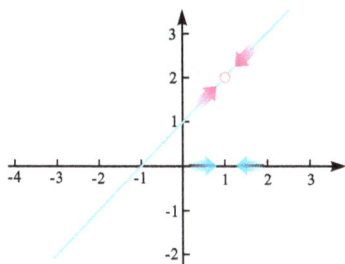

Figure 2.2.3: Behaviors of $f(x)$ when $x \to 1$.

equal to 1, $x - 1$ is never 0, and thus for all other x, $\frac{x^2-1}{x-1}$ is equal to $\frac{(x-1)(x+1)}{x-1} = x + 1$. Thus, we have

$$\lim_{x \to 1} \frac{x^2 - 1}{x - 1} = \lim_{x \to 1}(x + 1) = 2.$$

It is now clearer to see that when x tends to 1, $x + 1$ tends to 2, as shown in Figure 2.2.3.

The above examples have given us some visual and numerical approaches to the limit, so that we can make a guess in these cases. Computer programs or electronic calculators can also be used to evaluate limits, but one needs to be careful since they only have the capacity to deal with a certain range of numbers. For example, a calculator may work out

$$\lim_{x \to 0}\left(\frac{1}{x^2} - \frac{1}{(\sin x)^2} \right)$$

correctly, but it may not be able to work out

$$\lim_{x \to 0}\left(\frac{1}{x^{100}} - \frac{1}{(\sin x)^{100}} \right).$$

As x is very small, both $1/x^{100}$ and $1/(\sin x)^{100}$ may exceed the maximum number that the computer algebra can work with. Table 2.2.3 shows values of $\frac{1}{x^{100}} - \frac{1}{(\sin x)^{100}}$ for some very small values of x.

Table 2.2.3: Computer fails to evaluate the limit numerically.

	$1/x{^}100 - 1/(\sin(x)){^}100$
1	−31 339 022.21
0.5	−8.343 24E+31
0.2	−7.49E+69
0.1	−1.8143E+99
0.05	−5.3939E+128
0.001	−1.6667E+295
0.0001	#DIV/0!

You may be the next to invent a better computer, but we will need to know how the machine works before we make it, so we need to convince ourselves algebraically, in a mathematical way. First, let us clarify what we meant by saying "$f(x)$ is arbitrarily close to L, as close as we please." We need a rigorous definition of the limit. The following definition is credited to Cauchy and Weierstrass.

Definition 2.2.1. For a function $f(x)$ defined on some open interval containing a (but not necessarily at a itself), we say $\lim_{x \to a} f(x) = L$ if, given any number $\varepsilon > 0$, there is a corresponding number $\delta > 0$, such that

$$|f(x) - L| < \varepsilon \quad \text{whenever } 0 < |x - a| < \delta.$$

NOTE. "ϵ-δ definition of limit" is a formalization of the notion of limit. It was first given by Bernard Bolzano in 1817. Augustin-Louis Cauchy occasionally used ϵ-δ arguments in proofs. The definitive modern statement was ultimately provided by Karl Weierstrass. http://en.wikipedia.org/wiki/(ε,δ)-definition_of_limit

Since ε is any positive number, we can make it as small as we want. The definition says that, if $\lim_{x \to a} f(x) = L$, then, for any given number ε, no matter how small it is, we can always find a corresponding number $\delta > 0$, such that the distance between $f(x)$ and L will be less than ε as long as x is in the deleted δ-neighborhood of a. For instance, if $\varepsilon = 0.1$, we can make $f(x)$ be within 0.1 units of L by requiring that x be within the $\delta_{0.1}$ units of a; if $\varepsilon = 0.01$, we can make $f(x)$ be within 0.01 units of L by requiring that x be within $\delta_{0.01}$ units of a. This is the meaning of "as close as we please". Figure 2.2.4 illustrates the $\delta - \varepsilon$ definition of a limit.

Example 2.2.2 (Building blocks). Show that $\lim_{x \to a} c = c$ and $\lim_{x \to a} x = a$, where c is a constant and a is any real number.

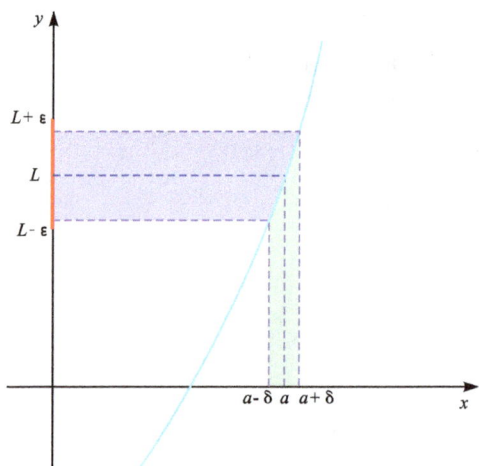

Figure 2.2.4: $\varepsilon - \delta$ definition of a limit.

Proof. Intuitively speaking, these facts are so obvious that they need no proof, but now we confirm them by using the ε-δ language.

(1) Let $f(x) = c$. Then given any positive number ε, we could choose δ to be any positive number, since

$$|f(x) - c| = |c - c| = 0 < \varepsilon, \quad \text{for all } x.$$

(2) Let $f(x) = x$. Then, given any positive number ε, we can choose $\delta = \varepsilon$, so that, whenever $0 < |x - a| < \delta$,

$$|f(x) - a| = |x - a| < \delta = \varepsilon.$$

By the formal definition of a limit, we have $\lim_{x \to a} c = c$ and $\lim_{x \to a} x = a$, as illustrated in Figure 2.2.5 and Figure 2.2.6. □

Example 2.2.3. Prove that $\lim_{x \to 0} |x| = 0$.

Solution. Given any number $\varepsilon > 0$, choose $\delta = \varepsilon$ so that, whenever $0 < |x - 0| < \delta$, we have

$$||x| - 0| = |x| = |x - 0| < \delta = \varepsilon.$$

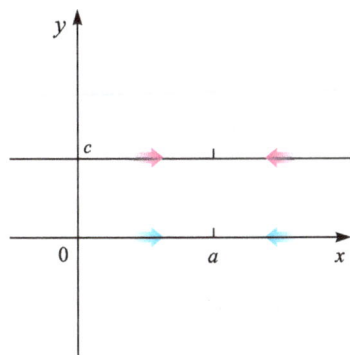

Figure 2.2.5: The limit of a constant function.

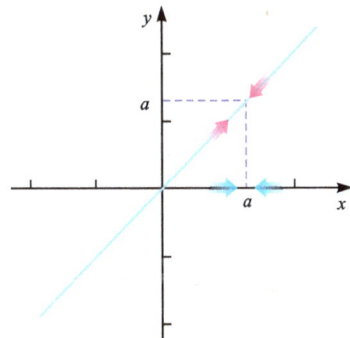

Figure 2.2.6: The limit of x when $x \to a$ is a.

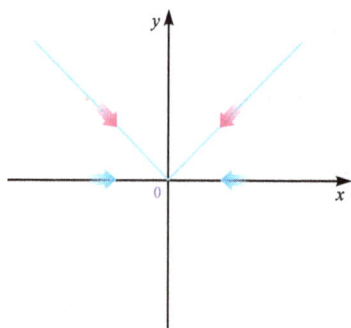

Figure 2.2.7: The limit of |x| when x → 0 is 0.

This means $\lim_{x \to 0} |x| = 0$. Figure 2.2.7 illustrates this limit.

Example 2.2.4. Prove that $\lim_{x \to 1} \frac{x^2-1}{x-1} = 2$.

Proof. Given any number $\varepsilon > 0$, in order to find a corresponding number δ such that

$$\left| \frac{x^2 - 1}{x - 1} - 2 \right| < \varepsilon \quad \text{whenever } 0 < |x - 1| < \delta,$$

we simplify the inequality to obtain

$$\left| \frac{x^2 - 1}{x - 1} - 2 \right| < \varepsilon \leftrightarrow \left| \frac{(x - 1)(x + 1)}{x - 1} - 2 \right| < \varepsilon$$
$$\leftrightarrow |(x + 1) - 2| < \varepsilon$$
$$\leftrightarrow |x - 1| < \varepsilon.$$

This means that, if $0 < |x - 1| < \varepsilon$, then $|\frac{x^2-1}{x-1} - 2| < \varepsilon$, so we can choose $\delta = \varepsilon$. Then

$$\left| \frac{x^2 - 1}{x - 1} - 2 \right| < \varepsilon \quad \text{when } |x - 1| < \delta = \varepsilon,$$

so, by the definition of the limit, we have $\lim_{x \to 1} \frac{x^2-1}{x-1} = 2$. □

Example 2.2.5. Show that $\lim_{x \to 2} x^2 = 4$.

Proof. For any given number $\varepsilon > 0$, in order for

$$|x^2 - 4| = |(x - 2)(x + 2)| < \varepsilon,$$

we need

$$|x - 2| < \frac{\varepsilon}{|x + 2|}.$$

Because $x \to 2$, we could assume $1 < x < 3$, i.e., $|x - 2| < 1$. Then $3 < |x + 2| < 5$. Now let the number δ be the minimum of 1 and $\frac{\varepsilon}{5}$, i.e., $\delta = \min(1, \frac{\varepsilon}{5})$. Then $\delta < 1$ and $\delta < \frac{\varepsilon}{5}$, so whenever $0 < |x - 2| < \delta$, we have

$$|x^2 - 4| = |x - 2||x + 2| < \frac{\varepsilon}{5} \times 5 = \varepsilon.$$

Thus $\lim_{x \to 2} x^2 = 4$ by the formal definition of a limit. $\square$

2.2.2 Properties of limits of functions

As x approaches a, can the limit of $f(x)$ be two different values? Of course, the answer is no. If $\lim_{x \to a} f(x)$ exists, it must be unique. Since, as $x \to a$, if $f(x) \to L$ and $f(x) \to M$, then, when x is sufficiently close to a, $f(x)$ can be made arbitrarily close to L and arbitrarily close to M, L and M must be sufficiently close to each other; as close as possible. Thus L must be equal to M.

Theorem 2.2.1 (Uniqueness). *If $\lim_{x \to a} f(x)$ exists, it must be unique.*

Proof. Intuitively, if $\lim_{x \to a} f(x) = L$, $\lim_{x \to a} f(x) = M$, then $f(x)$ can be made sufficiently close to L and M. Thus, L and M must be sufficiently close to each other. This indicates that L must be equal to M. A rigorous proof can be found in Section 2.7. $\square$

Another property is that $f(x)$ must be bounded on some neighborhood of $x = a$ (except possibly at a) if $\lim_{x \to a} f(x)$ exists. This is because the values of $f(x)$ will be close to the limit for all values of x in that deleted neighborhood of a.

Theorem 2.2.2 (Boundedness). *If $\lim_{x \to a} f(x) = L$, then there is a deleted neighborhood I of $x = a$ such that $f(x)$ is bounded on I.*

Proof. Intuitively, if $\lim_{x \to a} f(x) = L$, then $f(x)$ is close to L when x is close to a. Therefore, all values of $f(x)$ should be close to L when x is near a, except possibly at $x = a$. This means $f(x)$ must be bounded when x is near a. A rigorous proof can be found in Section 2.7. $\square$

NOTE. A deleted neighborhood of $x = a$ instead of a neighborhood of $x = a$ is necessary. For example, consider the *Dirac δ function*

$$f(x) = \begin{cases} 0, & \text{when } x \neq 0 \\ \infty, & \text{when } x = 0. \end{cases}$$

For this function, $\lim_{x \to 0} f(x) = 0$, but it is not bounded on any open interval containing $x = 0$. This function is a *generalized function* and it is used to model some abstractions such as point charges, point masses, and electron points.

If $\lim_{x \to a} f(x) = L > 0$, then, when x is sufficiently close to a, $f(x)$ will be sufficiently close to L. This means that, on some deleted neighborhood of a, we will have $f(x) > 0$.

Theorem 2.2.3. *If $\lim_{x \to a} f(x) = L > 0$, then there is a deleted neighborhood of $x = a$ such that $f(x) > 0$ for all x in that deleted neighborhood.*

Proof. A rigorous proof can be found in Section 2.7. □

From Theorem 2.2.3, we deduce the following result.

Corollary 2.2.4. *If $f(x) > 0$ for all x near a, except possibly at $x = a$, and $\lim_{x \to a} f(x)$ exists, then $\lim_{x \to a} f(x) \geqslant 0$.*

Proof. If $\lim_{x \to a} f(x) < 0$, then, as shown above, for all x sufficiently near a, $f(x) < 0$. This contradicts the assumption that $f(x) > 0$ for all x near a (except possibly $x = a$). □

NOTE. We cannot draw the conclusion that $\lim_{x \to a} f(x) > 0$. This is because, even if $f(x)$ is not 0 for all x near a, the limit $\lim_{x \to a} f(x)$ may be 0. For example, $|x| > 0$ for all x near 0, but $\lim_{x \to 0} |x| = 0$.

Corollary 2.2.5 (Order properties). *If $f(x) > g(x)$ for all x near a, except possibly at $x = a$, and both $\lim_{x \to a} f(x)$ and $\lim_{x \to a} g(x)$ exist, then $\lim_{x \to a} f(x) \geqslant \lim_{x \to a} g(x)$.*

Proof. Let $h(x) = f(x) - g(x)$ and the result follows from Corollary 2.2.4. □

2.2.3 Limit laws

In the previous section, we tried to use evaluation and graphs to guess the values of limits and determined the limits of some simple functions. In this section we show the following properties of limits. These are called the *limit laws* (or *limit rules*). These can be used to calculate the limits of more complicated functions from the known limits of simpler functions.

Limit laws

If c is a constant and the two limits $\lim_{x \to a} f(x)$ and $\lim_{x \to a} g(x)$ exist, then:

1. $\lim_{x \to a} [f(x) + g(x)] = \lim_{x \to a} f(x) + \lim_{x \to a} g(x)$ (sum rule);
2. $\lim_{x \to a} [f(x) - g(x)] = \lim_{x \to a} f(x) - \lim_{x \to a} g(x)$ (difference rule);
3. $\lim_{x \to a} [f(x)g(x)] = \lim_{x \to a} f(x) \lim_{x \to a} g(x)$ (product rule);
4. $\lim_{x \to a} \frac{f(x)}{g(x)} = \frac{\lim_{x \to a} f(x)}{\lim_{x \to a} g(x)}$, provided $\lim_{x \to a} g(x) \neq 0$ (quotient rule).

These rules show that the limit operations follow the basic mathematical operations in a natural way. It is easy to believe that these properties are true. For instance, the sum rule says the limit of a sum is the sum of the limits. This is because, if $f(x)$ tends to L and $g(x)$ tends to M as $x \to a$, it is reasonable to conclude that $f(x) + g(x)$ tends to $L + M$ as $x \to a$. This gives us an intuitive basis for believing that the sum rule is true.

Proofs of these limit laws can be found in Section 2.7.

Corollary 2.2.6. *If $\lim_{x \to a} f(x)$ exists, using law 3, we have the following:*
1. $\lim_{x \to a}[cf(x)] = c\lim_{x \to a} f(x)$, *where c is any constant (constant multiple rule);*
2. $\lim_{x \to a}[f(x)]^n = (\lim_{x \to a} f(x))^n$, *where n is a positive integer (power rule).*

Proof. For 1, let $g(x)$ in the product rule be $g(x) = c$. For 2, repeatedly using the product rule with $g(x) = f(x)$ gives the proof. □

Example 2.2.6. Find $\lim_{x \to 3}(3x^2 - 5x + 7)$.

Solution.

$$\lim_{x \to 3}(3x^2 - 5x + 7)$$
$$= \lim_{x \to 3} 3x^2 - \lim_{x \to 3} 5x + \lim_{x \to 3} 7 \quad \text{(sum/difference rule)}$$
$$= 3\lim_{x \to 3} x^2 - 5\lim_{x \to 3} x + \lim_{x \to 3} 7 \quad \text{(constant multiple rule)}$$
$$= 3\left(\lim_{x \to 3} x\right)^2 - 5\lim_{x \to 3} x + \lim_{x \to 3} 7 \quad \text{(power rule)}$$
$$= 3 \times 3^2 - 5 \times 3 + 7$$
$$= 19.$$

In fact, using a combination of the sum, difference, and power rules, we state a more general rule.

Theorem 2.2.7 (Direct substitution rule for polynomials). *For any polynomial*

$$f(x) = a_0 + a_1 x + a_2 x^2 + \cdots + a_n x^n \quad \text{where } a_0, a_1, \dots, a_n \text{ are constants,}$$

we have

$$\lim_{x \to c} f(x) = a_0 + a_1 c + a_2 c^2 + \cdots + a_n c^n, \quad \text{where c is a constant.}$$

Example 2.2.7. Find $\lim_{x \to 1} \frac{x^2 - 2x + 5}{x^3 + 5x - 7}$.

Solution.

$$\lim_{x \to 1} \frac{x^2 - 2x + 5}{x^3 + 5x - 7}$$

$$= \frac{\lim_{x \to 1}(x^2 - 2x + 5)}{\lim_{x \to 1}(x^3 + 5x - 7)} \quad \text{(quotient rule)}$$

$$= \frac{1^2 - 2 \times 1 + 5}{1^3 + 5 \times 1 - 7} \quad \text{(direct substitution rule)}$$

$$= -4.$$

Example 2.2.8. Find $\lim_{x \to 2} \frac{x^2 - 7x + 10}{x^2 + 5x - 14}$.

Solution. The limit of the denominator is $\lim_{x \to 2}(x^2 + 5x - 14) = 2^2 + 5(2) - 14 = 0$, so we cannot use the quotient rule that says "the limit of a quotient is the quotient of the limits". Instead, a standard approach is to factor the denominator and the numerator, if possible, and then simplify the expression by canceling factors. The fact that the denominator has limit zero as x approaches 2 suggests that 2 is a root of the denominator, so $x - 2$ is a factor of the denominator. It turns out that $x - 2$ is also a factor of the numerator. Thus we have

$$\lim_{x \to 2} \frac{x^2 - 7x + 10}{x^2 + 5x - 14} = \lim_{x \to 2} \frac{(x - 2)(x - 5)}{(x - 2)(x + 7)}.$$

The factor $x - 2$ can be canceled, because, for all $x \neq 2$ and $x \neq -7$, the function is equal to $(x - 5)/(x + 7)$. We have

$$\lim_{x \to 2} \frac{x^2 - 7x + 10}{x^2 + 5x - 14} = \lim_{x \to 2} \frac{x - 5}{x + 7} = \frac{2 - 5}{2 + 7} = -\frac{1}{3}.$$

Theorem 2.2.8 (Direct substitution rule for rational functions). *If c is a constant, for any rational function $\frac{P(x)}{Q(x)}$, where $P(x)$ and $Q(x)$ are two polynomials, if $Q(c) \neq 0$, then $\lim_{x \to c} \frac{P(x)}{Q(x)} = \frac{P(c)}{Q(c)}$.*

Proof. It follows from the quotient rule and the direct substitution rule for polynomials. □

NOTE. We will see later that, for most of the functions we know so far, the direct substitution rule holds. For example, $\lim_{x \to c} \cos x = \cos c$ and $\lim_{x \to c} \sqrt{x} = \sqrt{c}$ for any number $c > 0$.

Example 2.2.9. Show that $\lim_{x \to c} \cos x = \cos c$ for any constant c.

Proof. Given a number $\varepsilon > 0$, choose $\delta = \varepsilon$. Then, when $0 < |x - c| < \delta$, we have

$$|\cos x - \cos c| = \left| 2 \sin \frac{x + c}{2} \sin \frac{x - c}{2} \right| \leq 2 \left| \sin \frac{x - c}{2} \right|$$

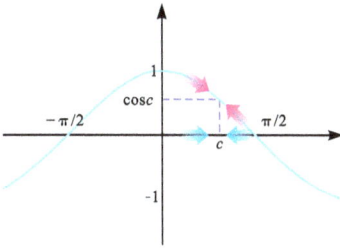

Figure 2.2.8: The limit of $\cos x$ when $x \to c$ is $\cos c$.

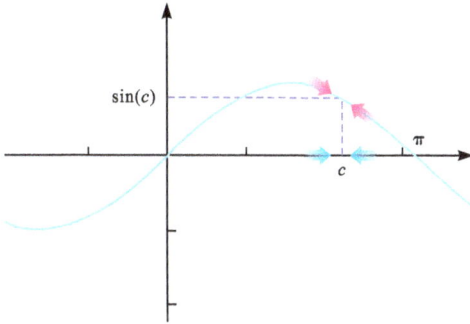

Figure 2.2.9: The limit of $\sin x$ when $x \to c$ is $\sin c$.

$$< 2 \left| \frac{x-c}{2} \right| = |x - c| < \varepsilon.$$

This means $\lim_{x \to c} \cos x = \cos c$. Figure 2.2.8 illustrates this limit. □

NOTES. 1. $|x| \geqslant |\sin x|$ for all x. This equality holds only when $x = 0$.

2. Similarly, we have the direct substitution rule for $\sin x$, that is, $\lim_{x \to c} \sin x = \sin c$ for any $c \in \mathbf{R}$, as illustrated in Figure 2.2.9.

Example 2.2.10. Show that $\lim_{x \to c} \sqrt{x} = \sqrt{c}$ for any constant $c > 0$.

Proof. Given a number $\varepsilon > 0$, choose $\delta = \sqrt{c}\varepsilon$. Then, if $0 < |x - c| < \delta$,

$$|\sqrt{x} - \sqrt{c}| = \left| \frac{(\sqrt{x} - \sqrt{c})(\sqrt{x} + \sqrt{c})}{\sqrt{x} + \sqrt{c}} \right| \quad \text{(conjugate pair)}$$

$$= \left| \frac{x - c}{\sqrt{x} + \sqrt{c}} \right| < \frac{|x - c|}{\sqrt{c}} < \varepsilon.$$

This means $\lim_{x \to c} \sqrt{x} = \sqrt{c}$ for any $c > 0$. Figure 2.2.10 illustrates this limit. □

There is also a limit law for function composition. It says that, if $g(x) \to b$ as $x \to a$ and $f(u) \to L$ as $u \to b$, $f(g(x)) \to L$ as $x \to a$.

Theorem 2.2.9 (Substitution rule). *Suppose that $u = g(x)$ is defined on some interval containing a, but not necessarily at a, $f(u)$ is defined on some interval containing b, but*

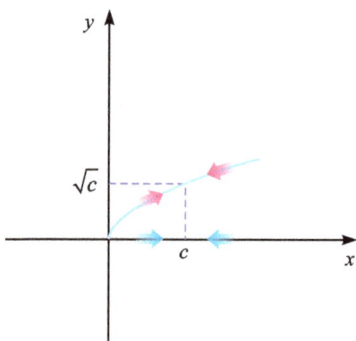

Figure 2.2.10: The limit of $\sqrt{x}$ when $x \to c$ is $\sqrt{c}$.

not necessarily at b, and $\lim_{x \to a} g(x) = b$ *and* $\lim_{u \to b} f(u) = L$. *Then*

$$\lim_{x \to a} f(g(x)) = \lim_{u \to b} f(u) = L.$$

Proof. See Section 2.7. □

This theorem enables us to make substitutions in evaluating limits. Intuitively, the theorem is reasonable, because, if x is close to a, $g(x) \to b$ and $f(g(x)) \to L$ as $g(x) \to b$.

Example 2.2.11. Find $\lim_{x \to 2} \sqrt{1 + x}$.

Solution.

$$\lim_{x \to 2} \sqrt{1 + x} = \lim_{u \to 3} \sqrt{u} \quad \text{(substitution } u = 1 + x, u \to 3 \text{ as } x \to 2\text{)}$$
$$= \sqrt{3} \quad \text{(direct substitution by Example 2.2.10).}$$

Also, we could write the above in the following way:

$$\lim_{x \to 2} \sqrt{1 + x} = \sqrt{\lim_{x \to 2}(1 + x)} = \sqrt{1 + \lim_{x \to 2} 2} = \sqrt{1 + 2} = \sqrt{3}.$$

Example 2.2.12. Find $\lim_{x \to 4} \frac{\sqrt{2x+1}-3}{x-4}$.

Solution. If we try to evaluate at $x = 4$, we obtain an indeterminate form, $0/0$. The key here is to find a hidden factor of $x - 4$ in the numerator that will cancel with this factor in the denominator. One technique is to *rationalize the numerator* by making use of the algebraic identity

$$(a - b)(a + b) = a^2 - b^2.$$

We multiply the numerator and denominator by $\sqrt{2x + 1} + 3$, so that the numerator becomes

$$(\sqrt{2x + 1})^2 - 3^2 = 2x - 8,$$

thus eliminating the square root. Here are the details:

$$\lim_{x \to 4} \frac{\sqrt{2x+1}-3}{x-4} = \lim_{x \to 4} \frac{(\sqrt{2x+1}-3)}{(x-4)} \frac{(\sqrt{2x+1}+3)}{(\sqrt{2x+1}+3)}$$

$$= \lim_{x \to 4} \frac{(\sqrt{2x+1})^2 - 3^2}{(x-4)(\sqrt{2x+1}+3)}$$

$$= \lim_{x \to 4} \frac{2x+1-9}{(x-4)(\sqrt{2x+1}+3)}$$

$$= \lim_{x \to 4} \frac{2(x-4)}{(x-4)\sqrt{2x+1}+3}$$

$$= \lim_{x \to 4} \frac{2}{\sqrt{2x+1}+3}$$

$$= \frac{\lim_{x \to 4} 2}{\lim_{x \to 4} \sqrt{2x+1} + \lim_{x \to 4} 3}$$

$$= \frac{2}{\sqrt{2 \times 4 + 1} + 3} = \frac{1}{3}.$$

2.2.4 One-sided limits

In the previous sections, our limits were determined by requiring x to approach the number a from both the left and the right. However, sometimes we wish to consider these as separate cases: x approaches a from the left ($x \to a^-$) and x approaches a from the right ($x \to a^+$). The corresponding limits are known as one-sided limits, the left-hand limit and the right-hand limit, respectively. Here is the notation:

Left-hand limit:

$$\lim_{x \to a^-} f(x) \text{ is the limit as } x \text{ approaches } a \text{ from the left.}$$

Right-hand limit:

$$\lim_{x \to a^+} f(x) \text{ is the limit as } x \text{ approaches } a \text{ from the right.}$$

Figure 2.2.11 illustrates the two one-sided limits.

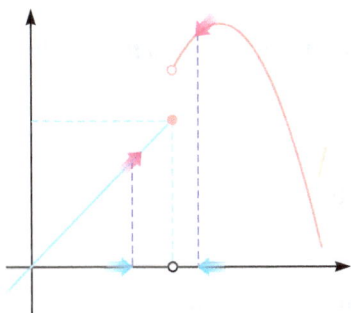

Figure 2.2.11: One-sided limits.

Example 2.2.13. Investigate the one-sided limits of the function as $x \to 0^+$ and $x \to 0^-$.

$$f(x) = \frac{|x|}{x} = \begin{cases} 1, & \text{when } x > 0 \\ -1, & \text{when } x < 0. \end{cases}$$

Solution. The left-hand limit, $\lim_{x \to 0^-} f(x)$, means x approaches 0 from the left, so we are only concerned with values of x that are less than 0. In this case $f(x) = -1$ is a constant function, so

$$\lim_{x \to 0^-} f(x) = \lim_{x \to 0^-} (-1) = -1.$$

Similarly we have $\lim_{x \to 0^+} f(x) = \lim_{x \to 0^+} (+1) = 1$. We indicate this situation symbolically by writing

$$\lim_{x \to 0^-} \frac{|x|}{x} = -1 \quad \text{and} \quad \lim_{x \to 0^+} \frac{|x|}{x} = 1.$$

Figure 2.2.12 illustrates these two limits.

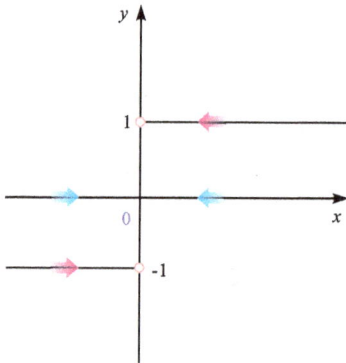

Figure 2.2.12: The limit of $\frac{|x|}{x}$ when $x \to 0$ does not exist.

Using the ε-δ language, the definitions of one-sided limits are given as follows.

Definition 2.2.2. The *left-hand limit* of $f(x)$ as x approaches a from the left is L, written as

$$\lim_{x \to a^-} f(x) = L,$$

if, for any given number $\varepsilon > 0$, there is a number $\delta > 0$, such that

$$|f(x) - L| < \varepsilon, \quad \text{whenever } a - \delta < x < a.$$

The *right-hand limit* of $f(x)$ as x approaches a from the right is L, written as

$$\lim_{x \to a^+} f(x) = L,$$

if, for any given number $\varepsilon > 0$, there is a number $\delta > 0$, such that

$$|f(x) - L| < \varepsilon, \quad \text{whenever } a < x < a + \delta.$$

By comparing the definition of limits and the definition of one-sided limits, we derive the following result.

Theorem 2.2.10. *A limit exists if and only if both one-sided limits exist and are equal. That is,*

$$\lim_{x \to a} f(x) = L \quad \Leftrightarrow \quad \lim_{x \to a^-} f(x) = \lim_{x \to a^+} f(x) = L.$$

This means that, if either of the one-sided limits does not exist or if the one-sided limits are not the same, the function has no limit. Thus, the function $f(x) = |x|/x$ has no limit as $x \to 0$, because $\lim_{x \to 0^-} f(x) = -1 \neq 1 = \lim_{x \to 0^+} f(x)$.

Example 2.2.14. For the greatest integer function $f(x) = [x]$, find $\lim_{x \to 1^-} f(x)$, $\lim_{x \to 1^+} f(x)$, and $\lim_{x \to 1} f(x)$.

Solution. The notation $x \to 1^-$ means x approaches 1 from the left. For these values of x, $[x] = 0$, so

$$\lim_{x \to 1^-} f(x) = \lim_{x \to 1^-} 0 = 0.$$

The notation $x \to 1^+$ means x approaches 1 from the right. For these values of x, $[x] = 1$, so

$$\lim_{x \to 1^+} f(x) = \lim_{x \to 1^+} 1 = 1.$$

Hence, $\lim_{x \to 1} f(x)$ does not exist, because $\lim_{x \to 1^-} f(x) \neq \lim_{x \to 1^+} f(x)$, as seen in Figure 2.2.13.

Example 2.2.15. Use the definition of a one-sided limit to prove that $\lim_{x \to 0^+} \sqrt{x} = 0$.

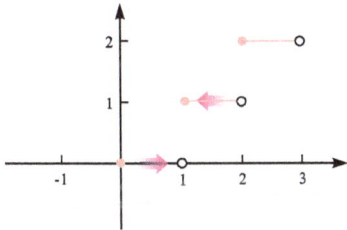

Figure 2.2.13: The limit of $[x]$ when $x \to 1$ does not exist.

Solution. Given $\varepsilon > 0$, let $\delta = \varepsilon^2$. If $0 < x < \delta$, then

$$|f(x) - 0| = |\sqrt{x} - 0| < \sqrt{\delta} = \sqrt{\varepsilon^2} = \varepsilon.$$

Then, according to the definition of the right-hand limit, $\lim_{x \to 0^+} \sqrt{x} = 0$.

Example 2.2.16. If

$$f(x) = \begin{cases} x^2 - 3x, & \text{if } x \leqslant 2 \\ x - 3, & \text{if } x > 2, \end{cases}$$

find $\lim_{x \to 2^+} f(x)$ and $\lim_{x \to 2^-} f(x)$. Does $\lim_{x \to 2} f(x)$ exist?

Solution. For the right-hand limit, we consider only $x > 2$. These values of x are greater than 2, which means $f(x) = x - 3$, so

$$\lim_{x \to 2^+} f(x) = \lim_{x \to 2^+} (x - 3).$$

Now, we proceed by using the limit theorems and the usual steps, to obtain

$$\lim_{x \to 2^+} f(x) = \lim_{x \to 2^+} (x - 3) = 2 - 3 = -1.$$

Similarly, for the left-hand limit $\lim_{x \to 2^-} f(x)$, we consider only $x < 2$. These values of x are to the left of 2, for which $f(x) = x^2 - 3x$, so

$$\lim_{x \to 2^-} f(x) = \lim_{x \to 2^-} (x^2 - 3x)$$
$$= \lim_{x \to 2^-} (x^2 - 3x) = 2^2 - 3(2) = -2.$$

Since the one-sided limits as $x \to 2$ do not agree, $\lim_{x \to 2} f(x)$ does not exist. Figure 2.2.14 illustrates these limits.

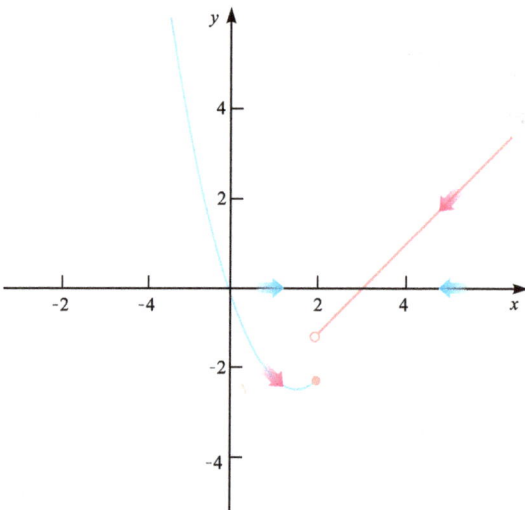

Figure 2.2.14: The limit of $f(x)$ in Example 2.2.16 when $x \to 2$ doesn't exist.

2.2.5 Limits involving infinity and asymptotes

If we investigate the limit

$$\lim_{x \to 0} \frac{1}{x^2}$$

as x approaches 0, the denominator gets closer and closer to zero, so the fraction $1/x^2$ gets larger and larger and is unbounded, as seen in Figure 2.2.15. It will not tend to any finite number. Therefore, the limit does not exist. However, since we introduced the two symbols ∞ and $-\infty$ in Chapter 1, we now use them to indicate a limit that does not exist in this unbounded way. For example, $\lim_{x \to a} f(x) = \infty$ means that $f(x)$ becomes larger and larger and can be made bigger than any given positive number as x approaches a. A formal definition is the following.

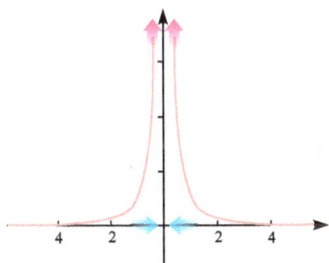

Figure 2.2.15: Behaviors of $y = \frac{1}{x^2}$ when $x \to 0$.

Definition 2.2.3. We say $\lim_{x \to a} f(x) = \infty$ if, for any given number $M > 0$ (no matter how large it is), we can find a number δ, such that

$$f(x) > M \quad \text{whenever } 0 < |x - a| < \delta.$$

Similarly, one can give a definition for $\lim_{x \to a} f(x) = -\infty$ stated as "when x approaches a, $f(x)$ becomes smaller and smaller and can be less than any given negative number when x is sufficiently close to a."

Example 2.2.17. Show that $\lim_{x \to 2} \frac{1}{(x-2)^2} = \infty$.

Proof. Given any positive number $M > 0$, choose $\delta = \frac{1}{\sqrt{M}}$. Then, whenever $0 < |x - 2| < \delta$,

$$\frac{1}{(x-2)^2} > \frac{1}{\delta^2} = \frac{1}{(\frac{1}{\sqrt{M}})^2} = M.$$

Hence, $\lim_{x \to 2} 1/(x-2)^2 = \infty$ by Definition 2.2.3. $\square$

If $f(x)$ is a quotient of two functions, then $f(x) \to \infty$ (or $-\infty$) usually implies that the denominator approaches 0 in the limiting process. This was shown in the example above, where $(x - 2) \to 0$, when $x \to 2$, so the reciprocal tends to infinity. In general, we have the following.

Theorem 2.2.11. *If $f(x)$ is not 0 near $x = a$, except possibly at a, then $\lim_{x \to a} |f(x)| = \infty$ if and only if $\lim_{x \to a} \frac{1}{f(x)} = 0$.*

Proof. "$\Longrightarrow$" Given any number $\varepsilon > 0$, since $\lim_{x \to a} |f(x)| = \infty$, we can find a number $\delta > 0$ such that, whenever $0 < |x - a| < \delta$,

$$|f(x)| > \frac{1}{\varepsilon}.$$

However, then

$$\left| \frac{1}{|f(x)|} - 0 \right| = \left| \frac{1}{f(x)} \right| = \frac{1}{|f(x)|} < \frac{1}{\frac{1}{\varepsilon}} = \varepsilon,$$

so $\lim_{x \to a} 1/f(x) = 0$.

"$\Longleftarrow$" Given any number $M > 0$, since $\lim_{x \to a} 1/f(x) = 0$, for the positive number $1/M$, we can find a number $\delta > 0$ such that, whenever $0 < |x - a| < \delta$, we have

$$\left| \frac{1}{f(x)} - 0 \right| < \frac{1}{M}.$$

This means $\frac{1}{|f(x)|} < \frac{1}{M}$, so $|f(x)| > M$ and therefore $\lim_{x \to a} |f(x)| = \infty$. □

Example 2.2.18. Investigate the following limits:
(a) $\lim_{x \to 0} \frac{2}{x^2}$; (b) $\lim_{x \to 0} \frac{1}{x}$.

Solution. (a) Since $\lim_{x \to 0} x^2 = 0$, we have

$$\lim_{x \to 0} \frac{2}{x^2} = 2 \lim_{x \to 0} \frac{1}{x^2} = 2 \times \infty = \infty.$$

(b) Since $\lim_{x \to 0} x = 0$, we know that $1/|x| \to \infty$. However, we need to be very careful here. When $x \to 0^+$, this means $\frac{1}{x} > 0$, so $\frac{1}{|x|} = \frac{1}{x}$. When $x \to 0^+$, $\frac{1}{x} \to \infty$, so we have

$$\lim_{x \to 0^+} \frac{1}{x} = \infty.$$

However, when $x \to 0^-$, $\frac{1}{|x|} = -\frac{1}{x}$. This means

$$\lim_{x \to 0^-} \frac{1}{x} = -\infty.$$

We conclude that $\lim_{x \to 0} \frac{1}{x}$ does not exist. Figure 2.2.16 illustrates this limit.

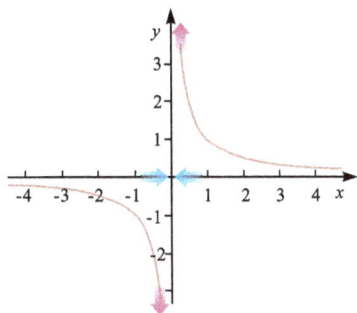

Figure 2.2.16: The limit of $y = \frac{1}{x}$ when $x \to 0$ does not exist.

Vertical asymptotes

There is a geometric significance for the limit $\lim_{x \to a^+} f(x) = \pm\infty$ or $\lim_{x \to a^-} f(x) = \pm\infty$. In this case, the line $x = a$ is a *vertical asymptote* to the graph of $f(x)$.

Example 2.2.19. Find all the vertical asymptotes for the function

$$f(x) = \frac{x+5}{x^2 - 2x - 3}.$$

Solution. To find vertical asymptotes, we need to determine when $f(x) \to \pm\infty$. This means we need to investigate when the denominator tends to 0. Factoring the denominator, we obtain

$$f(x) = \frac{x+5}{(x-3)(x+1)}.$$

When $x \to 3$, the denominator of $f(x)$ tends to 0, but the numerator tends to the finite number 8, which is not 0, so $f(x) \to \pm\infty$ when $x \to 3$. This is also true when $x \to -1$. This means $x = -1$ and $x = 3$ are two vertical asymptotes to the graph of $f(x)$. Figure 2.2.17 shows the graph of $f(x)$.

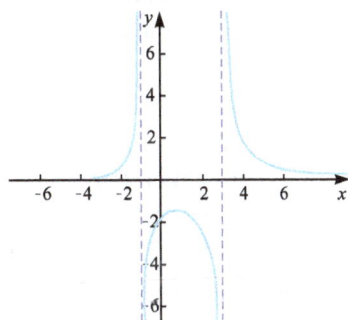

Figure 2.2.17: Graph of $f(x)$ in Example 2.2.19.

Example 2.2.20. Find all vertical asymptotes for $f(x) = \tan x$.

Solution. $\tan x = \frac{\sin x}{\cos x}$, so we need to determine when $\cos x = 0$. This happens when $x = k\pi + \frac{\pi}{2}$, $k \in \mathbf{Z}$, and in these cases $\sin x \neq 0$, so when $x \to k\pi + \frac{\pi}{2}$, $k \in \mathbf{Z}$, $\tan x \to \pm\infty$. Thus the vertical asymptotes to the graph of $\tan x$ occur exactly when $x = k\pi + \frac{\pi}{2}$, $k \in \mathbf{Z}$. Figure 2.2.18 shows the graph of $\tan x$.

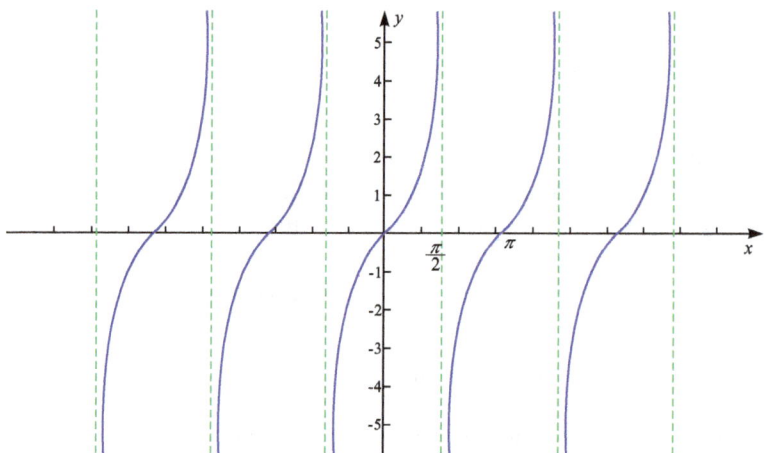

Figure 2.2.18: Graph of $y = \tan x$.

There is one more situation in which we use the infinity symbol with a limit. These are the "limits at infinity", denoted by

$$\lim_{x \to \infty} f(x) = L, \quad \lim_{x \to -\infty} f(x) = L \quad \text{or} \quad \lim_{x \to \pm\infty} f(x) = L.$$

This means that, when x gets sufficiently large (or small), the value of $f(x)$ gets arbitrarily close to the number L. For example, we can guess that the limit $\lim_{x \to \infty} \frac{1}{x} = 0$, since, as the denominator gets larger, the reciprocal gets smaller. A formal definition is given below.

NOTE. Sometimes, we use the notation $\pm\infty$ to mean positive infinity or negative infinity, so $\lim_{x \to \pm\infty} f(x)$ means $\lim_{x \to \infty} f(x)$ or $\lim_{x \to -\infty} f(x)$. This notation should only be used when the behavior of $f(x)$ is the same when $x \to \infty$ and when $x \to -\infty$. For example $\lim_{x \to +\infty} \frac{1}{x} = 0$ and $\lim_{x \to -\infty} \frac{1}{x} = 0$ can be rewritten as $\lim_{x \to \pm\infty} \frac{1}{x} = 0$. If the behavior of $f(x)$ as $x \to \infty$ is different from that when $x \to -\infty$, then we have to discuss $\lim_{x \to +\infty} f(x)$ and $\lim_{x \to -\infty} f(x)$ separately.

Figure 2.2.19 illustrates the limit $\lim_{x \to \pm\infty} f(x) = L$.

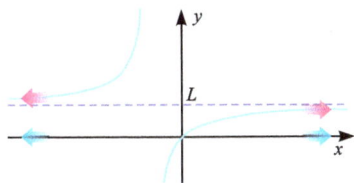

Figure 2.2.19: Limits at infinity.

Definition 2.2.4 (Limits at infinity).

1. The limit $\lim_{x\to\infty} f(x) = L$ if, for any number $\varepsilon > 0$, there exists a number $N > 0$ such that

$$|f(x) - L| < \varepsilon \quad \text{whenever } x > N.$$

2. The limit $\lim_{x\to-\infty} f(x) = L$ if, for any number $\varepsilon > 0$, there exists a number $N > 0$ such that

$$|f(x) - L| < \varepsilon \quad \text{whenever } x < -N.$$

3. The limit $\lim_{x\to\pm\infty} f(x) = L$ if, for any number $\varepsilon > 0$, there exists a number $N > 0$ such that

$$|f(x) - L| < \varepsilon \quad \text{whenever } |x| > N.$$

Example 2.2.21. Show that $\lim_{x\to\infty} \frac{1}{x} = 0$.

Solution. From Theorem 2.2.11, we know the denominator tends to infinity and the reciprocal tends to 0, but now we use Definition 2.2.4. Given a number $\varepsilon > 0$, we choose $N = \frac{1}{\varepsilon}$, so

$$\left|\frac{1}{x} - 0\right| = \frac{1}{x} < \frac{1}{\frac{1}{\varepsilon}} = \varepsilon, \quad \text{whenever } x > N.$$

Similarly, we can show that $\lim_{x\to-\infty} \frac{1}{x} = 0$, so sometimes we write $\lim_{x\to\pm\infty} \frac{1}{x} = 0$, as illustrated in Figure 2.2.20.

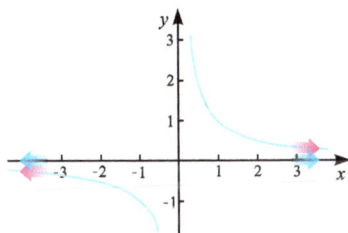

Figure 2.2.20: The limit of $y = \frac{1}{x}$ when $x \to \pm\infty$ is 0.

NOTE. Limit laws also work for limits involving infinity.

Example 2.2.22. Find $\lim_{x \to \infty} \frac{5x^2+x+7}{x^2+1}$.

Solution. We have

$$
\begin{aligned}
\lim_{x \to \infty} \frac{5x^2 + x + 7}{x^2 + 1} &= \lim_{x \to \infty} \frac{5 + \frac{1}{x} + \frac{7}{x^2}}{1 + \frac{1}{x^2}} \quad \text{(dividing by } x^2\text{)} \\
&= \frac{\lim_{x \to \infty} 5 + \lim_{x \to \infty} \frac{1}{x} + \lim_{x \to \infty} \frac{7}{x^2}}{\lim_{x \to \infty} 1 + \lim_{x \to \infty} \frac{1}{x^2}} \quad \text{(limit laws)} \\
&= \frac{5 + 0 + 0}{1 + 0} = 5.
\end{aligned}
$$

Horizontal asymptotes

There is also a geometric meaning for the limits $\lim_{x \to \infty} f(x) = b$ and $\lim_{x \to -\infty} f(x) = b$. In these cases, $y = b$ is a *horizontal asymptote* to the graph of the function $f(x)$. In Example 2.2.22, we know that $y = 5$ is a horizontal asymptote to the graph of the function. The graph of a function may have up to two horizontal asymptotes, as seen in the following example.

Example 2.2.23. Find the horizontal asymptotes for the function

$$ y = \tan^{-1} x \quad (\text{Note: } \tan^{-1} x = \arctan x.) $$

Solution. The graph of $y = \tan^{-1} x$ and the graph of $y = \tan x$ on $\left(-\frac{\pi}{2}, \frac{\pi}{2}\right)$ are reflections in $y = x$. In fact,

$$ \lim_{x \to \frac{\pi}{2}^-} \tan x = \lim_{x \to \frac{\pi}{2}^-} \frac{\sin x}{\cos x} = \infty \quad \text{and} \quad \lim_{x \to \frac{\pi}{2}^+} \tan x = \lim_{x \to \frac{\pi}{2}^+} \frac{\sin x}{\cos x} = -\infty. $$

$x = -\frac{\pi}{2}$ and $x = \frac{\pi}{2}$ are two vertical asymptotes of the graph of $y = \tan x$, so

$$ \lim_{x \to \infty} \tan^{-1} x = \frac{\pi}{2} \quad \text{and} \quad \lim_{x \to -\infty} \tan^{-1} x = -\frac{\pi}{2}. $$

Therefore, there are two horizontal asymptotes of the graph of $y = \tan^{-1} x$.

Another example of functions which have two horizontal asymptotes is

$$ f(x) = \begin{cases} \frac{x+1}{x-1}, & x < 0 \\ \frac{\sin x}{x}, & x > 0. \end{cases} $$

To the left and to the right, $y = 1$ and $y = 0$, respectively, are two horizontal asymptotes.

Slant asymptotes

Graphs of functions may have slant asymptotes. You already know the graph of the hyperbola

$$\frac{x^2}{a^2} - \frac{y^2}{b^2} = 1, \quad \text{where } a, b \text{ are nonzero constants.}$$

It has two slant asymptotes, $y = \frac{b}{a}x$ and $y = -\frac{b}{a}x$. A slant asymptote is a nonhorizontal, nonvertical line that the graph of a function gets arbitrary close to when x moves increasingly far to the left or to the right along the x-axis.

Definition 2.2.5. If there are constants m and b such that

$$\lim_{x \to \infty} (f(x) - mx - b) = 0 \quad \text{or} \quad \lim_{x \to -\infty} (f(x) - mx - b) = 0,$$

then $y = mx + b$ is a *slant asymptote* to the graph of the function $f(x)$.

But how do we find a slant asymptote if there is one? Notice that, if $y = mx + b$ is a slant asymptote, then

$$\lim_{x \to \pm\infty} (f(x) - mx - b) = 0,$$

so

$$b = \lim_{x \to \pm\infty} (f(x) - mx).$$

Thus

$$\lim_{x \to \pm\infty} \frac{f(x) - mx - b}{x} = 0.$$

Therefore,

$$\lim_{x \to \pm\infty} \left(\frac{f(x)}{x} - m - \frac{b}{x} \right) = 0.$$

Since $\lim_{x \to \pm\infty} \frac{b}{x} = 0$, we obtain

$$m = \lim_{x \to \pm\infty} \frac{f(x)}{x}.$$

In fact, we can prove the converse is also true. That is, if

$$\lim_{x \to \pm\infty} \frac{f(x)}{x} = m \quad \text{and} \quad \lim_{x \to \pm\infty} (f(x) - mx) = b,$$

then

$$\lim_{x \to \pm\infty} (f(x) - mx - b) = 0,$$

so $y = mx + b$ is a slant asymptote.

NOTES. 1. If $\lim_{x \to \pm\infty} \frac{f(x)}{x} = m$ and $\lim_{x \to \pm\infty} (f(x) - mx) = b$, then $y = mx + b$ is a slant asymptote to the graph of $f(x)$.
2. If $f(x)$ behaves differently when $x \to +\infty$ and $x \to -\infty$, then one has to investigate the two one-sided limits separately.

Example 2.2.24. Find any slant asymptotes for

$$f(x) = \sqrt{x^2 + 4x}.$$

Solution. Simplifying gives $f(x) = |x| \sqrt{1 + \frac{4}{x}}$, when $x > 0$ or $x \leqslant -4$. Therefore, $x \to \infty$ and $x \to -\infty$, $f(x)$ behaves differently. Thus we discuss the two cases separately instead of discussing $x \to \pm\infty$.

Since

$$\lim_{x \to \infty} \frac{f(x)}{x} = \lim_{x \to \infty} \frac{|x|}{x} \sqrt{1 + \frac{4}{x}} = \lim_{x \to \infty} \frac{x}{x} \sqrt{1 + \frac{4}{x}} = 1,$$

$$\lim_{x \to \infty} (f(x) - x) = \lim_{x \to \infty} \left(\sqrt{x^2 + 4x} - x \right)$$

$$= \lim_{x \to \infty} \frac{(\sqrt{x^2 + 4x} - x)(\sqrt{x^2 + 4x} + x)}{\sqrt{x^2 + 4x} + x}$$

$$= \lim_{x \to \infty} \frac{x^2 + 4x - x^2}{\sqrt{x^2 + 4x} + x} = \lim_{x \to \infty} \frac{4x}{\sqrt{x^2 + 4x} + x}$$

$$= \lim_{x \to \infty} \frac{4x}{x\sqrt{1 + \frac{4}{x}} + x}, x > 0,$$

$$= \lim_{x \to \infty} \frac{4}{\sqrt{1 + \frac{4}{x}} + 1} = 2.$$

The slant asymptote to the graph of $f(x)$ as $x \to \infty$ is $y = x + 2$.

Similarly, we find that $y = -x - 2$ is the slant asymptote to the graph of $f(x)$ as $x \to -\infty$. Figure 2.2.21 shows the graph of this function and its two slant asymptotes.

Example 2.2.25. Determine whether or not the graph of the function $f(x) = x^2 - \sqrt{x}$, $x > 0$, has a slant asymptote.

Solution. If a slant asymptote exists, then $\lim_{x \to \infty} \frac{f(x)}{x}$ must exist. However,

$$\lim_{x \to \infty} \frac{f(x)}{x} = \lim_{x \to \infty} \frac{x^2 - \sqrt{x}}{x} = \lim_{x \to \infty} \left(x - \frac{1}{\sqrt{x}} \right) = \infty - 0 = \infty,$$

so there is no slant asymptote to the graph of $f(x) = x^2 - \sqrt{x}$, as shown in Figure 2.2.22.

Sometimes, there is a simpler way to identify slant asymptotes for rational functions.

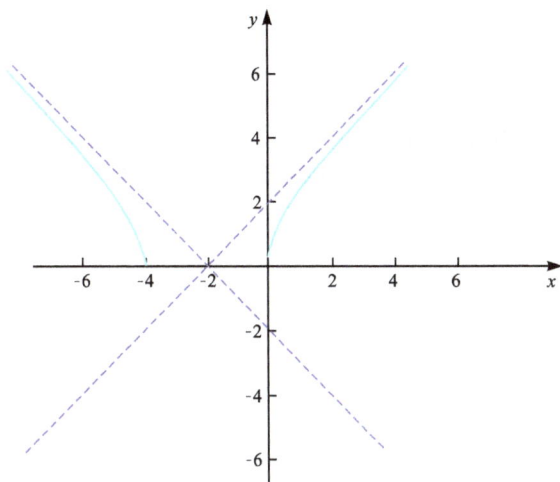

Figure 2.2.21: Graph of $f(x)$ in Example 2.2.24 and its asymptotes.

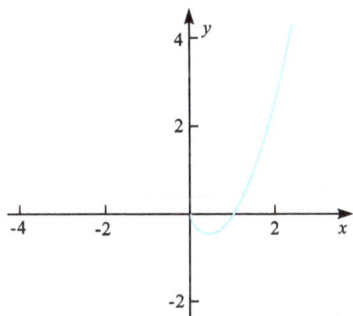

Figure 2.2.22: Graph of $f(x)$ in Example 2.2.25.

Example 2.2.26. Find all the asymptotes for

$$f(x) = \frac{2x^2 + 3x + 2}{2x + 1}$$

and find the x- and y-intercepts. Sketch the curve.

Solution. When $x \to -1/2$, $f(x) \to \pm\infty$, so $x = -1/2$ is the vertical asymptote. Since

$$\lim_{x \to \pm\infty} \frac{2x^2 + 3x + 2}{2x + 1} = \lim_{x \to \pm\infty} \frac{2x + 3 + \frac{2}{x}}{2 + \frac{1}{x}} = \pm\infty,$$

it has no horizontal asymptote. However, using long division, we have

$$f(x) = \frac{2x^2 + 3x + 2}{2x + 1} = x + 1 + \frac{1}{2x + 1},$$

so we have $\lim_{x\to\pm\infty}(f(x) - x - 1) = \lim_{x\to\pm\infty}\frac{1}{2x+1} = 0$. Therefore, $y = x + 1$ is the only slant asymptote.

In order to find the intercepts, we plug in $x = 0$ to obtain the y-intercept 2 and if we let $y = 0$, we have $2x^2 + 3x + 2 = 0$, which has no real solution, so there is no x-intercept. This means the graph of the function does not cross the x-axis. The graph is shown in Figure 2.2.23.

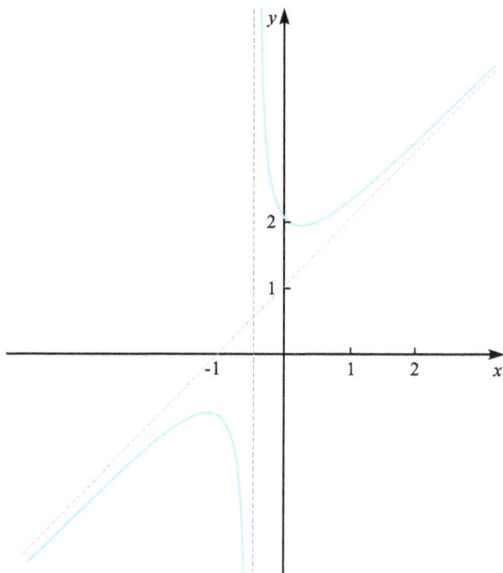

Figure 2.2.23: Graph of $f(x)$ in Example 2.2.26 and its asymptotes.

2.3 Limits of sequences

To better explore limits of functions, we now consider limits of sequences which are special functions with domain the set of natural numbers **N**.

2.3.1 Definitions and properties

A *sequence* is a list of numbers written in a definite order:

$$a_1, a_2, a_3, \ldots, a_n, \ldots.$$

The number a_1 is called the first term, a_2 is the second term and in general a_n is the nth term of the sequence. In many cases, a sequence is defined by giving a formula for the nth term that is valid for all values of n. Some examples of sequences are

(1)	1,	$\frac{1}{2}$,	$\frac{1}{3}$,	...,	$\frac{1}{n}$,	...
(2)	$\frac{1}{1\times2}$,	$\frac{1}{2\times3}$,	$\frac{1}{3\times4}$,	...,	$\frac{1}{n\times(n+1)}$,	...
(3)	2,	$\frac{1}{2}$,	$\frac{4}{3}$,	...,	$\frac{n+(-1)^{n-1}}{n}$,	...
(4)	$\frac{1}{2}$,	$\frac{2}{3}$,	$\frac{3}{4}$,	...,	$\frac{n}{n+1}$,	...

In (1) and (2), as n gets larger and larger, the sequence of numbers comes closer and closer to 0. In (3) and (4), as n gets larger and larger, we see that the sequence values come closer and closer to 1, even though no term is actually 1. In other words, intuitively speaking, we have

$$\lim_{n\to\infty}\frac{1}{n}=0, \quad \lim_{n\to\infty}\frac{1}{n\times(n+1)}=0, \quad \text{and} \quad \lim_{n\to\infty}\frac{n}{n+1}=1.$$

To analyze these limits, we use the results for limits of functions, since a sequence can be viewed as a special function with domain $\{1,2,3,...\}$, the set of positive integers, range $\{a_1,a_2,a_3,...\}$, and the rule f such that $f(n)=a_n$. Thus

$$\lim_{n\to\infty}a_n = \lim_{n\to\infty}f(n).$$

Therefore, in view of the definition of limits of functions $\lim_{x\to\infty}f(x)$, we have the precise definition of limits of sequences.

Definition 2.3.1. $\lim_{n\to\infty}a_n = L$ if and only if, for any given number $\varepsilon > 0$, there is a number N such that

$$|a_n - L| < \varepsilon \quad \text{whenever } n > N.$$

Example 2.3.1. Prove that the sequence $\{x_n\} = \{\frac{2n+1}{n}\}$ has limit 2.

Solution. It is easy to see that the sequence tends towards 2 as $n \to \infty$, so we can guess the limit of the sequence is 2. However, we now prove this by using the precise definition of the limit. For a given number $\varepsilon > 0$, we must find an N so that, whenever $n > N$,

$$|x_n - 2| = \left|\frac{2n+1}{n} - 2\right| < \varepsilon.$$

Equivalently, $\quad \dfrac{1}{n} < \varepsilon \quad$ or $\quad \dfrac{1}{\varepsilon} < n.$

Hence, if N is any number larger than $\frac{1}{\varepsilon}$, then $\frac{1}{\varepsilon} < n$ whenever $n > N$, so $\lim_{n\to\infty}\frac{2n+1}{n} = 2$.

Example 2.3.2. Show that $\lim_{n\to\infty}b^{\frac{1}{n}} = 1$ for any number $b > 1$.

Proof. Given a number $\varepsilon > 0$, we want to find a number N such that, when $n > N$, we have

$$\left|b^{\frac{1}{n}} - 1\right| < \varepsilon.$$

Since $b > 1$ and $b^{\frac{1}{n}} > 1$, the above inequality implies

$$b^{\frac{1}{n}} - 1 < \varepsilon,$$

so

$$b^{\frac{1}{n}} < 1 + \varepsilon.$$

Taking the natural logarithm on both sides gives

$$\frac{1}{n}\ln b < \ln(1 + \varepsilon),$$

so

$$n > \frac{\ln b}{\ln(1 + \varepsilon)}.$$

Now we can choose N to be any number that is larger than the number $\frac{\ln b}{\ln(1+\varepsilon)}$. According to the definition of limits, we now have

$$\lim_{n \to \infty} b^{\frac{1}{n}} = 1. \qquad \square$$

Although $\lim_{n \to \infty} f(n)$ is different from $\lim_{x \to \infty} f(x)$, the two limits must have some sort of connection. We notice that the difference between $x \to \infty$ and $n \to \infty$ is that x takes the value of every positive real number, but n only takes the values of all positive integers. The set of positive integers is a subset of the set of all positive real numbers. Hence, $\lim_{n \to \infty} f(n)$ must inherit something from $\lim_{x \to \infty} f(x)$. For example, if $\lim_{x \to \infty} f(x) = L$, then $\lim_{n \to \infty} f(n)$ must also be L, as shown in Figure 2.3.1. A nice theorem connecting the limit of a sequence and the limit of a function is due to Heine.

Theorem 2.3.1 (Heine). $\lim_{x \to a} f(x) = L$ *if and only if* $\lim_{n \to \infty} f(x_n) = L$ *for every sequence* $\{x_n\}$ *that converges to* a.

Proof. See Section 2.7. $\qquad \square$

Heinrich Eduard Heine (1821–1881) was a German mathematician. Heine became known for results on special functions and in real analysis. In particular, he authored an important treatise on spherical harmonics and Legendre functions. He also investigated basic hypergeometric series. He introduced the Mehler–Heine formula. http://en.wikipedia.org/wiki/Eduard_Heine#Selected_Works

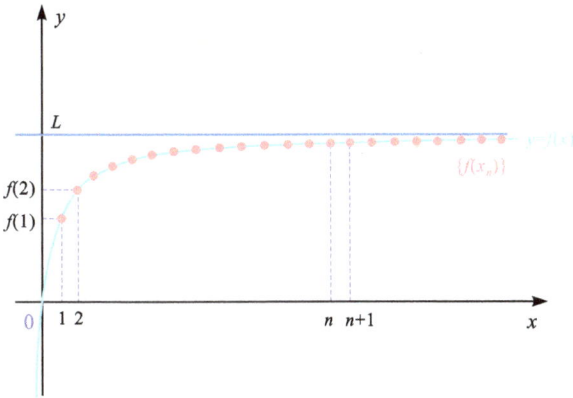

Figure 2.3.1: Limit of a sequence vs limit of a finction as $x \to \infty$.

This theorem says that $\lim_{x \to a} f(x) = L$, if and only if, for any sequence $\{x_n\}$ that converges to a, the corresponding sequence of function values $\{f(x_n)\}$ also converges to L. It may be hard to determine the existence of $\lim_{x \to a} f(x)$ by investigating all possible sequences $\{f(x_n)\}$ where $x_n \to a$. However, Heine's theorem enables us to determine the nonexistence of certain types of limits.

Example 2.3.3. Show that the Heaviside function

$$f(x) = \begin{cases} 0, & \text{for } x < 0 \\ 1, & \text{for } x > 0 \end{cases}$$

does not have a limit as $x \to 0$.

Solution. We choose two sequences, both of which converge to 0:

$\{x_n\} =$	$-1,$	$-\frac{1}{2},$	$-\frac{1}{3},$	$\cdots,$	$-\frac{1}{n},$	$\cdots$	(all terms are negative)
$\{x_n'\} =$	$1,$	$\frac{1}{2},$	$\frac{1}{3},$	$\cdots,$	$\frac{1}{n},$	$\cdots$	(all terms are positive)

Then we have

$$\lim_{n \to \infty} f(x_n) = \lim_{n \to \infty} f\left(\frac{-1}{n}\right) = 0 \quad \text{and} \quad \lim_{n \to \infty} f(x_n') = \lim_{n \to \infty} f\left(\frac{1}{n}\right) = 1.$$

Therefore, $\lim_{x \to 0} f(x)$ does not exist. Figure 2.3.2 illustrates this limit.

Example 2.3.4. Show that $\lim_{x \to 0} \sin \frac{1}{x}$ does not exist.

Proof. The function $f(x) = \sin(1/x)$ is undefined at 0, but the limit may, or may not, exist. Notice that $\sin \frac{1}{x}$ oscillates between -1 and 1 when x approaches 0, so we choose

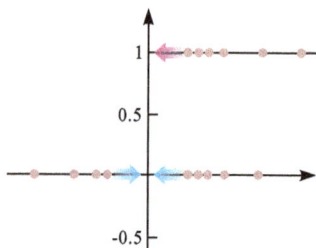

Figure 2.3.2: Use Heine's theorem to show a limit doesn't exist.

two sequences

$$\{x_n\} = \left\{\frac{1}{2n\pi}\right\} \quad \text{and} \quad \{x_n'\} = \left\{\frac{1}{2n\pi + \frac{\pi}{2}}\right\}.$$

Then we have

$$\lim_{n\to\infty} \sin\frac{1}{x_n} = \lim_{n\to\infty} \sin\frac{1}{\frac{1}{2n\pi}} = \lim_{n\to\infty} \sin 2n\pi = 0,$$

$$\lim_{n\to\infty} \sin\frac{1}{x_n'} = \lim_{n\to\infty} \sin\frac{1}{\frac{1}{2n\pi+\frac{\pi}{2}}} = \lim_{n\to\infty} \sin\left(2n\pi + \frac{1}{2}\pi\right) = 1.$$

Hence, by Theorem 2.3.1, we know that $\lim_{x\to0} \sin\frac{1}{x}$ does not exist. The graph of $y = f(x)$ is shown in Figure 2.3.3. $\qquad\square$

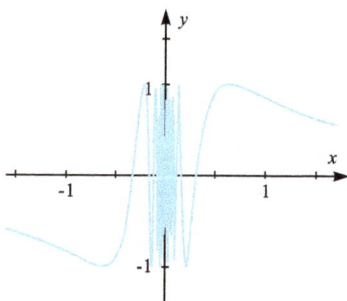

Figure 2.3.3: Graph of $y = \sin\frac{1}{x}$.

Example 2.3.5. Show that $\lim_{x\to\infty} \sin x$ does not exist.

Solution. The graph of the sine function oscillates infinitely many times between -1 and 1, as $x \to \infty$. Therefore, $\sin x$ does not tend to any single value, so $\lim_{x\to\infty} \sin x$ must not exist. We now confirm this analytically. We choose two sequences

$$\{x_n\} = \{2n\pi\} \quad \text{and} \quad \{x_n'\} = \left\{2n\pi + \frac{\pi}{2}\right\}.$$

Both sequences tend to infinity, but

$$\lim_{n\to\infty} \sin x_n = \lim_{n\to\infty} \sin 2n\pi = 0 \quad \text{and} \quad \lim_{n\to\infty} \sin x_n = \lim_{n\to\infty} \sin\left(2n\pi + \frac{\pi}{2}\right) = 1,$$

so $\lim_{x\to\infty} \sin x$ does not exist.

Some theorems on convergent sequences

Properties of convergent sequences are analogs of those for functions since a sequence can be regarded as a function. We state some useful theorems for convergent sequences. The proofs of them are similar to those for functions.

Theorem 2.3.2. *If the sequence $\{x_n\}$ is convergent, then it has a unique limit (there can only be one limit).*

Theorem 2.3.3. *If the sequence $\{x_n\}$ is convergent, then the sequence $\{x_n\}$ must be bounded.*

Theorem 2.3.4. *If $x_n > 0$ for sufficiently large n (there is a number N such that $x_n > 0$ for all $n > N$), then $\lim_{n\to\infty} x_n \geqslant 0$.*

NOTE. Even if each term of a sequence is not 0, the limit of the sequence can be 0. For example $\lim_{n\to\infty} \frac{1}{n} = 0$.

Monotone and bounded sequences theorem

Similar to functions, monotone sequences and bounded sequences are defined as follows.

Definition 2.3.2. A sequence $\{x_n\}$ is said to be:
1. *monotone increasing* (also called increasing) if $x_n \leqslant x_{n+1}$ for all n;
2. *monotone decreasing* (also called decreasing) if $x_n \geqslant x_{n+1}$ for all n;
3. *bounded above* if there is a number M, such that $x_n \leqslant M$ for all n;
4. *bounded below* if there is a number m, such that $x_n \geqslant m$ for all n;
5. *bounded* if it is bounded above and bounded below;
6. *monotone* if it is either increasing or decreasing.

For example, the sequence $\{\frac{1}{n}\}$, which is $1, \frac{1}{2}, \frac{1}{3}, \ldots, \frac{1}{n}, \ldots$, is monotone decreasing with an upper bound of 1 and a lower bound of 0. The sequence $\{\frac{n}{n+1}\}$, which is $\frac{1}{2}$, $\frac{2}{3}, \frac{3}{4} \ldots, \frac{n}{n+1}, \ldots$, is monotone increasing with an upper bound of 1 and a lower bound of $\frac{1}{2}$. Also, it is easy to see that, if an increasing sequence is bounded above, it must be bounded. Similarly, if a decreasing sequence is bounded below, then it must be bounded. The graph of the sequence $\{\frac{n}{n+1}\}$ is shown in Figure 2.3.4.

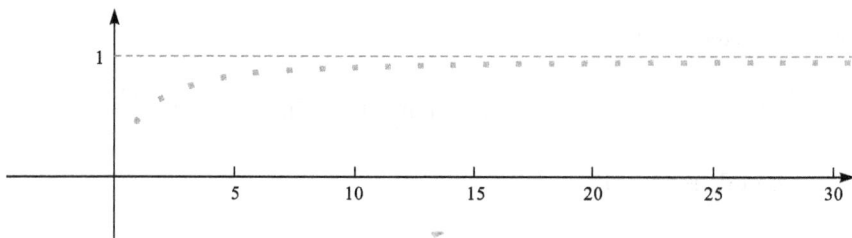

Figure 2.3.4: Graph of the sequence $\{\frac{n}{n+1}\}$.

There is a very important property of monotonic and bounded sequences that says they must converge. This is an existence theorem that tells us about a case where a limit exists, but tells us nothing about its value. Hence, the value of the limit must be found, or estimated, in some other way. The theorem can be explained in an obvious geometrical way. By the least upper bound property, if the sequence $\{x_n\}$ is increasing and has bound M, then there must be a least upper bound, β, such that $x_n \leqslant \beta$ for all n. The rising values of x_n cannot increase above β but eventually must become arbitrarily close to β (otherwise there would be a smaller least upper bound than β). Hence, β is the limit of the sequence. We now state a theorem about this.

Theorem 2.3.5. *A monotonic and bounded sequence must converge.*

Proof. See Section 2.7. □

Example 2.3.6. Show that the sequence $a_1 = \sqrt{2}$, $a_2 = \sqrt{2 + \sqrt{2}}$, $a_3 = \sqrt{2 + \sqrt{2 + \sqrt{2}}}$, ... converges. Then find $\lim_{n \to \infty} a_n$.

Proof. First, we notice that $a_n > 0$ and $a_n = \sqrt{2 + a_{n-1}}$. Now we prove by induction that the sequence $\{a_n\}$ is bounded above. For $n = 1$, $a_1 = \sqrt{2} < 2$. Now we assume $a_k < 2$ for $k \geqslant 2$. Then

$$a_{k+1} = \sqrt{2 + a_k} \leqslant \sqrt{2 + 2} = 2,$$

so by induction, we obtain

$$a_n < 2 \quad \text{for all integers } n \geqslant 1.$$

For $n \geqslant 2$ we have

$$\frac{a_n}{a_{n-1}} = \frac{\sqrt{2 + a_{n-1}}}{a_{n-1}} = \sqrt{\frac{2}{a_{n-1}^2} + \frac{1}{a_{n-1}}} \geqslant \sqrt{\frac{2}{2^2} + \frac{1}{2}} = 1,$$

so $a_n \geqslant a_{n-1}$. This means the sequence $\{a_n\}$ is increasing.

Now we have proved that the sequence is increasing and bounded above, so it has a limit.

To find the limit, we suppose $\lim_{n\to\infty} a_n = L$. Then $\lim_{n\to\infty} a_{n-1} = L$, so it follows from

$$a_n = \sqrt{2 + a_{n-1}}$$

that $L = \sqrt{2 + L}$. This means $L^2 - L - 2 = 0$. Solving for L, we have $L = -1$ (rejected, because all terms are positive) and $L = 2$, so $\lim_{n\to\infty} a_n = 2$. □

The "e" limit

In the following, we show that $\lim_{n\to\infty}(1 + \frac{1}{n})^n$ has a very unexpected value. We first use Theorem 2.3.5 to prove the existence of the limit of the corresponding sequence $\{x_n\} = \{(1 + \frac{1}{n})^n\}$. We graph the sequence $\{x_n\}$ and see how it looks.

Figure 2.3.5 shows the graph of the sequence $\{x_n\}$, for $n = 1, 2, \ldots 16$.

Table 2.3.1 shows numerical values of the sequence $\{x_n\}$ for some integers n.

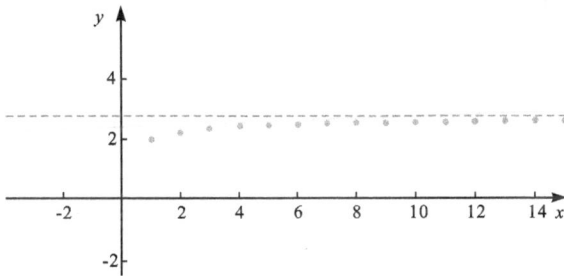

Figure 2.3.5: Graph of the sequence $\{(1 + \frac{1}{n})^n\}$.

Table 2.3.1: Select values of the sequence $\{(1 + \frac{1}{n})^n\}$.

n	$(1 + 1/n) \wedge n$
1	2
2	2.25
3	2.370 370 37
10	2.593 742 46
100	2.704 813 829
1000	2.716 923 932
10 000	2.718 145 927
100 000	2.718 268 237
1000 000	2.718 280 469
10 000 000	2.718 281 694
100 000 000	2.718 281 786

It seems, both from the graph and the table, that $\{(1+\frac{1}{n})^n\}$ is monotonic increasing and bounded above, so $\lim_{n\to\infty}(1+\frac{1}{n})^n$ should exist. Now we prove this by using the binomial expansion and by induction.

Theorem 2.3.6. *The limit of the sequence* $(1+\frac{1}{n})^n$ *as* $n \to \infty$ *exists.*

Proof. By the binomial expansion, we have

$$
\begin{aligned}
x_n &= \left(1+\frac{1}{n}\right)^n \\
&= 1 + n\cdot\frac{1}{n} + \frac{n(n-1)}{2!}\frac{1}{n^2} + \frac{n(n-1)(n-2)}{3!}\frac{1}{n^3} + \cdots \\
&\quad + \frac{n(n-1)(n-2)\cdots(n-(n-1))}{n!}\frac{1}{n^n} \\
&= 1 + 1 + \frac{1}{2!}\left(1-\frac{1}{n}\right) + \frac{1}{3!}\left(1-\frac{1}{n}\right)\left(1-\frac{2}{n}\right) + \\
&\quad \cdots + \frac{1}{n!}\left(1-\frac{1}{n}\right)\left(1-\frac{2}{n}\right)\cdots\left(1-\frac{n-1}{n}\right) \\
&\leqslant 1 + 1 + \frac{1}{2!}\left(1-\frac{1}{n+1}\right) + \frac{1}{3!}\left(1-\frac{1}{n+1}\right)\left(1-\frac{2}{n+1}\right) + \cdots \\
&\quad + \frac{1}{n!}\left(1-\frac{1}{n+1}\right)\left(1-\frac{2}{n+1}\right)\cdots\left(1-\frac{n-1}{n+1}\right) \\
&\quad + \frac{1}{(n+1)!}\left(1-\frac{1}{n+1}\right)\left(1-\frac{2}{n+1}\right)\cdots\left(1-\frac{n}{n+1}\right) \\
&= x_{n+1}.
\end{aligned}
$$

This shows that $\{x_n\}$ is a monotonic increasing sequence. Furthermore, it is bounded because

$$
\begin{aligned}
x_n &= 1 + 1 + \frac{1}{2!}\left(1-\frac{1}{n}\right) + \frac{1}{3!}\left(1-\frac{1}{n}\right)\left(1-\frac{2}{n}\right) + \cdots \\
&\quad + \frac{1}{n!}\left(1-\frac{1}{n}\right)\left(1-\frac{2}{n}\right)\cdots\left(1-\frac{n-1}{n}\right) \\
&\leqslant 1 + 1 + \frac{1}{2!} + \frac{1}{3!} + \cdots \frac{1}{n!} \\
&\leqslant 1 + 1 + \frac{1}{2} + \frac{1}{2^2} + \cdots + \frac{1}{2^{n-1}} \\
&= 1 + \frac{1-(\frac{1}{2})^n}{1-\frac{1}{2}} < 3.
\end{aligned}
$$

Hence, $0 < x_n < 3$ for all n, so $\{x_n\}$ is a monotonic and bounded sequence. By Theorem 2.3.5, it has a limit. □

NOTE. This limit, of course, is unique and is denoted by the letter e, thus

$$
\lim_{n\to\infty}\left(1+\frac{1}{n}\right)^n = e.
$$

As shown in the previous table, using a large value of n, we estimate the value of $e \approx$ 2.718 28, correct to five decimal places.

Leonhard Euler
(1707–1783) was a pioneering Swiss mathematician and physicist. He is considered to be the preeminent mathematician of the eighteenth century and one of the greatest mathematicians to have ever lived. He made many important discoveries in mathematics and physics. He also introduced much of the modern mathematical terminology and notation. A statement attributed to Pierre-Simon Laplace expresses Euler's influence on mathematics: "Read Euler, read Euler, he is the master of us all." http://en.wikipedia.org/wiki/Leonhard_Euler

In 1683, *Jacob Bernoulli* looked at the problem of compound interest and, in examining continuous compound interest, he tried to find the limit of $(1 + \frac{1}{n})^n$ as n tends to infinity. It is believed that *Euler* proved that e is an irrational number and introduced e as the base of the natural logarithm. *Euler* gave an approximation for e to 18 decimal places $e \approx 2.718\,281\,828\,459\,045\,235$ (O'Connor, J. J.; Robertson, E. F. "*The number e*". MacTutor History of Mathematics).

The number e is the base of the natural logarithm, $\ln x$, and it has a very special role in calculus, as we will see later.

There is a similar result for a monotone function defined on some left neighborhood of a point $x = a$, as stated without proof in the following theorem.

Theorem 2.3.7. *If the function $f(x)$ is defined on an interval (c, a) and it is monotone and bounded on (c, a), then $\lim_{x \to a^-} f(x)$ exists.*

2.3.2 Subsequences

A subsequence carries much of the information of its mother sequence. We first give the definition of a subsequence.

Definition 2.3.3. Let $\{x_n\}$ be a sequence. A *subsequence* $\{x_{n_k}\}$ is a sequence such that $x_{n_k} \in \{x_n\}$ and $n_k < n_{k+1}$ for all $k \in \mathbf{N}$.

This definition tells us that terms of a subsequence must retain the order of the terms in the parent sequence. For example, both
$$1, \frac{1}{3}, \frac{1}{5}, \dots, \frac{1}{2n+1}, \dots$$
and
$$1, \frac{1}{4}, \frac{1}{8}, \dots, \frac{1}{2^n}, \dots$$
are subsequences of the sequence $\{\frac{1}{n}\}$, which is $1, \frac{1}{2}, \frac{1}{3}, \dots, \frac{1}{n}, \dots$.

If a sequence $\{x_n\}$ approaches L as $n \to \infty$, then any subsequence of $\{x_{n_k}\}$ must have the same tendency since all terms in the subsequence are also terms of $\{x_n\}$ and the order of the terms is also retained as in $\{x_n\}$.

NOTE. The subscript n_k implies that $n_k \geq k$.

Theorem 2.3.8. *If $\{x_n\}$ converges to L, then any subsequence of $\{x_n\}$ also converges to L.*

Proof. If $\lim_{n\to\infty} x_n = L$ and $\{x_{n_k}\}$ is a subsequence of $\{x_n\}$, then, given a number $\varepsilon > 0$, there is a number, say, M, such that

$$|x_n - L| < \varepsilon \quad \text{whenever } n > M.$$

If we choose N to be any number greater than M, then, for every $k > N$, we have $n_k > N > M$ and

$$|x_{n_k} - L| < \varepsilon.$$

Thus, $\lim_{k\to\infty} x_{n_k} = L$. □

A *peak term* of a sequence is a term that is larger than all subsequent terms. That is, if x_m is a peak term of a sequence $\{x_n\}$, then, whenever $n > m$, we have $x_n < x_m$.

We now state the *Bolzano–Weierstrass theorem*, which is a fundamental theorem in analysis.

Theorem 2.3.9. *A bounded sequence $\{x_n\}$ of infinitely many real numbers must have a convergent subsequence.*

Proof. See Section 2.7. □

Karl Weierstrass (1815–1897) was a German mathematician cited as the "father of modern analysis". Despite leaving university without a degree, he studied mathematics and trained as a teacher, eventually teaching mathematics, physics, botany, and gymnastics. http://en.wikipedia.org/wiki/Karl_Weierstrass

2.4 Squeeze theorem and Cauchy's theorem

First we investigate an interesting limit. We have

$$\lim_{x\to 0}\left(x^2 \sin \frac{1}{x}\right).$$

What makes this limit interesting is that we cannot use the product limit laws which state that the limit of a product is the product of the limits, i.e.,

$$\lim_{x\to a}(f(x)g(x)) = \left(\lim_{x\to a} f(x)\right)\left(\lim_{x\to a} g(x)\right),$$

provided the limits exist.

Even though $\lim_{x \to 0} x^2 = 0$, $\lim_{x \to 0} \sin \frac{1}{x}$ does not exist, so

$$\lim_{x \to 0} x^2 \times \lim_{x \to 0} \sin \frac{1}{x} \text{ is not defined.}$$

Although you might think 0 times anything is 0, this is not always true. For example, if you evaluate

$$\lim_{x \to 0} x^2 \left(\frac{1}{x^4} \right) = \left(\lim_{x \to 0} x^2 \right) \times \left(\lim_{x \to 0} \frac{1}{x^4} \right),$$

then you have the form $0 \times$ something. It is obvious that the limit is not 0; instead, it is ∞.

However, notice that

$$-1 \leqslant \sin \frac{1}{x} \leqslant 1 \quad \text{for all } x \neq 0.$$

Hence, we have

$$-x^2 \leqslant x^2 \sin \frac{1}{x} \leqslant x^2 \quad \text{for all } x \neq 0.$$

When we let $x \to 0$, both $-x^2$ and x^2 approach 0 and therefore all values between them also approach 0, as shown in Figure 2.4.1.

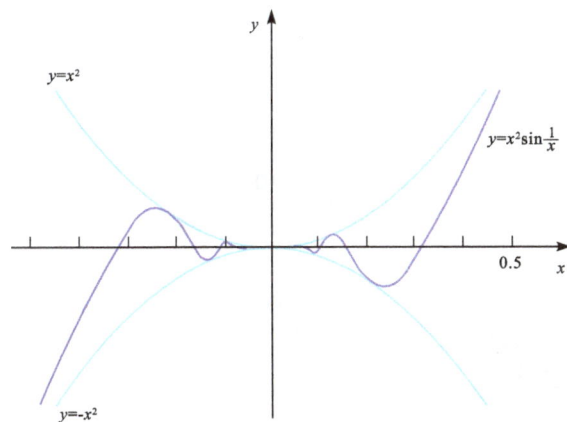

Figure 2.4.1: Graphs of $y = x^2$, $y = -x^2$ and $y = x^2 \sin \frac{1}{x}$.

The way we solved the above example by "sandwiching" or "squeezing" $x^2 \sin(1/x)$ between two other functions with known equal limits suggests that we can write down a principle about a general "sandwiching process". It is called the *sandwich theorem* or the *squeeze theorem*.

Theorem 2.4.1. *Suppose that $f(x) \leqslant g(x) \leqslant h(x)$ for all x close to a, except possibly for $x = a$. If $\lim_{x \to a} f(x) = \lim_{x \to a} h(x) = L$, then $\lim_{x \to a} g(x) = L$.*

Proof. See Section 2.7. $\qquad\qquad\qquad\qquad\qquad\qquad\qquad\qquad\qquad\qquad\qquad\square$

The squeeze theorem, of course, also works for limits of sequences which are special functions.

Example 2.4.1. Find $\lim_{n \to \infty} \frac{n}{2^n}$.

Solution. We have

$$0 \leqslant \frac{n}{2^n} = \frac{n}{(1+1)^n} = \frac{n}{1 + n + \frac{n(n-1)}{2} + \cdots} < \frac{n}{\frac{n(n-1)}{2}}$$

and

$$\lim_{n \to \infty} \frac{n}{\frac{n(n-1)}{2}} = \lim_{n \to \infty} \frac{2}{n-1} = 0,$$

so $\lim_{n \to \infty} \frac{n}{2^n} = 0$.

Example 2.4.2. Find $\lim_{n \to \infty} \frac{5^n}{n!}$.

Solution. We have

$$0 \leqslant \frac{5^n}{n!} = \frac{5 \times 5 \times 5 \times 5 \times 5}{1 \times 2 \times 3 \times 4 \times 5} \times \frac{5^{n-5}}{6 \times 7 \times 8 \times \cdots \times n} \leqslant \frac{5^5}{5!} \times \left(\frac{5}{6}\right)^{n-5}$$

and

$$\lim_{n \to \infty} \frac{5^5}{5!} \times \left(\frac{5}{6}\right)^{n-5} = \frac{5^5}{5!} \times \lim_{n \to \infty} \left(\frac{5}{6}\right)^{n-5} = 0,$$

so $\lim_{n \to \infty} \frac{5^n}{n!} = 0$.

Example 2.4.3. Find $\lim_{n \to \infty} \sqrt[n]{2^n + 5^n}$.

Solution. We have

$$5 = \sqrt[n]{5^n} \leqslant \sqrt[n]{2^n + 5^n} \leqslant \sqrt[n]{5^n + 5^n} \leqslant 5 \times 2^{\frac{1}{n}}.$$

Then

$$5 \leqslant \lim_{n \to \infty} \sqrt[n]{2^n + 5^n} \leqslant \lim_{n \to \infty} 5 \times 2^{\frac{1}{n}} = 5 \times \lim_{n \to \infty} 2^{\frac{1}{n}} = 5,$$

so $\lim_{n \to \infty} \sqrt[n]{2^n + 5^n} = 5$.

Example 2.4.4. Show that $\lim_{x \to a} f(x) = 0$ if and only if $\lim_{x \to a} |f(x)| = 0$.

Proof. "$\Longrightarrow$" If $\lim_{x \to a} f(x) = 0$, then, given any number $\varepsilon > 0$, we have $\delta > 0$ such that

$$|f(x) - 0| < \varepsilon \quad \text{whenever } 0 < |x - a| < \delta.$$

However, this also means $||f(x)| - 0| < \varepsilon$ whenever $0 < |x - a| < \delta$, so $\lim_{x \to a} |f(x)| = 0$.
"$\Longleftarrow$" Assume $\lim_{x \to a} |f(x)| = 0$. Then, by the inequalities

$$-|f(x)| \leq f(x) \leq |f(x)|$$

and the squeeze theorem, we have $\lim_{x \to a} f(x) = 0$. $\qquad \square$

Example 2.4.5. Show that $\lim_{n \to \infty} b^{\frac{1}{n}} = 1$ for $0 < b \leq 1$.

Proof. If $b = 1$, the statement is true. Now suppose $0 < b < 1$. Let $b = \frac{1}{a}$. Then $a > 1$ and

$$0 \leq |b^{\frac{1}{n}} - 1| = \left| \left(\frac{1}{a} \right)^{\frac{1}{n}} - 1 \right| = \left| \frac{1}{a^{\frac{1}{n}}} - 1 \right|$$

$$= \frac{|a^{\frac{1}{n}} - 1|}{a^{\frac{1}{n}}} \leq |a^{\frac{1}{n}} - 1| = a^{\frac{1}{n}} - 1.$$

Moreover,

$$\lim_{n \to \infty} 0 = 0 \quad \text{and} \quad \lim_{n \to \infty} (a^{\frac{1}{n}} - 1) = 0 \quad \text{by Example 2.3.2.}$$

So, by the Squeeze theorem, $\lim_{n \to \infty} (b^{\frac{1}{n}} - 1) = 0$. This means $\lim_{n \to \infty} b^{\frac{1}{n}} = 1$. $\qquad \square$

NOTE. Together with Example 2.3.2, we have $\lim_{n \to \infty} b^{\frac{1}{n}} = 1$ for any constant $b > 0$.

Example 2.4.6. Find $\lim_{x \to 0} \frac{\sin x}{x}$ (the function $\sin x$ requires x to be computed in radian angle measure, not degree angle measure).

Solution. Computing numerical data for small values of x suggests that $\lim_{x \to 0} (\sin x)/x = 1$, as does the graph of $(\sin x)/x$. To verify this by using the squeeze theorem we first need to find bounding functions.

Consider the unit circle with center $O(0,0)$ and radius 1. Let S be the point $(1,0)$ and label P on the upper half of the circle so that $\angle POS$ has (radian) measure x. Then P is the point $(\cos x, \sin x)$. From point P, draw a line perpendicular to the x-axis intersecting the axis at point $R(\cos x, 0)$. Let the line perpendicular to the x-axis at point S meet OP at point $Q(1, \tan x)$.

From Figure 2.4.2, we know that

$$\text{area}(\triangle ORP) \leq \text{area}(\text{sector } OSP) \leq \text{area}(\triangle OSQ).$$

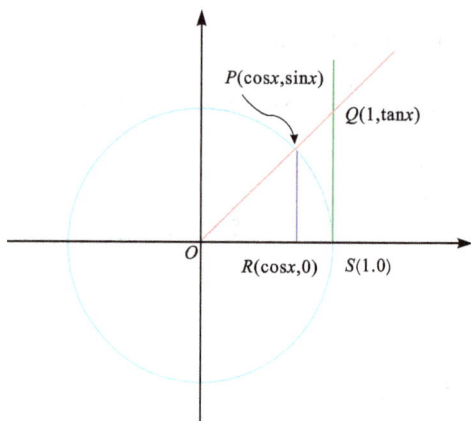

Figure 2.4.2: Diagram for Example 2.4.6.

By the formulas for area of a triangle and area of a sector, these inequalities become

$$\frac{1}{2}\cos x \sin x \leqslant \frac{1}{2}x \leqslant \frac{1}{2}\tan x.$$

Since x is a small positive number, $\sin x > 0$ and $\cos x > 0$. Multiplying the inequalities by $2/\sin x$ gives

$$0 < \cos x \leqslant \frac{x}{\sin x} \leqslant \frac{1}{\cos x}.$$

Taking reciprocals of the expression (thus reversing the direction of the inequalities) gives the inequalities

$$\cos x \leqslant \frac{\sin x}{x} \leqslant \frac{1}{\cos x}.$$

We know from Example 2.2.9 that

$$\lim_{x\to 0}\cos x = \cos 0 = 1 \quad \text{and} \quad \lim_{x\to 0}\frac{1}{\cos x} = \frac{1}{\lim_{x\to 0}\cos x} = 1.$$

Hence, by the squeeze theorem, we find the previous guess for the limit was correct:

$$\lim_{x\to 0}\frac{\sin x}{x} = 1.$$

Example 2.4.7. Find $\lim_{x\to 0}\frac{\tan x}{x}$.

Solution. We have

$$\lim_{x\to 0}\frac{\tan x}{x} = \lim_{x\to 0}\frac{\sin x}{x}\frac{1}{\cos x} = \lim_{x\to 0}\frac{\sin x}{x}\lim_{x\to 0}\frac{1}{\cos x} = 1 \times 1 = 1.$$

Example 2.4.8. Find $\lim_{x\to 0} \frac{1-\cos x}{x^2}$.

Solution. We have

$$
\lim_{x\to 0} \frac{1-\cos x}{x^2} = \lim_{x\to 0} \frac{2\sin^2 \frac{x}{2}}{x^2}
$$
$$
= \frac{1}{2}\left(\lim_{x\to 0} \frac{\sin\frac{x}{2}}{\left(\frac{x}{2}\right)}\right)^2
$$
$$
= \frac{1}{2}\left(\lim_{t\to 0} \frac{\sin t}{t}\right)^2
$$
$$
= \frac{1}{2}\cdot 1^2
$$
$$
= \frac{1}{2}.
$$

Example 2.4.9. Use the squeeze theorem to show that

$$
\lim_{x\to\infty}\left(1+\frac{1}{x}\right)^x = \lim_{x\to-\infty}\left(1+\frac{1}{x}\right)^x = e.
$$

Proof. For any positive number $x > 0$, there exists a unique integer n such that

$$
n \leqslant x < n+1.
$$

Therefore,

$$
\left(1+\frac{1}{n+1}\right)^n < \left(1+\frac{1}{n+1}\right)^x < \left(1+\frac{1}{x}\right)^x \leqslant \left(1+\frac{1}{n}\right)^{n+1}.
$$

Moreover,

$$
\lim_{n\to+\infty}\left(1+\frac{1}{n+1}\right)^n = \lim_{n\to+\infty}\left(1+\frac{1}{n+1}\right)^{n+1}\left(1+\frac{1}{n+1}\right)^{-1} = e
$$

and

$$
\lim_{n\to+\infty}\left(1+\frac{1}{n}\right)^{n+1} = \lim_{n\to+\infty}\left(1+\frac{1}{n}\right)^n\left(1+\frac{1}{n}\right)
$$
$$
= \lim_{n\to+\infty}\left(1+\frac{1}{n}\right)^n \lim_{n\to+\infty}\left(1+\frac{1}{n}\right) = e \times 1 = e.
$$

By the squeeze theorem, we obtain

$$
\lim_{x\to+\infty}\left(1+\frac{1}{x}\right)^x = \lim_{n\to+\infty}\left(1+\frac{1}{n}\right)^n = e.
$$

If x tends to negative infinity, let $x = -(t+1)$ and then $t \to +\infty$. Then

$$
\lim_{x\to-\infty}\left(1+\frac{1}{x}\right)^x = \lim_{t\to+\infty}\left(1-\frac{1}{t+1}\right)^{-(t+1)} = \lim_{t\to+\infty}\left(\frac{t+1-1}{t+1}\right)^{-(t+1)}
$$

$$= \lim_{t\to+\infty}\left(\frac{t}{t+1}\right)^{-(t+1)} = \lim_{t\to+\infty}\left(\frac{t+1}{t}\right)^{t+1}$$

$$= \lim_{t\to+\infty}\left(1+\frac{1}{t}\right)^{t+1} = \lim_{t\to+\infty}\left(1+\frac{1}{t}\right)^{t}\lim_{t\to+\infty}\left(1+\frac{1}{t}\right)$$

$$= e\times 1 = e.$$

Therefore, for $x\to-\infty$, we also have

$$\lim_{x\to-\infty}\left(1+\frac{1}{x}\right)^{x} = e.$$

Figure 2.4.3 shows the graph of the function $(1+\frac{1}{x})^{x}$. □

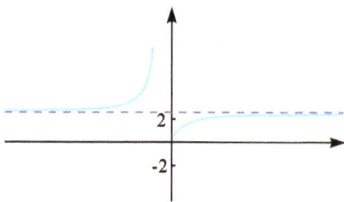

Figure 2.4.3: Graph of $y = (1+\frac{1}{x})^{x}$.

Example 2.4.10. Find $\lim_{x\to\infty}(1+\frac{2}{x})^{x}$.

Solution. We have

$$\lim_{x\to\infty}\left(1+\frac{2}{x}\right)^{x} = \lim_{x\to\infty}\left[\left(1+\frac{1}{\frac{x}{2}}\right)^{\frac{x}{2}}\right]^{2}.$$

Now, let $t = \frac{x}{2}$. Then $t\to\infty$ as $x\to\infty$, so

$$\lim_{x\to\infty}\left(1+\frac{2}{x}\right)^{x} = \lim_{x\to\infty}\left[\left(1+\frac{1}{\frac{x}{2}}\right)^{\frac{x}{2}}\right]^{2} = \lim_{t\to\infty}\left[\left(1+\frac{1}{t}\right)^{t}\right]^{2} = e^{2}.$$

In order to use our formal definition of limit, it is necessary for us to know the value L of the limit. For any particular value L, we can use the limit definition to test whether or not L is the limit of f as $x\to a$, but using the definition to show that L is not the limit of f does not tell us whether or not a limit actually exists as $x\to a$. A useful criterion for testing the existence of limits of functions is given by the next theorem.

Theorem 2.4.2 (Cauchy's theorem). *The limit* $\lim_{x\to a}f(x)$ *exists if and only if, for any given* $\varepsilon > 0$, *there exists a number* $\delta > 0$ *(that depends on* ε*) such that, for any* x' *and* x'' *satisfying* $0 < |x' - a| < \delta$ *and* $0 < |x'' - a| < \delta$, *we have*

$$|f(x') - f(x'')| < \varepsilon.$$

Augustin-Louis Cauchy (1789–1857) was a French mathematician who is reputed to be an early pioneer of analysis. He is also a profound and prolific mathematician who exercised a great influence over his contemporaries and successors. His writings cover the entire range of mathematics and mathematical physics. http://en.wikipedia.org/wiki/Augustin-Louis_Cauchy

Proof. "$\Longrightarrow$" If $\lim_{x\to a} f(x)$ exists, then there is a finite number L such that, for any given number $\varepsilon > 0$, there is a number $\delta > 0$ (corresponding to the number $\varepsilon/2$) such that

$$|f(x) - L| < \frac{\varepsilon}{2} \quad \text{whenever } 0 < |x - a| < \delta.$$

Choose any two numbers x' and x'' such that $0 < |x' - a| < \delta$ and $0 < |x'' - a| < \delta$. Then

$$\begin{aligned}
|f(x') - f(x'')| &= |f(x') - L - (f(x'') - L)| \\
&\leqslant |f(x') - L| + |f(x'') - L| \\
&< \frac{\varepsilon}{2} + \frac{\varepsilon}{2} = \varepsilon.
\end{aligned}$$

"$\Longleftarrow$" Not given here. $\qquad\qquad\qquad\qquad\qquad\qquad\qquad\qquad\qquad\qquad\qquad$ □

Theorem 2.4.3 (Sequence form of Cauchy's theorem). *The limit* $\lim_{n\to\infty} x_n$ *exists if and only if, for any given* $\varepsilon > 0$, *there exists an integer* $N > 0$ *(that depends on* ε*) such that, for any* n *and* m *satisfying* $n > N$ *and* $m > N$, *we have*

$$|x_n - x_m| < \varepsilon.$$

Example 2.4.11. Show that $\lim_{n\to\infty} s_n$ does not exist, where

$$s_n = 1 + \frac{1}{2} + \frac{1}{3} + \cdots + \frac{1}{n}.$$

Solution. If the limit exists, then, for the number $\varepsilon = \frac{1}{2}$, there is an integer $N > 0$ such that

$$|s_n - s_m| < \frac{1}{2} \quad \text{for all } n > N \text{ and } m > N.$$

If we choose $n = 4N$ and $m = 2N$, then

$$\begin{aligned}
|s_n - s_m| &= \left| 1 + \frac{1}{2} + \cdots + \frac{1}{4N} - \left(1 + \frac{1}{2} + \frac{1}{3} + \cdots + \frac{1}{2N} \right) \right| \\
&= \left| \frac{1}{2N+1} + \frac{1}{2N+2} + \cdots + \frac{1}{4N} \right| \\
&= \frac{1}{2N+1} + \frac{1}{2N+2} + \cdots + \frac{1}{4N} \\
&\geqslant \frac{1}{4N} + \frac{1}{4N} + \cdots + \frac{1}{4N} \\
&= \frac{4N}{4N} \quad \frac{(2N+1)+1}{ } = \frac{1}{2}.
\end{aligned}$$

This is a contradiction, so $\lim_{n\to\infty} s_n$ does not exist.

2.5 Infinitesimal functions and asymptotic functions

One of the great ideas derived from calculus is the one called "infinitesimal". Calculus was originally called the infinitesimal analysis. We first give the definition of an infinitesimal function.

In the following definitions, the letter a could be a finite real number or $\pm\infty$. The connection between infinitesimal functions and limits is given by the following theorem.

Definition 2.5.1. If $\lim_{x \to a} \alpha(x) = 0$, then we say $\alpha(x)$ is infinitesimal as $x \to a$.

Theorem 2.5.1. $\lim_{x \to a} f(x) = L$ *if and only if the difference* $\alpha(x) = f(x) - L$ *is infinitesimal as* $x \to a$.

Proof. This is because

$$\lim_{x \to a} \alpha(x) = 0 \quad \Longleftrightarrow \quad \lim_{x \to a}(f(x) - L) = 0$$

$$\Longleftrightarrow \quad \lim_{x \to a} f(x) - \lim_{x \to a} L = 0 \quad \Longleftrightarrow \quad \lim_{x \to a} f(x) = L. \qquad \square$$

Example 2.5.1. Since $\lim_{x \to \infty} \frac{2x}{x+7} = 2$, $\frac{2x}{x+7} - 2 = \frac{-14}{x+7} \to 0$, as $x \to \infty$.

Corollary 2.5.2. *If* $\lim_{x \to a} \frac{f(x)}{g(x)} = L$, *then* $f(x) = L \cdot g(x) + \alpha \cdot g(x)$, *where* $\alpha \to 0$ *as* $x \to a$.

Proof. This follows directly from Theorem 2.5.1. $\qquad \square$

Definition 2.5.2. We say that two functions $\alpha(x)$ and $\beta(x)$ are *asymptotic* as $x \to a$ if $\lim_{x \to a} \frac{\beta(x)}{\alpha(x)} = 1$, and we write $\beta(x) \sim \alpha(x)$, as $x \to a$.

From Examples 2.4.6, 2.4.7, and 2.4.8, we know that

$$\text{as } x \to 0, \quad \sin x \sim x, \quad \tan x \sim x, \quad 1 - \cos x \sim \frac{x^2}{2}.$$

Figure 2.5.1, Figure 2.5.2 and Figure 2.5.3 show graphs of these functions.

Example 2.5.2. Show that $\arcsin x \sim x$ as $x \to 0$.

Proof. By the definition of inverse functions, we have

$$x = \sin(\arcsin x) = x \quad \text{for } -\frac{\pi}{2} \leqslant x \leqslant \frac{\pi}{2}.$$

Then from above, $\sin x \sim x$ as $x \to 0$, which means

$$x = \sin(\arcsin x) \sim \arcsin x \quad \text{as } x \to 0.$$

Figure 2.5.4 shows this pair of asymptotic functions. $\qquad \square$

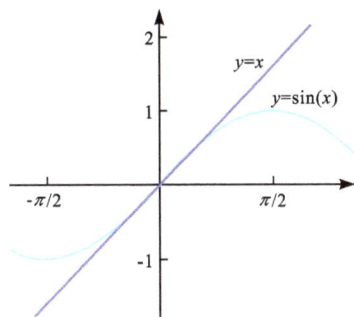

Figure 2.5.1: Graphs of $y = x$ and $y = \sin x$.

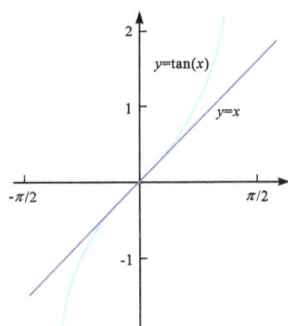

Figure 2.5.2: Graphs of $y = x$ and $y = \tan x$.

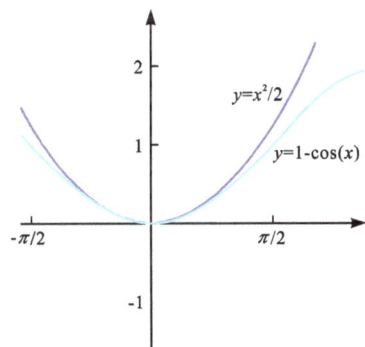

Figure 2.5.3: Graphs of $y = \frac{x^2}{2}$ and $y = 1 - \cos x$.

Example 2.5.3. Prove that $\sqrt[n]{1 + x} - 1 \sim \frac{1}{n} x$ as $x \to 0$.

Solution. Recall the factorization

$$(a^n - b^n)$$
$$= (a - b)(a^{n-1} + a^{n-2}b + a^{n-3}b^2 + \cdots + a^2 b^{n-3} + ab^{n-2} + b^{n-1}).$$

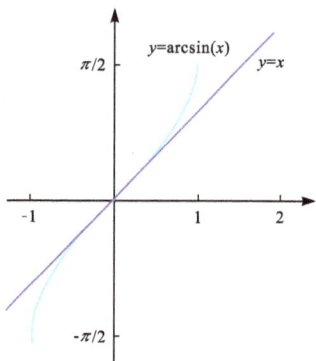

Figure 2.5.4: Graphs of $y = x$ and $y = \arcsin x$.

Let $a = \sqrt[n]{1+x}$ and $b = 1$. Consider

$$\lim_{x \to 0} \frac{\sqrt[n]{1+x} - 1}{\frac{1}{n}x}$$

$$= \lim_{x \to 0} \frac{a - b}{\frac{1}{n}x}$$

$$= \lim_{x \to 0} \frac{(\sqrt[n]{1+x})^n - 1^n}{\frac{x}{n}[\sqrt[n]{(1+x)^{n-1}} + \sqrt[n]{(1+x)^{n-2}} + \cdots + \sqrt[n]{(1+x)} + 1]}$$

$$= \lim_{x \to 0} \frac{1 + x - 1}{\frac{x}{n}[\sqrt[n]{(1+x)^{n-1}} + \sqrt[n]{(1+x)^{n-2}} + \cdots + \sqrt[n]{(1+x)} + 1]}$$

$$= \lim_{x \to 0} \frac{1}{\frac{1}{n}[\sqrt[n]{(1+x)^{n-1}} + \sqrt[n]{(1+x)^{n-2}} + \cdots + \sqrt[n]{(1+x)} + 1]}$$

$$= \frac{1}{\frac{1}{n}(1 + 1 + \cdots + 1)} = \frac{1}{\frac{n}{n}} = 1.$$

As a result, we show the graphs of $y = \frac{x}{3}$ and $y = \sqrt[3]{1+x} - 1$ in Figure 2.5.5.

Example 2.5.4. Show that $x^3 + 7x^2 - x + 10 \sim x^3$ as $x \to \pm\infty$.

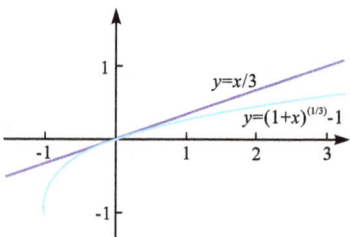

Figure 2.5.5: Graphs of $y = \sqrt[3]{1+x} - 1$ and $y = \frac{x}{3}$.

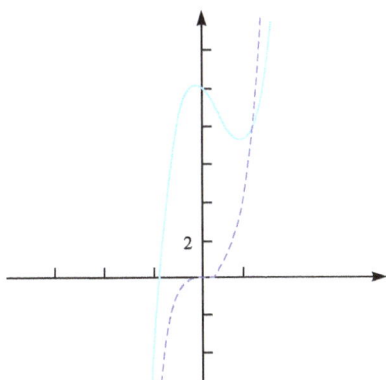

Figure 2.5.6: Graphs of $y = x^3$ and $y = x^3 - 2x^2 - x + 10$ in a small viewing window.

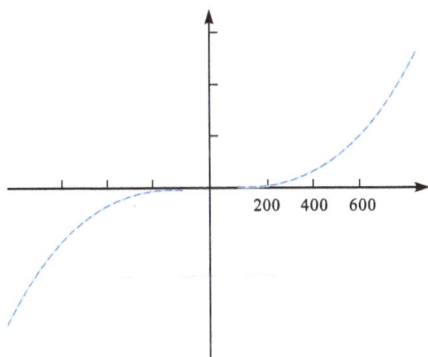

Figure 2.5.7: Graphs of $y = x^3$ and $y = x^3 - 2x^2 - x + 10$ in a large viewing window.

Proof.

$$\lim_{x \to \pm\infty} \frac{x^3 - 2x^2 - x + 10}{x^3} = \lim_{x \to \pm\infty} \left(1 - \frac{2}{x} - \frac{1}{x^2} + \frac{10}{x^3}\right) = 1.$$

Figure 2.5.6 and Figure 2.5.7 show the graphs of the two functions in different viewing windows. □

NOTE. Sometimes $a_n x^n$ is called the dominant term of the polynomial $P(x) = a_0 + a_1 x + \cdots + a_n x^n$, $a_n \neq 0$, because $P(x) \sim a_n x^n$ as $x \to \pm\infty$.

Using asymptotic functions to evaluate limits

When evaluating limits, we can replace the numerator, the denominator, or both of a quotient by corresponding asymptotic functions, provided that the limits exist.

Theorem 2.5.3. *If $\alpha(x) \sim \tilde{\alpha}(x)$ and $\beta(x) \sim \tilde{\beta}(x)$ as $x \to a$ and all of them are nonzero, then*

$$\lim_{x \to a} \frac{\alpha(x)}{\beta(x)} = \lim_{x \to a} \frac{\tilde{\alpha}(x)}{\beta(x)} = \lim_{x \to a} \frac{\alpha(x)}{\tilde{\beta}(x)} = \lim_{x \to a} \frac{\tilde{\alpha}(x)}{\tilde{\beta}(x)}.$$

Proof. For the first equality,

$$\lim_{x \to a} \frac{\alpha(x)}{\beta(x)} = \lim_{x \to a} \frac{\alpha(x)}{\beta(x)} \lim_{x \to a} \frac{\tilde{\alpha}(x)}{\alpha(x)} = \lim_{x \to a} \frac{\alpha(x)}{\beta(x)} \frac{\tilde{\alpha}(x)}{\alpha(x)} = \lim_{x \to a} \frac{\tilde{\alpha}(x)}{\beta(x)}.$$

Using similar ratios, one can deduce the remaining equalities. □

Example 2.5.5. Find $\lim_{x \to 0} \frac{1 - \cos x}{2x^2}$.

Solution. First note that $1 - \cos x \sim \frac{x^2}{2}$ as $x \to 0$. Then

$$\lim_{x \to 0} \frac{1 - \cos x}{2x^2} = \lim_{x \to 0} \frac{\frac{x^2}{2}}{2x^2} = \lim_{x \to 0} \frac{1}{4} = \frac{1}{4}.$$

Example 2.5.6. Find $\lim_{x \to 0} \frac{(1 + x^2)^{\frac{1}{3}} - 1}{\cos x - 1}$.

Solution. Observe that, as $x \to 0$, $1 - \cos x \sim \frac{1}{2} x^2$ and

$$(1 + x^2)^{\frac{1}{3}} - 1 \sim \frac{1}{3} x^2$$

by Example 2.5.3. Therefore,

$$\lim_{x \to 0} \frac{(1 + x^2)^{\frac{1}{3}} - 1}{\cos x - 1} = \lim_{x \to 0} \frac{\frac{1}{3} x^2}{-\frac{1}{2} x^2} = -\frac{2}{3}.$$

Example 2.5.7. Find $\lim_{x \to 0} \frac{\tan x - \sin x}{x^3}$.

Solution.

$$\lim_{x \to 0} \frac{\tan x - \sin x}{x^3} = \lim_{x \to 0} \frac{\frac{\sin x}{\cos x} - \sin x}{x^3} = \lim_{x \to 0} \frac{\sin x}{x^3} \left(\frac{1}{\cos x} - 1 \right)$$

$$= \lim_{x \to 0} \frac{\sin x}{x^3} \frac{1 - \cos x}{\cos x} = \lim_{x \to 0} \frac{x}{x^3} \frac{\frac{x^2}{2}}{\cos x}$$

$$= \frac{1}{2} \cdot \lim_{x \to 0} \frac{1}{\cos x} = \frac{1}{2}.$$

NOTE. This example shows that, if the function is a product of other functions, we can replace one or more of the "factors" by asymptotic functions, provided that all the limits involved exist.

Definition 2.5.3. If $\lim_{x \to a} \frac{f(x)}{g(x)} = 0$ as $x \to a$, then we say that $f(x)$ is a *negligible function* with respect to $g(x)$ as x tends to a. We denote this by $f(x) = o(g(x))$ (referred to as the "*small o notation*").

Intuitively speaking, if $f(x) = o(g(x))$, then $f(x)$ is much smaller than $g(x)$ as $x \to a$.

Example 2.5.8. Both $x^2 = o(x)$ and $x^2 - x^3 = o(2x - x^2)$ as x tends to 0, since

$$\lim_{x \to 0} \frac{x^2}{x} = \lim_{x \to 0} x = 0 \quad \text{and} \quad \lim_{x \to 0} \frac{x^2 - x^3}{2x - x^2} = \lim_{x \to 0} \frac{x - x^2}{2 - x} = \frac{0 - 0}{2 - 0} = 0.$$

Theorem 2.5.4. *As $x \to a$, $f(x) \sim g(x)$ if and only if $f(x) = g(x) + o(f(x))$.*

This theorem says that, if $f(x)$ and $g(x)$ are asymptotic functions as $x \to a$, the difference between $f(x)$ and $g(x)$ is a negligible function with respect to each of them. We leave the proof as an exercise.

Example 2.5.9. As $x \to 0$, $\sin x \sim x$, so we have $\sin x - x = o(x)$. Table 2.5.1 shows the values of $\sin x$ and the difference between x and $\sin x$ for some small values of x.

Table 2.5.1: Selected values of $\sin x$ and $x - \sin x$ for $x \approx 0$.

x	$\sin(x)$	$x - \sin(x)$
0.1	0.099 833 416 646 828	0.000 166 583 353 172
0.01	0.009 999 833 334 167	0.000 000 166 665 833
0.009	0.008 999 878 500 492	0.000 000 121 499 508
0.005	0.004 999 979 166 693	0.000 000 020 833 307
0.002	0.001 999 998 666 667	0.000 000 001 333 333
0.0001	0.000 100 000 000 000	0.000 000 000 000 000

2.6 Continuous and discontinuous functions

2.6.1 Continuity and discontinuity

The mathematical definition of a *continuous function* corresponds closely to the meaning of the word "continuity" in everyday language. We think of a continuous process as one that takes place gradually, without interruption or abrupt large changes. For example, the outside temperature changes continuously, the height of a child grows continuously, etc. The graph of a continuous function is a continuous curve with no holes, that you can draw without lifting your pencil from the page.

Figure 2.6.1 is the graph of a function defined on $[0, 4]$. It is not continuous at $x = 1$ and $x = 2$, because the curve jumps abruptly from one value to another at these places. The formal definition of continuity is given below.

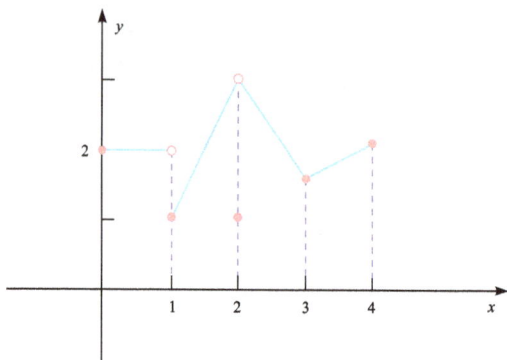

Figure 2.6.1: Continuity and discontinuity.

Definition 2.6.1. A function $y = f(x)$ is *continuous* at a point $x = a$, provided $\lim_{x \to a} f(x) = f(a)$.

NOTES. The definition is implicitly placing three requirements on the function f:
1. a is in the domain of $f(x)$, so that $f(a)$ is defined;
2. the function $f(x)$ is defined for x in a δ neighborhood of $x = a$ and $\lim_{x \to a} f(x)$ exists (as a finite real number);
3. $\lim_{x \to a} f(x) = f(a)$ (from this, we also have $\lim_{x \to a}(f(x) - f(a)) = 0$ or equivalently $\lim_{\Delta x \to 0} \Delta y = 0$).

In other words, f is continuous at a, if it is defined at a and $f(x) \to f(a)$ as $x \to a$. Thus, a continuous function f at $x = a$ has the property that, at $x = a$, small changes in x produce only small changes in $f(x)$. In fact, the change in $f(x)$ can be kept as small as we please by keeping the change in x sufficiently small, since $\lim_{\Delta x \to 0} \Delta y = 0$.

Figure 2.6.1 shows the graph of a function $f(x)$. This function f is not continuous at $x = 1$ since $\lim_{x \to 1} f(x)$ does not exist. This function is also not continuous at $x = 2$ since $\lim_{x \to 2} f(x)$ is not equal to $f(2)$.

If a function $f(x)$ is not continuous at $x = a$, we say that $f(x)$ is discontinuous at $x = a$ or $f(x)$ has a *discontinuity* at $x = a$. There are four types of common discontinuities. For example, in Figure 2.6.1, we say that f has jump discontinuity at $x = 1$ and f has a removable discontinuity at $x = 2$. Figure 2.6.2 shows some types of discontinuities. We now give a formal definition of these types of discontinuities.

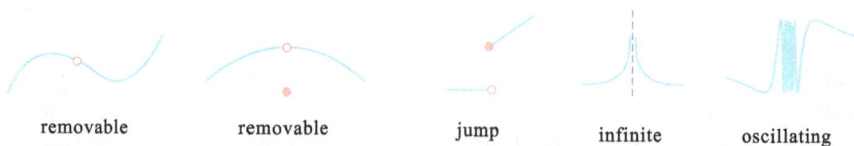

| removable | removable | jump | infinite | oscillating |

Figure 2.6.2: Some types of discontinuities.

Definition 2.6.2. If f is discontinuous at $x = a$, then at $x = a$:

1. f has a *removable discontinuity* if $\lim_{x \to a} f(x)$ exists but $f(a) \neq \lim_{x \to a} f(x)$;
2. f has a *jump discontinuity* if both $\lim_{x \to a^+} f(x)$ and $\lim_{x \to a^-} f(x)$ exist but $\lim_{x \to a^+} f(x) \neq \lim_{x \to a^-} f(x)$;
3. f has an *infinite discontinuity* if either $\lim_{x \to a^-} f(x)$ or $\lim_{x \to a^+} f(x)$ is ∞ or $-\infty$;
4. f has an *oscillating discontinuity* if $f(x)$ oscillates as $x \to a$.

Example 2.6.1. The function $f(x) = \frac{x-1}{x^2-1}$ has a discontinuity at $x = 1$. Write an extended function so that f is continuous at $x = 1$.

Solution. Since $f(x)$ is undefined at $x = 1$, it must be discontinuous at $x = 1$ and

$$\lim_{x \to 1} \frac{x-1}{x^2-1} = \lim_{x \to 1} \frac{x-1}{(x-1)(x+1)} = \lim_{x \to 1} \frac{1}{x+1} = \frac{1}{2}.$$

Because the limit exists but the function value at $x = 1$ does not, f must have a removable discontinuity there.

We now define

$$g(x) = \begin{cases} \frac{x-1}{x^2-1}, & \text{if } x \neq 1 \\ \frac{1}{2}, & \text{if } x = 1. \end{cases}$$

Notice that $g(x)$ is now continuous at $x = 1$. The discontinuity at $x = 1$ has therefore been removed. Figure 2.6.3 shows the graph of this function.

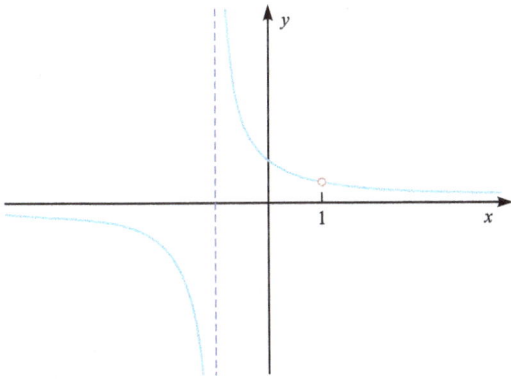

Figure 2.6.3: Graph of $f(x)$ in Example 2.6.1.

NOTE. The functions f and g also have a discontinuity at $x = -1$. However, this discontinuity cannot be removed; it is an infinite discontinuity (there is a vertical asymptote at $x = -1$).

Since the continuity of a function at point $x = a$ involves a limit, it is sensible to explore the significance of the two corresponding one-sided limits.

Definition 2.6.3. A function f is *continuous from the right* at $x = a$ if both $f(a)$ and $\lim_{x \to a^+} f(x)$ exist and

$$\lim_{x \to a^+} f(x) = f(a).$$

Similarly, f is *continuous from the left* at $x = a$ if both $f(a)$ and $\lim_{x \to a^-} f(x)$ exist and

$$\lim_{x \to a^-} f(x) = f(a).$$

For example, from Figure 2.6.1, we know that the function $f(x)$ is continuous from the right at $x = 1$ and $f(x)$ is continuous from the left at $x = 4$. Example 2.2.10 shows that the function $\sqrt{x}$ is continuous from the right at $x = 0$.

Obviously, if a function is continuous from the left at $x = a$ and also continuous from the right at $x = a$, then it must be continuous at $x = a$, since

$$\lim_{x \to a^+} f(x) = f(a) \quad \text{and} \quad \lim_{x \to a^-} f(x) = f(a) \quad \Longleftrightarrow \quad \lim_{x \to a} f(x) = f(a).$$

Thus,

$f(x)$ is continuous at $x = a$ if and only if

$f(x)$ is continuous from the left and from the right at $x = a$.

Evaluating limits of a continuous function is easy since, if f is continuous at $x = a$, $\lim_{x \to a} f(x) = f(a)$. This means that one can find $\lim_{x \to a} f(x)$ simply by substituting $x = a$ in $f(x)$ to find the limit $f(a)$.

Direct substitution rule for limits of a function at its continuities
If $f(x)$ is continuous at $x = a$, then $\lim_{x \to a} f(x) = f(a)$.

2.6.2 Continuous functions

Definition 2.6.4. If f is continuous at each point a in an interval $\mathbf{I}$, then f is said to be continuous on the interval $\mathbf{I}$.

Theorem 2.2.7 and Example 2.2.9 show that the polynomial functions $\cos x$ and $\sin x$ are continuous at every point $a \in (-\infty, \infty)$. We now show that several other basic functions are also continuous on their domains (we would expect this from their graphs).

Example 2.6.2. The exponential function $f(x) = b^x$, $b > 0$, $b \neq 1$, is continuous on its domain $(-\infty, \infty)$.

Proof. See Section 2.7. □

Building continuous functions

In Section 2.2.3, it was shown how the limit of a function $f(x)$ whose formula is an algebraic combination of basic functions could be computed from the limits of those much simpler basic functions. These same results apply to continuous functions. That is, continuous functions could be added, subtracted, multiplied, divided (provided the denominator is not 0), and composed to form new continuous functions. You can probably understand and appreciate these on the intuitive level, as well as construct simple proofs for limit theorems.

Theorem 2.6.1. *If $f(x)$ and $g(x)$ are continuous at $x = a$, then the following functions are also continuous at $x = a$:*

1. $(f + g)(x) = f(x) + g(x)$ *(sum rule);*
2. $(f - g)(x) = f(x) - g(x)$ *(difference rule);*
3. $(cf)(x) = cf(x)$ *(constant multiple rule);*
4. $(fg)(x) = f(x)g(x)$ *(product rule);*
5. $(\frac{f}{g})(x) = \frac{f(x)}{g(x)}$, *provided $g(a) \neq 0$ (quotient rule).*

Proof. Each of the five parts of this theorem follows from the corresponding limit laws from Section 2.2.3. For example, we give the proof of number 1. Since $f(x)$ and $g(x)$ are continuous at a, we have

$$\lim_{x \to a} f(x) = f(a) \quad \text{and} \quad \lim_{x \to a} g(x) = g(a).$$

Therefore,

$$\lim_{x \to a}[f(x) + g(x)] = \lim_{x \to a} f(x) + \lim_{x \to a} g(x) = f(a) + g(a) = (f + g)(a).$$

This shows that $f + g$ is continuous at a. □

Example 2.6.3. Each of the six basic trigonometric functions, $\sin x$, $\cos x$, $\tan x$, $\cot x$, $\sec x$, and $\csc x$, is continuous on its own domain.

Solution. We proved that $\sin x$ and $\cos x$ are continuous on their domains in Example 2.2.9. For the remaining functions, $\tan x$, $\cot x$, $\sec x$, and $\csc x$, we have

$$\tan x = \frac{\sin x}{\cos x}, \quad \cot x = \frac{\cos x}{\sin x}, \quad \sec x = \frac{1}{\cos x}, \quad \text{and} \quad \csc x = \frac{1}{\sin x}.$$

Then, because the functions $\sin x$ and $\cos x$ are continuous on **R** and because the remaining trigonometric functions are quotients of 1, $\cos x$, and $\sin x$, by Theorem 2.6.1, these four trigonometric functions are continuous on their domains.

Continuity of inverse functions

If $y = f(x)$ is a one-to-one function, then it has an inverse function, $x = f^{-1}(y)$. We claim that, if $y = f(x)$ is continuous, $x = f^{-1}(y)$ (or $y = f^{-1}(x)$ in the more traditional notation) is also continuous. For example, if $y = 2x + 1$, then $x = (y - 1)/2$ is its inverse (in this case you should be aware that x is now the dependent variable and y is the independent variable). Both of these functions are polynomials and therefore both are continuous on **R**.

Because the graphs of a function $y = f(x)$ and its inverse function $y = f^{-1}(x)$ are reflections across the line $y = x$, if f is continuous, then f^{-1} must also be continuous. Observe, however, that if you reverse the roles of the x- and y-axes in your graph, the graph of $x = f^{-1}(y)$ will be identical to the graph of $y = f(x)$ in the traditional xy-plane. Therefore, we have the following theorem (the proof is omitted here).

Theorem 2.6.2. *If $y = f(x)$ is a one-to-one continuous function defined on D with range R, then its inverse $y = f^{-1}(x)$ is also continuous on R.*

Example 2.6.4. Determine the intervals on which a logarithmic function and the six inverse trigonometric functions are continuous.

Solution. A logarithmic function $y = \log_b x$ ($b > 0$, $b \neq 1$) is the inverse of the exponential function $y = b^x$. Such an exponential function is continuous and one-to-one on the whole real line with range $y > 0$. Therefore, its inverse function is continuous on its entire domain $\{x | x > 0\}$.

The six trigonometric functions are also continuous and are one-to-one on their restricted domains. Therefore, the inverse trigonometric functions are all continuous on their domains. That is to say:
(1) $y = \arcsin x$ is continuous on $[-1, 1]$;
(2) $y = \arccos x$ is continuous on $[-1, 1]$;
(3) $y = \arctan x$ is continuous on $(-\infty, \infty)$;
(4) $y = \text{arccot}\, x$ is continuous on $(-\infty, \infty)$;
(5) $y = \text{arcsec}\, x$ is continuous on $(-\infty, -1] \cup [1, \infty)$;
(6) $y = \text{arccsc}\, x$ is continuous on $(-\infty, -1] \cup [1, \infty)$.

Continuity of composite functions

Another way of combining continuous functions f and g to form new continuous functions is to form the composite function $(f \circ g)(x) = f(g(x))$, in which we substitute the formula $g(x)$ in place of the variable of $f(x)$. Intuitively, if x is close to a, then $g(x)$ is close to b and, since f is continuous at b, $g(x)$ is close to b, so $f(g(x))$ is close to $f(b)$. The continuity properties of the composite function are given in the following theorem.

Theorem 2.6.3. *If $u = g(x)$ is continuous at $x = a$, $g(a) = b$, and f is continuous at $u = b$, then $f(g(x))$ is continuous at $x = a$. That is,*

$$\lim_{x \to a} f(g(x)) = f(b) = f(g(a)).$$

Proof. Since $f(u)$ is continuous at b, for any given number $\varepsilon > 0$, we can find a number δ_1 such that

$$|f(g(x)) - f(g(a))| = |f(u) - f(b)| < \varepsilon \quad \text{whenever } |u - b| < \delta_1.$$

However, $\lim_{x \to a} g(x) = b$ means that there exists a number δ such that

$$|g(x) - b| < \delta_1 \quad \text{whenever } |x - a| < \delta.$$

By the definition of a limit, $\lim_{x \to a} f(g(x)) = f(b)$ and $f(g(x))$ is continuous at a. $\quad\square$

NOTE. If $g(x)$ is continuous at $x = a$, $g(a) = b$, and $f(x)$ is continuous at $x = b$, then

$$\lim_{x \to a} f(g(x)) = f\left(\lim_{x \to a} g(x)\right) = f(g(a)).$$

Example 2.6.5. Find $\lim_{x \to \pi} 3^{\sin x}$.

Solution. Since 3^x and $\sin x$ are continuous and $x = \pi$ is a point in the domain of $3^{\sin x}$, we have

$$\lim_{x \to \pi} 3^{\sin x} = 3^{\lim_{x \to \pi} \sin x} = 3^{\sin \pi} = 3^0 = 1.$$

Example 2.6.6. Find $\lim_{x \to 0} \frac{e^x - 1}{x}$.

Solution. Let $u = e^x - 1$. Then $x = \ln(u + 1)$ and $u \to 0$ as $x \to 0$. Then

$$\lim_{x \to 0} \frac{e^x - 1}{x} = \lim_{u \to 0} \frac{u}{\ln(u + 1)} = \lim_{u \to 0} \frac{1}{\frac{1}{u}\ln(u + 1)}$$

$$= \lim_{u \to 0} \frac{1}{\ln(u + 1)^{\frac{1}{u}}} = \frac{1}{\ln \lim_{u \to 0}(u + 1)^{\frac{1}{u}}} = \frac{1}{\ln e} = 1.$$

NOTE. From this example we see that $e^x - 1 \sim x$ as $x \to 0$ (or $e^x - 1 \approx x$ for small x).

Example 2.6.7. A *power function* is defined as $y = x^u$, where $u \in \mathbf{R}$, $u \neq 0$. For some u, the domain of a power function is all real numbers, but for other u, the domain is restricted to positive numbers, $x > 0$. The domain must be determined in each individual case. Show that $y = x^u$ is continuous for $x > 0$.

Proof. This is because $y = x^u = e^{u \ln x}$. The exponential function and the logarithm are both continuous when $x > 0$, so $y = x^u$ is continuous on $x > 0$. $\quad\square$

Recall that an *elementary function* is a function built from a finite number of exponentials, logarithms, constants, power functions, trigonometric functions, and their inverses through composition and combinations using the four elementary operations $(+, -, \times, \div)$. Now, in view of Theorem 2.6.1, Theorem 2.6.2, and Theorem 2.6.3, we conclude as follows.

Theorem 2.6.4. *Each elementary function is continuous on its domain.*

Example 2.6.8. Find all x-values for which $f(x) = x^3 + e^x \sin x$ is continuous.

Solution. By Theorem 2.6.4, x^3, e^x, and $\sin x$ are continuous for all real x-values. Therefore, the function $f(x) = x^3 + e^x \sin x$ is also continuous for all real x-values.

Example 2.6.9. For the function $f(x) = \log_2(\sin x)$, find all x-values for which f is continuous.

Solution. By Theorem 2.6.4, the function $\log_2 x$ is continuous when $x > 0$ and $\sin x$ is continuous for all real x, so $f(x)$ is continuous when

$$\sin x > 0.$$

Solving this inequality gives

$$2n\pi + 0 < x < 2n\pi + \pi \quad \text{for all } n \in \mathbf{N}.$$

This means the function $f(x) = \log_2 \sin x$ is continuous on these open intervals. The graph of $f(x)$ is shown in Figure 2.6.4.

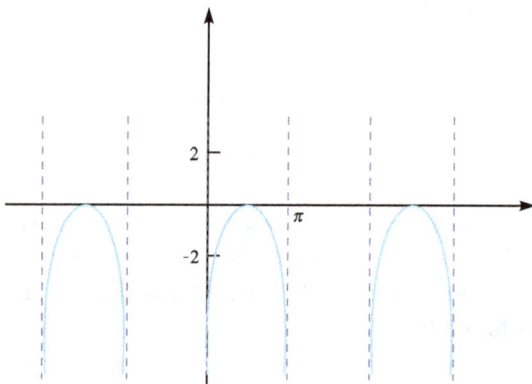

Figure 2.6.4: Graph of the function in Example 2.6.9.

Example 2.6.10. Find

$$\lim_{x\to 0}(2+\tan x)^{\ln(1+e^{\sin x^2})}.$$

Solution. This is a limit of an elementary function when $x \to 0$ and $x = 0$ is a point in its domain, so by the direct substitution rule we have

$$\lim_{x\to 0}(2+\tan x)^{\ln(1+e^{\sin x^2})} = (2+0)^{\ln(1+e^{\sin 0^2})}$$
$$= 2^{\ln(1+e^0)}$$
$$= 2^{\ln(1+1)}$$
$$= 2^{\ln 2}.$$

Example 2.6.11. Classify the discontinuity for the function

$$f(x) = \frac{2^{\frac{1}{x}}-1}{2^{\frac{1}{x}}+1}.$$

Solution. This function is an elementary function defined for all x except 0, so the only discontinuity of $f(x)$ occurs at $x = 0$. Then

$$\lim_{x\to 0^+}\frac{2^{\frac{1}{x}}-1}{2^{\frac{1}{x}}+1}\overset{t=\frac{1}{x}}{=}\lim_{t\to\infty}\frac{2^t-1}{2^t+1}=\lim_{t\to\infty}\frac{1-\frac{1}{2^t}}{1+\frac{1}{2^t}}=\frac{1-0}{1+0}=1,$$

$$\lim_{x\to 0^-}\frac{2^{\frac{1}{x}}-1}{2^{\frac{1}{x}}+1}=\lim_{t\to-\infty}\frac{2^t-1}{2^t+1}=\lim_{t\to\infty}\frac{0-1}{0+1}=-1.$$

This discontinuity is a jump discontinuity. Figure 2.6.5 shows the graph of $f(x)$.

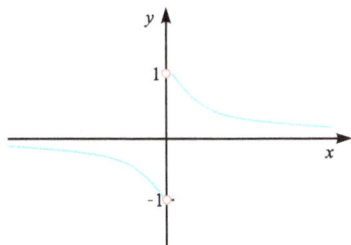

Figure 2.6.5: Graph of $f(x)$ in Example 2.6.11.

2.6.3 Theorems on continuous functions

A number of results involving continuous functions are very important in the development of calculus, including the extreme value theorem and the intermediate value theorem. First we give the definitions of global extrema.

Definition 2.6.5. If f has domain D, then:

1. f has an *absolute maximum* (also called *global maximum*) at $x = a \in D$ if $f(x) \leqslant f(a)$ for every $x \in D$;
2. f has an *absolute minimum* (also called *global minimum*) at $x = a \in D$ if $f(x) \geqslant f(a)$ for every $x \in D$.

NOTE. An absolute maximum can occur at more than one x-value in the domain of f. The same is true for absolute minima, as shown in Figure 2.6.8. Also, global extrema may occur at end points or at some interior points of an interval as shown in Figure 2.6.6 and Figure 2.6.7.

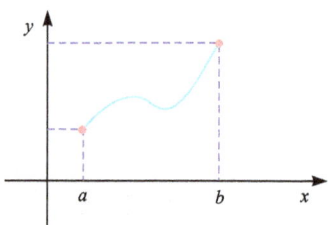

Figure 2.6.6: Global extrema may occur at endpoints.

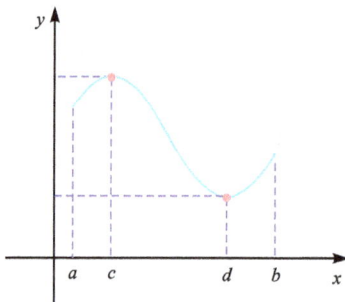

Figure 2.6.7: Absolute maximum and absolute minimum.

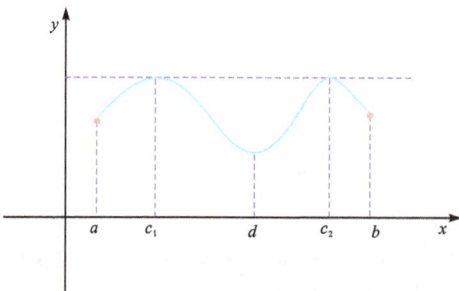

Figure 2.6.8: A function may obtain its global extrema at more than one point.

If you sketch the graph of a continuous function f on a closed interval $[a,b]$, then you find that the curve is bounded above and below by two horizontal lines. The curve will always have a global maximum and a global minimum somewhere in the interval. This is because the two points $A(a,f(a))$ and $B(b,f(b))$ are fixed and the graph which is a continuous curve goes from point A to point B. No matter how many turns or oscillations there are along the curve, the path cannot increase or decrease without bound as the graph moves from A to B. The highest points and the lowest points reached are the global extremum points. This observation is confirmed by the following two theorems.

Theorem 2.6.5 (Boundedness theorem). *If f is continuous on a closed interval $[a,b]$, then f must be bounded.*

Proof. See Section 2.7. □

Furthermore, we conclude that there must be both an absolute maximum and an absolute minimum somewhere in the closed interval.

Theorem 2.6.6 (Extreme value theorem). *If f is continuous on a closed interval $[a,b]$, then f must have an absolute minimum value $f(c)$ and an absolute maximum value $f(d)$ at some points c and d in $[a,b]$. That is,*

$$f(c) \leqslant f(x) \leqslant f(d) \quad \text{for all } x \in [a,b].$$

Proof. See Section 2.7. □

If a function is not continuous or is continuous only on an open interval, then the extreme value theorem may not hold. For example, consider the two functions $f(x)$ and $g(x)$:

$$f(x) = \begin{cases} x+1, & 0 \leqslant x \leqslant 2 \\ 4 - (x-2)^2, & 2 < x \leqslant 4 \end{cases} \quad \text{and} \quad g(x) = \frac{1}{x} \quad \text{for } x \in (0.5, 3).$$

Graphs of $f(x)$ and $g(x)$ are shown in Figure 2.6.9 and Figure 2.6.10 respectively.

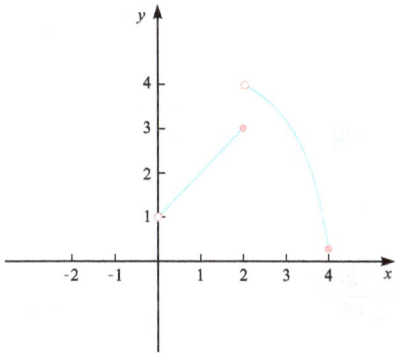

Figure 2.6.9: A function may not have global extrema if it is not continuous.

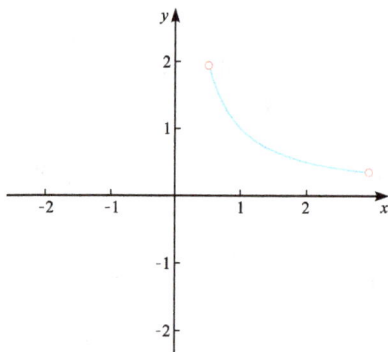

Figure 2.6.10: A function may not have global extrema if the interval is not closed.

The function $f(x)$ is defined on the closed interval $[0, 4]$, but it is not continuous on the interval and it has no global maximum value. The function $g(x)$ is continuous at each point in the interval $(0.5, 3)$, but the interval is not closed. This function has no global extrema in the interval.

NOTE. The extreme value theorem was originally proved by Bernard Bolzano in the 1830s in a work called *Function Theory*, but the work remained unpublished until 1930. Bolzano's proof consisted of showing that a continuous function on a closed interval was bounded and then showing that the function attained a maximum and a minimum value. His proof involved what is known today as the Bolzano–Weierstrass theorem. http://en.wikipedia.org/wiki/Extreme_value_theorem

Now we discuss another theorem on continuous functions defined on a closed interval, named the intermediate value theorem. If your height at birth was 50 cm and you are now 180 cm tall, then there must have been a time when your height was exactly 170 cm, since you grew continuously and your height could not jump from one point to another. Similarly, if the temperature changes continuously from negative to positive, then you would believe that there must have been an instant when the temperature was exactly 0.

Now let us go back to mathematics and explore this in terms of continuous functions. If the graph of a continuous function has one end below the x-axis and the other end above the x-axis, then the curve must cross the x-axis at least once. This is stated as the following theorem.

Theorem 2.6.7 (Bolzano's theorem). *If $f(x)$ is continuous on a closed interval $[a, b]$, and $f(a) \times f(b) < 0$, then there is at least a value $c \in [a, b]$ such that $f(c) = 0$.*

Proof. See Section 2.7. □

NOTE. This theorem was first proved by Bernard Bolzano in 1817. Augustin-Louis Cauchy provided a proof in 1821. http://en.wikipedia.org/wiki/Intermediate_value_theorem

Bernhard Bolzano
(1781–1848) was a Bohemian mathematician, logician, and philosopher.
http://en.wikipedia.org/wiki/Bernard_Bolzano

This theorem can be extended in an obvious way. For a continuous function f and any number m between $f(a)$ and $f(b)$, there must be at least one number c between a and b such that $f(c) = m$. We can prove this by leaving the shape of the curve untouched but shifting the axes up or down by m units.

Theorem 2.6.8 (Intermediate value theorem). *If $f(x)$ is continuous on a closed interval $[a,b]$ and m is any number between $f(a)$ and $f(b)$, then there is at least one value $c \in [a,b]$ such that $f(c) = m$.*

Proof. If $f(a) < m < f(b)$, then let $g(x) = f(x) - m$. Then $g(x)$ is also continuous on $[a,b]$. Since

$$g(a) = f(a) - m < 0$$

and $g(b) = f(b) - m > 0$, by Bolzano's theorem, there is a value $c \in [a,b]$ such that

$$g(c) = 0.$$

This means $f(c) = m$.

If $f(b) < m < f(a)$, then we let $g(x) = m - f(x)$. By an argument similar to the one above, we can also find a number c such that $f(c) = m$.

The intermediate value theorem states that a continuous function takes on every intermediate value between the function values $f(a)$ and $f(b)$. This means that any horizontal line $y = m$ between $y = f(a)$ and $y = f(b)$ must intersect the graph of $y = f(x)$ at least one x-value in the interval $[a,b]$, as seen in Figure 2.6.11 and Figure 2.6.12. If there is no point of intersection between the line $y = m$ and the graph of $y = f(x)$, then part of the graph of $y = f(x)$ will be below $y = m$ and some part of the graph of $y = f(x)$ will be above $y = m$, so the graph of $y = f(x)$ must jump over the line $y = m$ at some point and there would be some holes or breaks in the graph of $y = f(x)$. For some functions, an intermediate value m will be attained more than once, as in Figure 2.6.13.

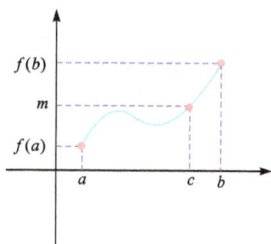

Figure 2.6.11: There is a "c" such that $f(c) = m$ in case that $f(a) < f(b)$.

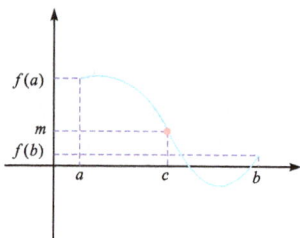

Figure 2.6.12: There is a "c" such that $f(c) = m$ in case that $f(a) > f(b)$.

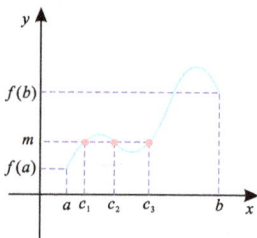

Figure 2.6.13: There may be more than one c such that $f(c) = m$.

For the intermediate value theorem to hold, the function f must be continuous on a closed interval. The theorem is not true in general for discontinuous functions, even if the function is defined on a closed interval. For example, consider the two functions

$$f(x) = \begin{cases} x, & 0 \leqslant x \leqslant 2 \\ x+2, & 2 < x \leqslant 4 \end{cases} \quad \text{and} \quad g(x) = x, 0 \leqslant x < 1.$$

The function $f(x)$ is not continuous on $[0, 4]$, because $f(x)$ cannot take every value between $f(2)$ and $f(4)$ as shown in Figure 2.6.14. The function $g(x)$ is not defined on a closed interval, so $g(1)$ does not exist. □

Example 2.6.12. Show that the equation $4x^3 + 3x - 2 = 0$ has at least one root between $x = 0$ and $x = 2$.

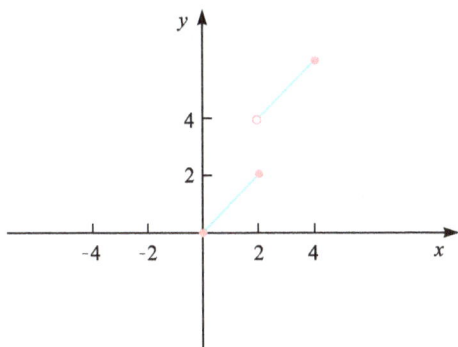

Figure 2.6.14: No intermediate values for a discontinuous function.

Proof. Let $f(x) = 4x^3 + 3x - 2$ and look for a number c between $a = 0$ and $b = 2$ such that $f(c) = 0$. We try

$$f(0) = 0 - 0 - 2 = -2, \quad \text{so } f(0) < 0,$$
$$f(2) = 32 + 6 - 2 = 36, \quad \text{so } f(2) > 0.$$

Since f is continuous on $[0, 2]$, it follows from Bolzano's Theorem that $f(x)$ must have at least one root $x = c$ in the interval $(0, 2)$. ☐

NOTE. In fact, we can locate a root more precisely by subdividing it into smaller intervals and using the intermediate value theorem again. For example, because $f(1) = 5 > 0$, we know there must be a root in $(0, 1)$.

Example 2.6.13. Show that the equation $\sin x = x - 1$ has a real root.

Proof. Let $f(x) = x - 1 - \sin x$. Then we must prove that there is a number c such that $f(c) = 0$. Note that

$$f(0) = 0 - 1 - 0 = -1 < 0,$$

while

$$f(\pi) = \pi - 1 - \sin \pi = \pi - 1 > 0.$$

Since $f(x)$ is continuous on $[0, \pi]$ (in fact on all of **R**), we may apply Bolzano's theorem to conclude the existence of some c in $(0, \pi)$ such that $f(c) = c - 1 - \sin c = 0$, and c is therefore a root of the equation $\sin x = x - 1$. ☐

Example 2.6.14. Every polynomial of odd degree has a real root.

Proof. Let $f(x) = a_0 + a_1x + \cdots + a_nx^n$, where n is odd. Assuming that $a_n > 0$ (a similar proof holds if $a_n < 0$), we have

$$\lim_{x \to +\infty} (a_0 + a_1x + \cdots + a_nx^n) = \lim_{x \to +\infty} \left[\left(\frac{a_0}{x^n} + \frac{a_1}{x^{n-1}} + \cdots + a_n \right) x^n \right]$$
$$= a_n \lim_{x \to +\infty} x^n = +\infty.$$

Similarly,

$$\lim_{x \to -\infty} (a_0 + a_1x + \cdots + a_nx^n) = a_n \lim_{x \to -\infty} x^n = -\infty.$$

Since $f(x)$ is continuous, there must be a sufficiently large x-value, b, such that $0 < f(b)$. Similarly there must be a negative x-value a in the other direction for which $f(a) < 0$. Hence, by Bolzano's theorem, there is at least one c with $a < c < b$ such that $f(c) = a_0 + a_1c + \cdots + a_nc^n = 0$, where c is the required root. $\quad\square$

Example 2.6.15. If $f(x)$ is continuous on the interval $[a, b]$ and

$$a < x_1 < x_2 < \cdots < x_n < b,$$

then show that there exists a number $c \in [a, b]$ such that

$$f(c) = \frac{f(x_1) + f(x_2) + \cdots + f(x_n)}{n}.$$

Proof. Let

$$f(x_m) = \min\{f(x_1), f(x_2), \ldots, f(x_n)\} \quad \text{and}$$
$$f(x_M) = \max\{f(x_1), f(x_2), \ldots, f(x_n)\}.$$

Then

$$f(x_m) \leqslant \frac{f(x_1) + f(x_2) + \cdots + f(x_n)}{n} \leqslant f(x_M).$$

Since $f(x)$ is continuous on $[a, b]$ and the interval $[x_m, x_M]$ (or $[x_M, x_m]$) is a subset of $[a, b]$, we know that $f(x)$ is also continuous on the closed interval $[x_m, x_M]$ (or $[x_M, x_m]$). Following from the intermediate value theorem, there exists a number $c \in [x_m, x_M]$ (or $[x_M, x_m]$) such that

$$f(c) = \frac{f(x_1) + f(x_2) + \cdots + f(x_n)}{n}. \quad\square$$

2.6.4 Uniform continuity

We know that the function $y = \frac{1}{x}$ is continuous at each point in $(0,1)$. However, from the graph of the function $y = \frac{1}{x}$, we know that the value of y changes more and more rapidly as x approaches 0. This means that, even if we keep the change in x, Δx, constant, the change in y, Δy, which depends on the specific x-value in $(0,1)$, can be larger than any given number. This is different from other continuous functions, like $y = \sin x$; if Δx is small, Δy is also small at each point x. We now give the definition of uniform continuity.

Definition 2.6.6 (Uniform continuity). Let $f(x)$ be defined on an interval **I**. Then f is *uniformly continuous* on **I**, if, for any given positive number ε, there exists a number δ (which depends only on ε) such that, for every two points $a, b \in I$, we have $|f(a) - f(b)| < \varepsilon$ whenever $|a - b| < \delta$.

Notice that the number δ is independent of the choices of values of a and b. That is, f is uniformly continuous if it is possible to guarantee that $f(a)$ and $f(b)$ will be as close to each other as we please by requiring only that the distance $|a - b|$ between a and b is less than a sufficiently small value δ that does not depend on the values a or b.

Example 2.6.16. Prove that $y = \frac{1}{x}$ is continuous on $(0,1]$, but not uniformly continuous on $(0,1]$.

Solution. The function $y = \frac{1}{x}$ is obviously continuous on $(0,1]$, since it is an elementary function (a power function) defined on $(0,1]$. We prove by contradiction that the function is not uniformly continuous on $(0,1]$.

For a given small number $0 < \varepsilon < 1$, if $y = \frac{1}{x}$ is uniformly continuous on $(0,1]$, there must be a number δ such that $|f(a) - f(b)| < \varepsilon$ whenever $|a - b| < \delta$. Now we choose two numbers $a = \frac{1}{n}$ and $b = \frac{1}{n+1}$, where n is a positive integer. For a sufficiently large n, the value of $|a - b|$ will be smaller than δ since $|a - b| = |\frac{1}{n} - \frac{1}{n+1}| = \frac{1}{n(n+1)} \to 0$ as $n \to \infty$. However, for these numbers, we have

$$|f(a) - f(b)| = \left| \frac{1}{\frac{1}{n}} - \frac{1}{\frac{1}{n+1}} \right| = |n - (n+1)| = 1 > \varepsilon.$$

This contradicts the definition of uniform continuity. Therefore, $y = \frac{1}{x}$ is not uniformly continuous on $(0,1]$.

A uniformly continuous function must be continuous, but a continuous function may not be uniformly continuous. A useful theorem to determine whether or not a continuous function is uniformly continuous is given below without a proof. The theorem is due to Eduard Heine and Georg Cantor.

Theorem 2.6.9 (Heine–Cantor theorem). *If $f(x)$ is continuous on a closed interval $[a,b]$, then $f(x)$ is also uniformly continuous on $[a,b]$.*

2.7 Some proofs in Chapter 2

Proof of Theorem 2.2.1. Assume $\lim_{x\to a} f(x) = L$ and $\lim_{x\to a} f(x) = M$, where $L \neq M$. Then $|L - M| = d \neq 0$. Now choose $\varepsilon = d/2 > 0$. By the definition of a limit, there exist two numbers, δ_1 and δ_2, such that

$$|f(x) - L| < \frac{d}{2} \quad \text{whenever } 0 < |x - a| < \delta_1,$$

$$|f(x) - M| < \frac{d}{2} \quad \text{whenever } 0 < |x - a| < \delta_2.$$

Then, when $|x - a| < \min\{\delta_1, \delta_2\}$,

$$|L - M| = |L - f(x) + f(x) - M|$$
$$\leqslant |L - f(x)| + |f(x) - M|$$
$$< \frac{d}{2} + \frac{d}{2} = d.$$

This is a contradiction to the assumption that $|L - M| = d$. $\square$

Proof of Theorem 2.2.2. Since $\lim_{x\to a} f(x) = L$, for any given number $\varepsilon > 0$, there is a corresponding δ such that

$$|f(x) - L| < \varepsilon \quad \text{whenever } 0 < |x - a| < \delta.$$

Now choose $\varepsilon = 1$. Then there is a δ such that $|f(x) - L| < 1$ whenever $0 < |x - a| < \delta$. This means

$$-1 < f(x) - L < 1 \quad \text{whenever } 0 < |x - a| < \delta,$$

so $L - 1 < f(x) < L + 1$ whenever $0 < |x - a| < \delta$.

This means that $f(x)$ is bounded on the deleted neighborhood $(a, a - \delta) \cup (a, a + \delta)$ of $x = a$. $\square$

Proof of Theorem 2.2.3. Since $L > 0$, $\frac{L}{2} > 0$. Because $\lim_{x\to a} f(x) = L$, for a given number $\varepsilon = \frac{L}{2}$, there is a corresponding $\delta > 0$, such that

$$|f(x) - L| < \frac{L}{2} \quad \text{whenever } 0 < |x - a| < \delta.$$

This means

$$-\frac{L}{2} < f(x) - L < \frac{L}{2} \quad \text{whenever } 0 < |x - a| < \delta,$$

so

$$L - \frac{L}{2} < f(x) < \frac{L}{2} + L \quad \text{whenever } 0 < |x - a| < \delta.$$

Therefore, $f(x) > L - \frac{L}{2} = \frac{L}{2} > 0$ for all x in the deleted δ neighborhood $0 < |x - a| < \delta$ of $x = a$. $\square$

Proof (Limit law 1). Assume $\lim_{x \to a} f(x) = L$ and $\lim_{x \to a} g(x) = M$, where both L and M are two real numbers. Then, given any number $\varepsilon > 0$, we have δ_1 and δ_2 such that

$$|f(x) - L| < \frac{\varepsilon}{2} \quad \text{whenever } 0 < |x - a| < \delta_1 \quad \text{and}$$

$$|g(x) - M| < \frac{\varepsilon}{2} \quad \text{whenever } 0 < |x - a| < \delta_2.$$

Now choose δ to be the smaller one of the two numbers δ_1 and δ_2, that is, $\delta = \min\{\delta_1, \delta_2\}$. Then, whenever $0 < |x - a| < \delta$, we have $0 < |x - a| < \delta_1$ and $0 < |x - a| < \delta_2$, and $|(f(x) + g(x)) - (L + M)| \leq |f(x) - L| + |g(x) - M| < \frac{\varepsilon}{2} + \frac{\varepsilon}{2} = \varepsilon$, so

$$\lim_{x \to a} (f(x) + g(x)) = L + M.$$

Therefore,

$$\lim_{x \to a} (f(x) + g(x)) = \lim_{x \to a} f(x) + \lim_{x \to a} g(x).$$

This completes the proof. $\square$

Proof (Limit law 3). Since $\lim_{x \to a} f(x) = L$ and $\lim_{x \to a} g(x) = M$, where both L and M are two real numbers, given any positive number $\varepsilon > 0$, the number $\frac{\varepsilon}{|L| + |M| + 1} > 0$ and we can find δ_1 and δ_2 such that

$$|f(x) - L| < \frac{\varepsilon}{|L| + |M| + 1} \quad \text{whenever } 0 < |x - a| < \delta_1 \quad \text{and}$$

$$|g(x) - M| < \frac{\varepsilon}{|L| + |M| + 1} \quad \text{whenever } 0 < |x - a| < \delta_2.$$

Furthermore, for the number $\varepsilon = 1$, we can find a number δ_3 such that

$$|g(x) - M| \leq 1 \quad \text{whenever } 0 < |x - a| < \delta_3.$$

This means

$$|g(x)| - |M| \leq |g(x) - M| \leq 1,$$

so

$$|g(x)| \leq |M| + 1 \quad \text{whenever } 0 < |x - a| < \delta_3.$$

We choose δ to be the smallest one of the three numbers δ_1, δ_2, and δ_3, since $x - a$ must satisfy the three conditions simultaneously. Let $\delta = \min\{\delta_1, \delta_2, \delta_3\}$. Then, whenever $0 < |x - a| < \delta$, we have $0 < |x - a| < \delta_1$, $0 < |x - a| < \delta_2$, and $0 < |x - a| < \delta_3$. Then

$$
\begin{aligned}
|f(x)g(x) - LM| &= |f(x)g(x) - Lg(x) + Lg(x) - LM| \\
&\leqslant |g(x)||f(x) - L| + |L||g(x) - M| \\
&< (|M| + 1)\frac{\varepsilon}{|L| + |M| + 1} + |L|\frac{\varepsilon}{|L| + |M| + 1} \\
&= \varepsilon.
\end{aligned}
$$

This implies that, for any given number $\varepsilon > 0$, we have found a number $\delta > 0$ such that

$$
|f(x)g(x) - LM| < \varepsilon \quad \text{whenever } 0 < |x - a| < \delta,
$$

so

$$
\lim_{x \to a}[f(x) \times g(x)] = LM = \lim_{x \to a} f(x) \times \lim_{x \to a} g(x).
$$

This completes the proof. ☐

The proofs of the remaining limit laws are left to the reader.

Proof of the substitution rule. Given any number $\varepsilon > 0$, since $\lim_{u \to b} f(u) = L$, there is a number $\delta_1 > 0$ such that

$$
|f(u) - L| < \varepsilon \quad \text{whenever } 0 < |u - b| < \delta_1.
$$

On the other hand, we have $\lim_{x \to a} g(x) = b$, so for the number $\delta_1 > 0$, there is a number δ_2 such that

$$
|g(x) - b| < \delta_1 \quad \text{whenever } 0 < |x - a| < \delta_2.
$$

Let $\delta = \min\{\delta_1, \delta_2\}$. Then, whenever $0 < |x - a| < \delta$, we have $|g(x) - b| = |u - b| < \delta_1$. Then

$$
|f(g(x)) - L| = |f(u) - L| < \varepsilon. \qquad ☐
$$

Proof of Heine's theorem. "$\Longrightarrow$" Given a number $\varepsilon > 0$, since $\lim_{x \to a} f(x) = L$, there is $\delta > 0$ such that

$$
|f(x) - L| < \varepsilon \quad \text{whenever } 0 < |x - a| < \delta.
$$

Thus, if $\lim_{n \to \infty} x_n = a$, this means that x_n approaches a but is not equal to a, so for the number δ, there is a number $N > 0$ such that

$$
0 < |x_n - a| < \delta \quad \text{whenever } n > N.
$$

In this case,

$$|f(x_n) - L| < \varepsilon.$$

We conclude that $\lim_{n \to \infty} f(x_n) = L$. □

We leave the proof of the necessary condition to the reader.

Proof of the monotonic and bounded sequence theorem. We suppose that a sequence $\{x_n\}$ is a monotonic increasing and bounded above sequence. We show that it must converge. Since the set $\{x_1, x_2, \ldots, x_n, \ldots\}$ is bounded above, it must have the supremum β according to Axiom 1.2.1 in Chapter 1. That is, $\beta = \sup\{x_n\}$. We claim that $x_n \to \beta$. Given any number $\varepsilon > 0$, $\beta - \varepsilon$ is not the supremum of $\{x_n\}$, so there must be a term, say, x_N, such that $x_N > \beta - \varepsilon$. However, the sequence $\{x_n\}$ is increasing, so

$$x_n > x_N \quad \text{for all } n > N.$$

This means that

$$\beta - x_n < \beta - x_N < \varepsilon, \quad \text{for all } n > N.$$

Because $\beta > x_n$ for all n, $|x_n - \beta| < \varepsilon$, for all $n > N$. This implies

$$\lim_{n \to \infty} x_n = \beta.$$ □

The proof of the case that the sequence is decreasing and bounded below is analogous. In this case, the sequence converges to the infimum of the set $\{x_n | n = \mathbf{N}\}$.

Proof of the Bolzano and Weierstrass theorem. We first prove that the sequence has a monotonic subsequence. If there are infinitely many peak terms of $\{x_n\}$, then these terms, in the same order as they appear in $\{x_n\}$, consist of a monotone decreasing subsequence. Otherwise, there must be a finite number of peak terms. Let x_N be the last peak term of $\{x_n\}$, while $n_1 = N + 1$. Then x_{n_1} is not a peak term. Therefore, there must be a term x_{n_2} after x_{n_1} such that $x_{n_2} > x_{n_1}$. However, x_{n_2} itself is not a peak term, so there exists a term x_{n_3} after x_{n_2} such that $x_{n_3} > x_{n_2}$. Repeating this process, we will obtain the sequence $\{x_{n_k}\}$:

$$x_{n_1} < x_{n_2} < x_{n_3} < \cdots.$$

This is a subsequence of $\{x_n\}$ and it is monotone increasing. Therefore, we conclude that the sequence $\{x_n\}$ has a monotonic subsequence. If the sequence $\{x_n\}$ is also bounded, so is its subsequence, so any bounded sequence $\{x_n\}$ must have a bounded monotonic subsequence and, by the monotonic and bounded convergence theorem, this subsequence is convergent. □

Proof of the squeeze theorem. Given a number $\varepsilon > 0$. Since $\lim_{x \to a} f(x) = L$, there is a number δ_1 such that

$$|f(x) - L| < \varepsilon \quad \text{whenever } 0 < |x - a| < \delta_1.$$

This means

$$L - \varepsilon < f(x) < L + \varepsilon \quad \text{whenever } 0 < |x - a| < \delta_1.$$

Similarly, there is a number $\delta_2 > 0$ such that

$$L - \varepsilon < h(x) < L + \varepsilon \quad \text{whenever } 0 < |x - a| < \delta_2.$$

Now choose $\delta = \min\{\delta_1, \delta_2\}$, so $0 < |x - a| < \delta$ implies that both $0 < |x - a| < \delta_1$ and $0 < |x - a| < \delta_2$ hold. In this case we have

$$L - \varepsilon < f(x) \leqslant g(x) \leqslant h(x) < L + \varepsilon.$$

This means

$$L - \varepsilon < g(x) < L + \varepsilon \quad \text{whenever } 0 < |x - a| < \delta,$$

so

$$|g(x) - L| < \varepsilon \quad \text{whenever } 0 < |x - a| < \delta.$$

Thus, by the definition of a limit, we have $\lim_{x \to a} g(x) = L$. $\qquad\square$

Proof of the continuity of an exponential function. To prove that $\lim_{x \to a} b^x = b^a$ ($b > 0$, $b \neq 1$) at any point $a \in (-\infty, \infty)$, we investigate

$$b^x - b^a = b^a(b^{x-a} - 1).$$

Then

$$\lim_{x \to a}(b^x - b^a) = \lim_{x \to a} b^a(b^{x-a} - 1)$$
$$= b^a \lim_{t \to 0}(b^t - 1) \quad \text{(substitution: } t = x - a\text{)}.$$

Hence, we need only to consider the limit $\lim_{t \to 0}(b^t - 1)$. Now we only need to be concerned with those values of t that are close to 0, for instance, $-1 < t < 1$. We can find an integer n for each of these t such that

$$|t| < \frac{1}{n} \quad \text{or equivalently} \quad \frac{1}{n} < t < \frac{1}{n},$$

so we have

$$b^{-\frac{1}{n}} < b^t < b^{\frac{1}{n}}, \quad \text{for } b > 1, b^x \text{ is increasing,} \quad \text{or}$$

$b^{\frac{1}{n}} < b^t < b^{-\frac{1}{n}}$ in the case that $0 < b < 1$, b^x is decreasing.

From Example 2.3.2, Example 2.4.4, and the squeeze theorem, we have

$$\lim_{n\to\infty} b^{\frac{1}{n}} = 1, \ \lim_{n\to\infty} b^{-\frac{1}{n}} = \lim_{n\to\infty}\left(\frac{1}{b}\right)^{\frac{1}{n}} = 1, \quad \text{so } \lim_{t\to 0} b^t = 1.$$

Therefore, we have

$$\lim_{x\to a}(b^x - b^a) = b^a \lim_{t\to 0}(b^t - 1) = 0.$$

This means that $\lim_{x\to a} b^x = b^a$, so b^x is continuous at a.   □

Proof of the boundedness theorem. Assume f is unbounded, say, unbounded above on $[a,b]$. This means, for any number M, we can find a number x in $[a,b]$ such that $f(x) > M$, so for $M = 1, 2, 3, \ldots, n, \ldots$, we can find a sequence $x_1, x_2, x_3, \ldots, x_n, \ldots$ such that each term is in $[a,b]$ and

$$f(x_1) > 1, f(x_2) > 2, f(x_3) > 3, \ldots, f(x_n) > n, \ldots.$$

This means $\lim_{n\to\infty} f(x_n) = \infty$. However, the sequence $\{x_n\}$ is bounded because $a \leqslant x_n \leqslant b$. Therefore, there is a convergent subsequence, say, $\{x_{n_k}\}$, such that $x_{n_k} \to c$ as $k \to \infty$. Because the function f is continuous at c, as a result of Heine's theorem we must have $\lim_{k\to\infty} f(x_{n_k}) = f(c)$. This contradicts $\lim_{k\to\infty} f(x_{n_k}) = \infty$, since $\{x_{n_k}\}$ is a subsequence of $\{x_n\}$, so the function f must be bounded above.

A similar argument can be used to prove that f must be bounded below. Therefore, a continuous function f must be bounded on the closed interval $[a,b]$.   □

Proof of the extreme value theorem. We prove the case that there is a number $d \in [a,b]$ such that $f(d)$ is the maximum value.

Since f is continuous on $[a,b]$, by the boundedness theorem it must be bounded. This means the set

$$S = \{f(x) \mid x \in [a,b]\}$$

is a bounded set. Of course, S is nonempty. Then, by the *least upper bound property*, we know that there is a number M such that $M = \sup S$. This implies that $f(x) \leqslant M$ for all x in $[a,b]$. We show that there is a number d in $[a,b]$ such that $f(d) = M$. If there is no x such that $f(x) = M$, then the function $\frac{1}{M-f(x)}$ must be continuous on $[a,b]$, so by Theorem 2.6.5 it must also be bounded. Given any number $\varepsilon > 0$, since M is the least upper bound of S, $M - \varepsilon$ is not the least upper bound. Thus there is an x such that $f(x) > M - \varepsilon$ and $\varepsilon > M - f(x)$, so

$$\frac{1}{M - f(x)} > \frac{1}{\varepsilon}.$$

Since ε can be any positive number, this means $\frac{1}{M-f(x)}$ can be made as large as we please, so $\frac{1}{M-f(x)}$ is unbounded. This contradicts the previous statement that $\frac{1}{M-f(x)}$ is bounded, so there must be a point d such that $f(d) = M$.

Similarly, there must be a point c such that $f(c) = m$ is the minimum value. □

Proof of Bolzano's theorem. We first prove the case where $f(a) < 0$ and $f(b) > 0$.

Let S be the set of all x in $[a, b]$ such that $f(x) \leq 0$. S is nonempty since $a \in S$ and S is bounded above by b. Hence, by the least upper bound property of real numbers, $c = \sup S$ exists. That is, c is the least number such that it is greater than or equal to every number in S. Furthermore, for any number x between b and c, we have $f(x) > 0$.

We now show that $f(c) = 0$. If $f(c) \neq 0$, assume $f(c) > 0$. Since f is continuous, we have

$$\lim_{x \to c} f(x) = f(c) > 0.$$

By Theorem 2.2.3, there is a number $\delta > 0$ such that

$$f(x) > 0 \quad \text{whenever } c - \delta < x < c + \delta,$$

so $\sup S \leq c - \delta$, which contradicts the assumption that $c = \sup S$.

Similarly, if $f(c) < 0$, then there is also a number $\delta > 0$ such that

$$f(x) < 0, \quad \text{whenever } c - \delta < x < c + \delta.$$

Then $\sup S$ must be at least $c + \delta$, which also contradicts the assumption that $c = \sup S$. Thus $f(c) = 0$.

For the case where $f(a) > 0$ and $f(b) < 0$, let $g(x) = -f(x)$, which is also continuous on $[a, b]$. Then there must be a c such that $g(c) = 0$ and this c, of course, satisfies $f(c) = 0$. □

2.8 Exercises

1. Find the average rate of change of each of the following functions on the indicated interval:

   (a) $f(x) = \tan x, [0, \frac{\pi}{4}]$;   (b) $g(t) = x^2 - 2x, [-2, 2]$;   (c) $r(\theta) = \sin 2\theta, [0, 2\pi]$.

2. Part of the graph of a piecewise defined function $f(x)$ is shown below. Find (a) $\lim_{x \to -3} f(x)$, (b) $\lim_{x \to 0} f(x)$, (c) $\lim_{x \to 3} f(x)$, and (d) $\lim_{x \to 4} f(x)$, or state they do not exist.

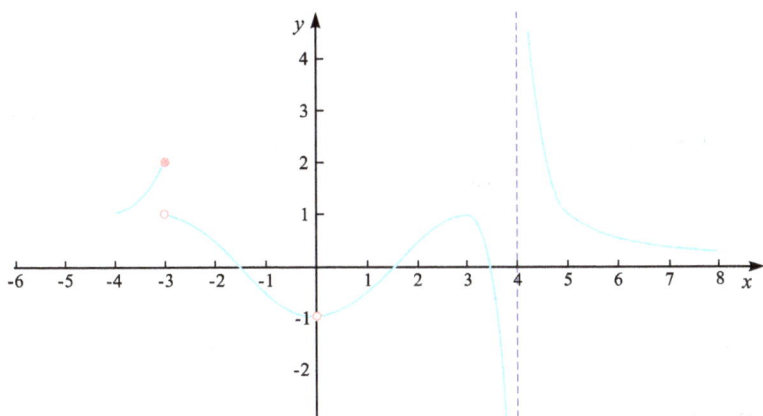

Question 2

3. Assume that $\lim_{x \to a} f(x) = -8$, $\lim_{x \to a} g(x) = 0$, and $\lim_{x \to a} h(x) = 5$. Find each of the following limits, if it exists:

   (a) $\lim_{x \to a} (f(x) - 2h(x))$; (b) $\lim_{x \to a} \frac{x^2 f(x) + g(x)}{h(x)}$;

   (c) $\lim_{x \to a} (g(x) - f(x))^3$; (d) $\lim_{x \to a} h(x) \sqrt[3]{f(x)}$.

4. Evaluate each of the following limits:

   (a) $\lim_{x \to 3} (3x^2 - 2x + 7)$; (b) $\lim_{x \to 2} \frac{x^2 - 4}{x + 2}$;

   (c) $\lim_{x \to -1} \frac{x^2 - 2x - 3}{x + 1}$; (d) $\lim_{x \to a} \frac{\sqrt{x+a} - \sqrt{2a}}{x - a}$, $a > 0$;

   (e) $\lim_{h \to 0} \frac{(x+h)^2 - x^2}{h}$; (f) $\lim_{\Delta x \to 0} \frac{\sqrt{x + \Delta x} - \sqrt{x}}{\Delta x}$;

   (g) $\lim_{t \to 0} \frac{\frac{1}{(1+t^2)} - 1}{t}$; (h) $\lim_{x \to \frac{\pi}{2}} (\cos^2 x - \sqrt{2\pi x})$.

5. (**Marginal cost**) The cost (in dollars) of producing x units of a certain commodity is modeled by $C(x) = 500 + 2x + 0.5x^2$.

   (a) Find the average rate of change of C with respect to x when the production level is changed from $x = 50$ to $x = 100$.

   (b) Find the instantaneous rate of change of C with respect to x when $x = 80$ (this rate is called the **marginal cost**).

6. Choose δ so that $|f(x)| < \frac{1}{1000}$, when $|x| < \delta$, for the following functions:

   (a) $f(x) = 2x$; (b) $f(x) = \sin 3x$; (c) $f(x) = x \cos x$; (d) $f(x) = x^2$.

7. Use the precise definition of a limit to show that:

   (a) $\lim_{x \to 2} (x + 1) = 3$; (b) $\lim_{x \to c} \sin x = \sin c$ for any $c \in \mathbf{R}$;

   (c) $\lim_{x \to -1} x^2 = 1$; (d) $\lim_{x \to 2} x^3 = 8$.

8. Using the function given in question 2, find (a) $\lim_{x \to 3^-} f(x)$, (b) $\lim_{x \to 3^+} f(x)$, (c) $\lim_{x \to 4^+} f(x)$, and (d) $\lim_{x \to 4^-} f(x)$. (e) Does $\lim_{x \to 3} f(x)$ exist? Explain.

9. Given

$$f(x) = \begin{cases} x^2 - 1, & \text{when } x < 0 \\ \sqrt{x}, & \text{when } 0 \leqslant x < 4 \\ [x], & x \geqslant 4, \end{cases}$$

find (a) $\lim_{x\to 5} f(x)$, (b) $\lim_{x\to 4} f(x)$, and (c) $\lim_{x\to 0} f(x)$ or explain why they do not exist.

10. The formal definition of "$f(x) \to L$ as $x \to -\infty$" is as follows: for any $\varepsilon > 0$, there is an $M > 0$, such that _ _ _ _ _ $< \varepsilon$ when $x < -M$. (a) Fill in the blank. (b) Give an example in which $f(x) \to 5$ as $x \to -\infty$.

11. Find the asymptotes and intercepts for the following functions and graph them (exact turning points are not required):

(a) $f(x) = \frac{x+2}{x^2-2x-3}$; (b) $f(x) = \frac{x^2+2x+1}{x^2-2x-3}$; (c) $g(x) = \frac{x}{x^2+1}$;

(d) $g(x) = \frac{x^2+x+1}{x-1}$; (e) $h(x) = \frac{x^2-3x-4}{x+2}$; (f) $f(x) = \frac{\sqrt{2x^2+1}}{3x-5}$;

(g) $f(t) = \frac{t^3}{t^2+3t-10}$.

12. (**Air pollution**) Assume the population size p (in thousands of people) of a certain community t years from now is modeled by $p(t) = 2000 - \frac{17}{1+4e^{-t}}$. Furthermore, suppose the average level of carbon monoxide in the air is modeled by $c(p) = 0.5\sqrt{p^2 - p + 17}$. What happens to the level of pollution c in the long run?

13. Find the limit for each of the following sequences:

(a) $\lim_{n\to\infty} \frac{2014n}{n^2+1}$; (b) $\lim_{n\to\infty} \frac{n}{\sqrt{1+n^2}}$;

(c) $\lim_{n\to\infty}(1 + \frac{(-1)^{n+1}}{\sqrt{n}})$; (d) $\lim_{n\to\infty} \frac{5^{n+1}}{7^n}$;

(e) $\lim_{n\to\infty}(\frac{1}{1\cdot 2} + \frac{1}{2\cdot 3} + \cdots + \frac{1}{n\cdot(n+1)})$;

(f) $\lim_{n\to\infty} \frac{n^2+n-1}{3n^2-n+1}$; (g) $\lim_{n\to\infty} \frac{3^n+6^n}{4^n+5^n}$;

(h) $\lim_{n\to\infty}(\frac{1^2}{n^3} + \frac{3^2}{n^3} + \cdots + \frac{(2n-1)^2}{n^3})$;

(i) $\lim_{n\to\infty} \frac{1+a+a^2+\cdots+a^n}{1+b+b^2+\cdots+b^n}$ ($|a| < 1, |b| < 1, ab \neq 0$).

14. Use Heine's theorem to explain why the following limits do not exist:

(a) $\lim_{x\to 0} \cos\frac{1}{x}$; (b) $\lim_{x\to 0} \sin\frac{1}{x^2}$; (c) $\lim_{x\to\infty} \frac{1}{x} \tan x$; (d) $\lim_{x\to 0} x\tan\frac{1}{x}$.

15. For each of the following sequences, determine a_5 and then the limit L. After which N is $|a_n - L| < \frac{1}{100}$? You may use a calculator.

(a) $1, -\frac{1}{2}, \frac{1}{3}, -\frac{1}{4}, \ldots$; (b) $\frac{1}{2}, \frac{2}{4}, \frac{3}{8}, \ldots, \frac{n}{2^n}, \ldots$; (c) $2, 2.2, 2.22, 2.222, \ldots$.

16. Use the monotonic and bounded theorems to prove the convergence of the following sequences:

(a) $\{x_n\} = \{\frac{1}{3+1} + \frac{1}{3^2+1} + \cdots + \frac{1}{3^n+1}\}$;

(b) $\{a_n\} = \{(1 - \frac{1}{2})(1 - \frac{1}{4}) \cdots (1 - \frac{1}{2^n})\}$.

17. (**Prove by induction**) Assume $x_1 = \frac{2}{3}$ and $x_{n+1} = \frac{2}{3-x_n}$. Prove that $x_n = \frac{2^{n+1}-2}{2^{n+1}-1}$ and find $\lim_{n\to\infty} x_n$.

18. If $x_{n+1} = \frac{1}{2}(x_n + \frac{a}{x_n})$, $n = 1,2,3\ldots$, $x_1 > 0$, $a > 0$, prove $\{x_n\}$ is monotonic and bounded and find $\lim_{n\to\infty} x_n$.

19. Suppose $x_1 > 0$ and $x_{n+1} = \frac{c(1+x_n)}{c+x_n}$, where $c > 1$.

(a) Show that, when $x_1 > \sqrt{c}$, $\{x_n\}$ is deceasing and bounded below.

(b) Show that, when $x_1 < \sqrt{c}$, $\{x_n\}$ is increasing and bounded above.

(c) Find $\lim_{n\to\infty} x_n$.

20. Show that $\frac{b^{n+1}-a^{n+1}}{b-a} \leqslant (n+1)b^n$, for $b \geqslant a > 0$.
    (a) Let $a = 1 + 1/(n+1)$ and $b = 1 + 1/n$. Show that $(1 + \frac{1}{n})^n$ is monotonically increasing.
    (b) Let $a = 1$, let $b = 1 + 1/2n$, and deduce that $(1 + 1/2n)^{2n} < 4$.
    (c) Prove that $\{(1 + \frac{1}{n})^n\}$ converges.

21. Give a counterexample for each of the following statements to show that it is false:
    (a) if $x_n \to L$ and $y_n \to L$, then $x_n/y_n \to 1$;
    (b) if $a_n < 0$ and $a_n \to L$, then $L < 0$;
    (c) assume $x_n y_n \to 0$; $\begin{cases} \text{(i)} & \text{if } x_n \text{ is bounded, then } y_n \to 0 \\ \text{(ii)} & \text{if } x_n \text{ is unbounded, then } y_n \text{ must be bounded.} \end{cases}$

22. Use the "e" limit $\lim_{n\to\infty}(1 + \frac{1}{n})^n = e$ to evaluate each of the following limits:
    (a) $\lim_{n\to\infty}(1 + \frac{1}{n+1})^n$;
    (b) $\lim_{n\to\infty}(1 + \frac{1}{n^2})^n$;
    (c) $\lim_{m\to 1^-} m^{\frac{2}{1-m}}$;
    (d) $\lim_{n\to\infty}(1 + \frac{5}{n})^{-2n}$;
    (e) $\lim_{n\to\infty}(\frac{1}{1\cdot 2} + \frac{1}{2\cdot 3} + \cdots + \frac{1}{n\cdot(n+1)})^n$; (f) $\lim_{n\to\infty}(\frac{n+x}{n-1})^n$;
    (g) $\lim_{n\to\infty}(1 - \frac{3}{n})^{3n}$.

23. Use the squeeze theorem to find each of the following limits:
    (a) $\lim_{n\to\infty} \frac{1}{n^2+1} + \frac{1}{n^2+2} + \cdots + \frac{1}{n^2+n}$;
    (b) $\lim_{x\to\infty} \frac{(\arctan x)^2}{x}$;
    (c) $\lim_{x\to\infty} \frac{\sin x + 2\cos \frac{1}{x}}{x}$;
    (d) $\lim_{n\to\infty} \frac{n^2}{3^n}$;
    (e) $\lim_{n\to\infty} \sqrt[n]{a^n + b^n + c^n}$, $a > b > c > 0$; (f) $\lim_{x\to 0} x[\frac{1}{x}]$;
    (g) $\lim_{x\to 0} x \sin \frac{1}{x}$;
    (h) $\lim_{n\to\infty} \frac{100^n}{n!}$.

24. For a sequence $\{a_n\}$, show that $\lim_{n\to\infty} a_n = L$ if and only if $\lim_{n\to\infty} a_{2n} = L$ and $\lim_{n\to\infty} a_{2n+1} = L$.

25. Use the "e" limit $\lim_{x\to\pm\infty}(1 + \frac{1}{x})^x = e$ to evaluate each of the following limits:
    (a) $\lim_{x\to\infty}(1 - \frac{1}{x})^{x+2}$; (b) $\lim_{x\to\infty}(\frac{3x+2}{3x-1})^x$;
    (c) $\lim_{u\to\infty}(\frac{u}{1+u})^{\sqrt{u}}$; (d) $\lim_{x\to 0}(1 + x)^{\frac{2}{x}}$.

26. (**Compounded interest**) Assume P dollars are invested at an annual interest rate r and the **future value** (accumulated value) in the account after t years is $M(t)$.
    (a) Show that, if the interest is compounded n times per year, then $M(t) = P(1 + \frac{r}{n})^{nt}$.
    (b) Show that, if the interest is **compounded continuously**, then $M(t) = Pe^{rt}$.
    (c) Suppose $\$100\,000$ is invested at an annual interest rate of 7%. Compute the future value in the account after 10 years if the interest is compounded (i) quarterly, (ii) monthly, and (iii) continuously.

27. (**Installment loan**) Assume Jack has borrowed an amount of $\$x_0$ from a bank at a monthly interest rate of r. He is supposed to pay the bank a fixed amount of $\$B$ each month. If he owes the bank $\$x_n$ at the end of the nth month, then:
    (a) show that $x_n = (1 + r)x_{n-1} - B$, $n = 1, 2, \ldots$;
    (b) find $\lim_{n\to\infty} x_n$;
    (c) find the number B such that Jack pays off the loan by the end of the mth month.

28. **(Stolz theorem)** Assume the sequence $\{y_n\}$ increases, $\lim y_n = +\infty$, and

$$\lim_{n\to\infty} \frac{x_{n+1} - x_n}{y_{n+1} - y_n} = l, \quad \text{where } \{x_n\} \text{ is a sequence.}$$

Then $\lim_{n\to\infty} \frac{x_n}{y_n} = l$.

Use Stolz's theorem to show that:

(a) $\lim_{n\to\infty} \frac{x_1 + x_2 + \cdots + x_n}{n} = a$, if $\lim_{n\to\infty} x_n = a$; (b) $\lim_{n\to\infty} (n!)^{\frac{1}{n^2}} = 1$.

29. **(Fibonacci sequence)** Fibonacci is considered to be the greatest European mathematician of the middle ages. He was born in Pisa, Italy in approximately AD 1175. He introduced the sequence of numbers named after him in his book of 1202 called Liber Abaci (Book of the Abacus). The Fibonacci sequence is

$$\{1, 1, 2, 3, \ldots, F_{n+2} = F_{n+1} + F_n, \ldots\}.$$

(a) Find the first 15 terms of the Fibonacci sequence.

(b) Let $a_n = \frac{F_{n+1}}{F_n}$. Then:

   (i) determine the first 5 terms of the sequence $\{a_n\}$;

   (ii) show that $a_{n+1} = 1 + \frac{1}{a_n}$ and $a_{n+2} = 2 - \frac{1}{1+a_n}$ and deduce that the sequence $\{a_n\}$ is bounded;

   (iii) either by considering $a_{n+2} - a_n$ or any other method, show that the sequence $\{a_{2k-1}\}$ is increasing and the sequence $\{a_{2k}\}$ is decreasing.

(c) Show that $\lim_{n\to\infty} \frac{F_{n+1}}{F_n} = \frac{1+\sqrt{5}}{2}$.

30. If $\sqrt{1 + ax^2} - 1$ and $\sin^2 x$ are asymptotic as $x \to 0$, find the constant a.

31. If $\sqrt{ax + b} - 2$ and x are asymptotic as $x \to 0$, find the constants a and b.

32. Use asymptotic functions to evaluate each of the following limits:

(a) $\lim_{x\to 0} \frac{\tan 3x}{2x}$; (b) $\lim_{x\to 0^+} \frac{\sin(x^n)}{(\sin x)^m}$, $m, n > 0$;

(c) $\lim_{x\to 0} \frac{\tan x - \sin x}{\sin^3 x}$; (d) $\lim_{x\to 0} \frac{\sin x - \tan x}{(\sqrt[3]{1+x^2}-1)(\sqrt{1+\sin x}-1)}$;

(e) $\lim_{x\to 0} \frac{1 - \cos x}{x \sin x}$; (f) $\lim_{x\to 0} \frac{\sin ax}{\tan bx}$, $b \neq 0$;

(g) $\lim_{x\to\infty} \frac{\sqrt{x+1} - \sqrt{x}}{\sin \sqrt{\frac{1}{x}}}$; (h) $\lim_{n\to\infty} \frac{n^3 \sqrt[n]{2}(1 - \cos \frac{1}{n^2})}{\sqrt{n^2+1}-n}$.

33. Show that $f(x) \sim g(x)$ as $x \to a$ if and only if $f(x) = g(x) + o(f(x))$ as $x \to a$.

34. Assume both $f(x)$ and $g(x)$ are infinitesimal when $x \to a$ and m is a positive integer. If $\lim_{x\to a} \frac{f(x)}{g^m(x)} = c \neq 0$, then we say that the infinitesimal function $f(x)$ has order m with respect to the other infinitesimal function $g(x)$ as $x \to a$.

(a) Show that $\sqrt{1 + x^3} - 1$ is negligible with respect to x as $x \to 0$.

(b) Find the order of $\sqrt{1 + x^3} - 1$ with respect to x as $x \to 0$.

35. Show that $o(x^2) + o(x^3) = o(x^2)$ and $o(x^2) - o(x^3) = o(x^2)$ as $x \to 0$.

36. Find the value of a such that each of the following functions is continuous:

(a) $f(x) = \begin{cases} x^2 - 1, & x < 3 \\ 2ax, & x \geq 3; \end{cases}$ (b) $f(x) = \begin{cases} 2x + 3, & x \leq 2 \\ ax + 1, & x > 2; \end{cases}$

(c) $f(x) = \begin{cases} e^x, & x < 0 \\ a + \ln(1 + x), & x \geq 0; \end{cases}$ (d) $f(x) = \begin{cases} e^{\frac{1}{x}}, & x < 0 \\ 0, & x = 0 \\ x^a \sin \frac{1}{x}, & x > 0. \end{cases}$

37. Locate and classify the discontinuities for the function whose graph is given in question 2.

38. Locate and classify the discontinuities for each of the following functions:

(a) $f(x) = \frac{\sin x}{|x|}$;

(b) $f(x) = \frac{1}{x^2 - 3x + 2}$;

(c) $g(t) = \frac{\sin t}{t^2 - 1}$;

(d) $t(\theta) = \frac{1}{1 - e^{\frac{\theta}{1-\theta}}}$;

(e) $f(x) = \lim_{n \to \infty} \frac{nx}{nx^2 + 1}$, $-1 \leqslant x < 1$;

(f) $k(y) = y \cos \frac{1}{y}$;

(g) $f(x) = [2x] - [x]$;

(h) $h(x) = \frac{x^2 - x}{|x|(x^2 - 1)}$;

(i) $f(x) = \lim_{t \to \infty} \frac{1 - x e^{tx}}{x + e^{tx}}$.

39. If $\lim_{x \to a} f(x) = L$, $\lim_{x \to a} g(x) = M$, both L and M are finite numbers, and $LM \neq 0$, prove that $\lim_{x \to a} (f(x))^{g(x)} = L^M$. Then find $\lim_{x \to 2} (x^2 + 1)^{\sin \frac{\pi}{x}}$.

40. Evaluate each of the following limits:

(a) $\lim_{x \to \pi} \frac{\sin(e^{\pi - x})}{x^2}$;

(b) $\lim_{x \to 0} \frac{\ln(1+x)}{\sin x}$;

(c) $\lim_{x \to 0} \frac{\arctan x}{\ln(1 + \sin x)}$;

(d) $\lim_{x \to 0} \frac{x(1 - \cos x)}{(1 - e^x) \sin x^2}$;

(e) $\lim_{y \to 0} \frac{3^y - 1}{y}$;

(f) $\lim_{\Delta x \to 0} \frac{e^{x + \Delta x} - e^x}{\Delta x}$;

(g) $\lim_{n \to \infty} (\frac{\sqrt[n]{a} + \sqrt[n]{b}}{2})^n$;

(h) $\lim_{n \to \infty} (\ln(n + 2) - \ln n)$;

(i) $\lim_{x \to 0} \frac{(\sin x) \tan(\ln(1+x))}{x \arcsin x}$;

(j) $\lim_{n \to \infty} (1 + \frac{2}{n} + \frac{3}{n^2})^n$;

(k) $\lim_{n \to \infty} [\tan(\frac{\pi}{4} + \frac{1}{n})]^n$.

41. Is it true that a continuous function that is never 0 on an interval never changes sign on that interval? Explain.

42. (**Dirichlet function**) Show that the function

$$D(x) = \begin{cases} 1, & \text{if } x \text{ is rational} \\ 0, & \text{if } x \text{ is irrational} \end{cases}$$

is discontinuous at every point. Explore the continuity of the function $y = x^2 D(x)$ at $x = 0$ (the Dirichlet function can be written analytically as $D(x) = \lim_{m \to \infty} \lim_{n \to \infty} \cos^{2n}(m! \pi x)$).

43. (**Riemann function**) Explore the continuity of the Riemann function (also called the Dirichlet ruler function)

$$R(x) = \begin{cases} \frac{1}{q}, & \text{if } x = \frac{p}{q} \text{ is a rational number in lowest terms} \\ 0, & \text{otherwise.} \end{cases}$$

44. Prove that the equation $x = a \sin x + b$ ($a > 0$, $b > 0$) has at least one positive root that is not more than $a + b$.

45. If $f(x)$ is continuous on $[0, 1]$, $f(0) = 0$, and $f(1) = 1$, then show that, for all positive integers n, there is a number c such that $0 < c < 1$ and $(f(c))^n + c = 1$.

46. Assume $f(x)$ is continuous on $[0, 2a]$ and $f(0) = f(2a)$. Show that there is a number $x_0 \in [0, a]$ such that $f(x_0 + a) = f(x_0)$.

47. Prove that the function $f(x) = 2^x - x^2$ has a negative root.

48. Suppose $P(a, b)$ is a point on the parabola $y = x^2$ and O is the origin. The perpendicular bisector of the line segment OP meets the y-axis at point $Q(0, c)$. Find $\lim_{a \to 0} c$ or explain why it does not exist.

49. (**Fixed point**) If $f(x)$ is continuous on $[0, 1]$ and $0 < f(x) < 1$, then show that there is a number $c \in (0, 1)$ such that $f(c) = c$ (this number c is called a fixed point of the function $f(x)$).

50. (**Clock hands**) Explain why there is a time when the hour hand and the minute hand of a clock coincide in each hour.

51. (**Body weight**) If you weighed 50 pounds at the beginning of year 2014 and by the end of 2014 you weighed 59 pounds, explain why there was a time point on which your weight was exactly 54 pounds.

52. (**Stabilizing a wobbly table**) A table whose four legs are equal in length was put on an uneven floor and it was unstable initially. By slightly rotating the table, it can be stabilized on the floor. Set up a mathematical model for this situation and explain.

53. Consider the function $f(x) = \begin{cases} 2 - x, & x < 0 \\ 2 + \sqrt{x}, & x \geq 0. \end{cases}$ Does $f(x)$ have a maximum and a minimum value on $[-2, 4]$? How does this example illustrate the extreme value theorem?

54. If $g(x) = \begin{cases} -x, & x \leq 0 \\ \frac{1}{x}, & x > 0, \end{cases}$ does $g(x)$ have a maximum value and a minimum value on $[-2, 4]$? Does this example contradict the extreme value theorem? Why or why not?

55. If $h(x) = \begin{cases} -x, & x \leq 0 \\ x^2 + 1, & x > 0, \end{cases}$ does $h(x)$ have a maximum value and a minimum value on $[-2, 4]$? Is $h(x)$ continuous on $[-2, 4]$? Does this example contradict the extreme value theorem? Why or why not?

56. Assume, for all $x \in (-\infty, \infty)$, $f(x^2) = f(x)$. If f is continuous at $x = 0$ and $x = 1$, then prove that $f(x)$ is constant.

3 The derivative

In this chapter, you will learn about:
- *the definition of the derivative;*
- *how to find derivatives using differentiation rules;*
- *linearization and the differentials.*

The derivative of a function $y = f(x)$ is another function, derived from $f(x)$, that gives information about how the function $f(x)$ changes, compared with changes in the independent variable x. Differential calculus studies methods for finding derivatives and applications of the derivatives in areas of mathematics, sciences, engineering, and many other disciplines.

3.1 Derivative of a function at a point

3.1.1 Instantaneous rates of change and derivatives revisited

As seen in Chapters 1 and 2, the average rate of change (average velocity) of a car moving with variable speed during a specified time interval $[t_1, t_2]$ is given by

$$\bar{v} = \frac{\Delta s}{\Delta t} = \frac{s(t_2) - s(t_1)}{t_2 - t_1}.$$

But what is the velocity of the particle at any given moment in time? How is this "instantaneous" velocity defined? That is, how can you measure the velocity of that car at a particular moment in time when the car's velocity is continually changing?

In order to study this problem mathematically, let us suppose that an object is moving along a straight line with variable velocity and its displacement s (from some fixed origin point) is given, as a function of time t, by $s = f(t)$. We will try to find a formula for the velocity of the object at a particular time t_0. If Δt is a small positive or negative time increment, then at time $t_0 + \Delta t$, the object has displacement $f(t_0 + \Delta t)$, so the average velocity during the time interval $(t_0, t_0 + \Delta t)$ is the change in displacement divided by the change in time:

$$\frac{f(t_0 + \Delta t) - f(t_0)}{\Delta t}.$$

It is reasonable to expect that, as Δt gets smaller and smaller, the average velocity will become closer and closer to the velocity at the particular time t_0. Hence, we define the instantaneous velocity of the object at time t_0 to be

$$\lim_{\Delta t \to 0} \frac{\Delta s}{\Delta t} = \lim_{\Delta t \to 0} \frac{\text{change in displacement } s}{\text{change in time } t}$$

https://doi.org/10.1515/9783110527780-003

$$= \lim_{\Delta t \to 0} \frac{f(t_0 + \Delta t) - f(t_0)}{\Delta t}$$

$$= \text{instantaneous velocity at time } t_0,$$

provided that this limit exists.

Now we consider a more general form. For the function $y = f(x)$, assume $M(x_0, y_0)$ is a point on the graph C of f. We choose a small positive or negative x-increment Δx which takes us to a nearby point $N(x_0 + \Delta x, y_0 + \Delta y)$ on C. The change in y is Δy, which is

$$\Delta y = f(x_0 + \Delta x) - f(x_0).$$

The *secant* line MN is the straight line joining M and N, as shown in Figure 3.1.1 and Figure 3.1.2. Notice that the slope m of the secant line MN is given by

$$m = \frac{\Delta y}{\Delta x} = \frac{f(x_0 + \Delta x) - f(x_0)}{\Delta x}.$$

It is reasonable to expect that, if N approaches M along the curve, the secant line will become closer and closer to the tangent line that touches the curve C at M. Also, the slope of the secant line MN will approach the slope of the tangent line. Hence, we define the *slope of the tangent line* at M to be the limit (if the limit exists) of the slope

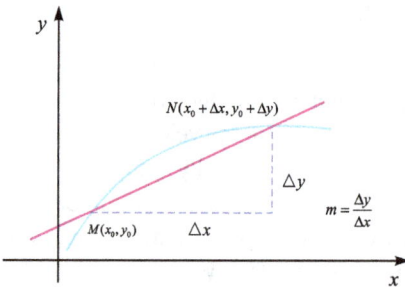

Figure 3.1.1: Secant line with positive slope.

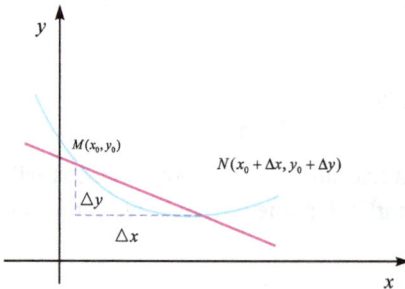

Figure 3.1.2: Secant line with negative slope.

m of the secant line as $\Delta x \to 0$, so we have

$$\lim_{\Delta x \to 0} \frac{f(x_0 + \Delta x) - f(x_0)}{\Delta x}$$

$$= \lim_{\Delta x \to 0} \frac{\Delta y}{\Delta x}$$

$$= \text{the slope of the tangent at } M(x_0, y_0).$$

Now, we have encountered the same type of limits twice even though they arose from different backgrounds. We will see that limits of the form

$$\lim_{\Delta x \to 0} \frac{f(x_0 + \Delta x) - f(x_0)}{\Delta x}$$

arise in many scientific fields whenever we want to calculate an instantaneous rate of change, such as a rate of a reaction in chemistry, or a marginal cost in economics. Since this type of limit occurs so frequently, it is worthy of being given a special name and notation. It is called the *derivative* of the function f at M and it is denoted by $f'(x_0)$.

Definition 3.1.1. The *derivative of a function at a point* $(x_0, f(x_0))$ is written as $f'(x_0)$ and is defined by

$$f'(x_0) = \lim_{\Delta x \to 0} \frac{f(x_0 + \Delta x) - f(x_0)}{\Delta x}.$$

The process of finding the derivative is called *differentiation*.

NOTES. The *derivative of a function at a point* $M(x_0, y_0)$ is:
1. the slope of the tangent line to the graph of f at M;
2. the instantaneous rate of change of $f(x)$ with respect to x at M.

There are many different notations for the derivative, or the rate of change of a function. The notation used often depends on how the function f is presented or on how the derivative is to be used. When a function is defined by an equation $y = f(x)$, the following are all notations for the derivative of f (or y) with respect to x at $x = x_0$:

$$f'(x_0), \frac{dy}{dx}\bigg|_{x=x_0}, \frac{df}{dx}\bigg|_{x=x_0}, \frac{d}{dx}f(x)\bigg|_{x=x_0},$$

$$D_x f(x_0), \dot{y}(x_0), y'(x_0), \quad \text{and} \quad (f(x))'_{x=x_0}.$$

Example 3.1.1. Find the derivative of $f(x) = x^2$ at $x = 2$.

Solution. We have

$$f'(2) = \lim_{\Delta x \to 0} \frac{f(2 + \Delta x) - f(2)}{\Delta x} = \lim_{\Delta x \to 0} \frac{(2 + \Delta x)^2 - 2^2}{\Delta x}$$

$$= \lim_{\Delta x \to 0} \frac{4 + 4\Delta x + (\Delta x)^2 - 4}{\Delta x} = \lim_{\Delta x \to 0} \frac{4\Delta x + (\Delta x)^2}{\Delta x}$$

$$= \lim_{\Delta x \to 0} (4 + \Delta x)$$

$$= 4,$$

so the derivative of $f(x) = x^2$ at $x = 2$ is 4.

Example 3.1.2. Find an equation of the tangent line to the graph of $f(x) = \frac{1}{x}$ at $(1,1)$.

Solution. First find the slope of the tangent line. We have

$$f'(1) = \lim_{\Delta x \to 0} \frac{f(1 + \Delta x) - f(1)}{\Delta x} = \lim_{\Delta x \to 0} \frac{\frac{1}{1 + \Delta x} - \frac{1}{1}}{\Delta x}$$

$$= \lim_{\Delta x \to 0} \frac{\frac{1 - (1 + \Delta x)}{1 + \Delta x}}{\Delta x}$$

$$= \lim_{\Delta x \to 0} \frac{-\Delta x}{\Delta x (1 + \Delta x)}$$

$$= \lim_{\Delta x \to 0} -\frac{1}{1 + \Delta x}$$

$$= -1.$$

This means the derivative of $f(x)$ at $x = 1$ is -1, so the slope of the line tangent to the graph of $f(x)$ at $x = 1$ is -1. Therefore, the point-slope form of the tangent line at $(1,1)$ is

$$y - 1 = -1 \times (x - 1),$$

which is the same as $y = -x + 2$ (slope-intercept form). Figure 3.1.3 shows the graph of $f(x)$ and its tangent at $x = 1$.

Example 3.1.3. A ball is dropped from rest from the top of a building. In the subsequent motion under gravity the displacement of the ball with respect to time is given

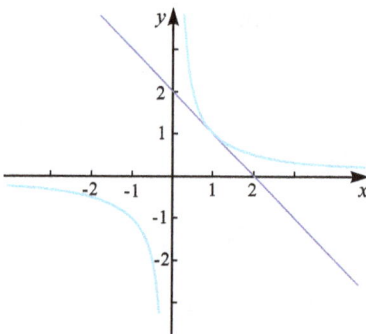

Figure 3.1.3: Graphs of $y = \frac{1}{x}$ and $y = -x + 2$.

by $s = \frac{1}{2}gt^2$, where g is a constant called gravitational acceleration. Find the velocity of the ball at any time t. Assume any air resistance is negligible.

Solution. The velocity at time t is the derivative of $s(t)$ at t. We have

$$v(t) = s'(t) = \lim_{\Delta t \to 0} \frac{s(t + \Delta t) - s(t)}{\Delta t}$$

$$= \lim_{\Delta t \to 0} \frac{\frac{1}{2}g(t + \Delta t)^2 - \frac{1}{2}gt^2}{\Delta t}$$

$$= \frac{1}{2}g \lim_{\Delta t \to 0} \frac{(t + \Delta t)^2 - t^2}{\Delta t}$$

$$= \frac{1}{2}g \lim_{\Delta t \to 0} \frac{t^2 + 2t \cdot \Delta t + (\Delta t)^2 - t^2}{\Delta t}$$

$$= \frac{1}{2}g \lim_{\Delta t \to 0} \frac{2t \cdot \Delta t + (\Delta t)^2}{\Delta t}$$

$$= \frac{1}{2}g \lim_{\Delta t \to 0} (2t + \Delta t)$$

$$= \frac{1}{2}g \times 2t$$

$$= gt,$$

so the instantaneous velocity at any time t is gt. For example, at $t = 1\,\text{sec}$, the velocity is $g \times 1 \approx 9.8\,\text{m/sec}$.

Some equivalent limits for finding derivatives at a point

We simplify (at least it looks simpler) the limit

$$\lim_{\Delta x \to 0} \frac{f(x_0 + \Delta x) - f(x_0)}{\Delta x}$$

by replacing Δx by h and x_0 by a to obtain

$$\lim_{\Delta x \to 0} \frac{f(x_0 + \Delta x) - f(x_0)}{\Delta x} = \lim_{h \to 0} \frac{f(a + h) - f(a)}{h}.$$

If we let $x = a + h$, then $h = x - a$. When $h \to 0$, then $x \to a$, so we have

$$\lim_{h \to 0} \frac{f(a + h) - f(a)}{h} = \lim_{x \to a} \frac{f(x) - f(a)}{x - a}.$$

Therefore, the derivative of a function f at the point $x = a$ (where $y = f(a)$) can also be given by the following two limits:

$$f'(a) = \lim_{h \to 0} \frac{f(a + h) - f(a)}{h} \quad \text{or}$$

$$f'(a) = \lim_{x \to a} \frac{f(x) - f(a)}{x - a}.$$

Example 3.1.4. Find the derivative of the function $f(x) = mx + b$ at $x = a$.

Solution. From the definition of the derivative, we have

$$\begin{aligned} f'(a) &= \lim_{h \to 0} \frac{f(a+h) - f(a)}{h} \\ &= \lim_{h \to 0} \frac{m(a+h) + b - (ma + b)}{h} \\ &= \lim_{h \to 0} \frac{ma + mh + b - ma - b}{h} \\ &= \lim_{h \to 0} \left(\frac{mh}{h} \right) \\ &= m, \end{aligned}$$

so

$$\frac{d}{dx}(mx + b)\Big|_{x=a} = m.$$

NOTE. At each point on the straight line $y = mx + b$, the derivative is the same as the slope m of the line.

Example 3.1.5. Find $\frac{dy}{dx}\big|_{x=1}$ for the function $y = 1 + \sqrt{x}$ and find the line tangent to the graph of y at $(1, 2)$.

Solution. The derivative of $f(x) = 1 + \sqrt{x}$ at $x = 1$ is given by

$$\begin{aligned} \frac{dy}{dx}\Big|_{x=1} &= \lim_{h \to 0} \frac{f(1+h) - f(1)}{h} \\ &= \lim_{h \to 0} \frac{(1 + \sqrt{1+h}) - (1 + \sqrt{1})}{h} \\ &= \lim_{h \to 0} \frac{\sqrt{1+h} - 1}{h} \\ &= \lim_{h \to 0} \frac{(\sqrt{1+h} - 1)(\sqrt{1+h} + 1)}{h(\sqrt{1+h} + 1)} \\ &= \lim_{h \to 0} \frac{h}{h(\sqrt{1+h} + 1)} \\ &= \lim_{h \to 0} \frac{1}{(\sqrt{1+h} + 1)} \\ &= \frac{1}{2}. \end{aligned}$$

Hence, the slope of the tangent line at $(1, 2)$ is $f'(1) = \frac{1}{2}$, so an equation of the tangent line is

$$y - 2 = \frac{1}{2}(x - 1) \quad \text{or} \quad y = \frac{1}{2}x + \frac{3}{2}.$$

The graph of this curve and its tangent line are shown in Figure 3.1.4.

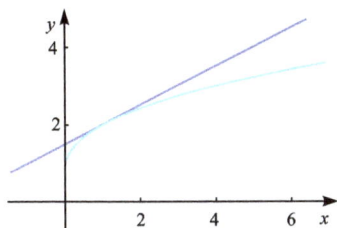

Figure 3.1.4: Graphs of $y = 1 + \sqrt{x}$ and $y = \frac{1}{2}x + \frac{3}{2}$.

Example 3.1.6. Find the derivative of $f(x) = x^3$ at $x = 2$ and find the normal line to the graph of $f(x)$ at $x = 2$.

Solution. We use the form

$$f'(2) = \lim_{x \to 2} \frac{f(x) - f(2)}{x - 2} = \lim_{x \to 2} \frac{x^3 - 2^3}{x - 2}$$

$$= \lim_{x \to 2} \frac{(x - 2)(x^2 + 2x + 4)}{x - 2}$$

$$= \lim_{x \to 2} (x^2 + 2x + 4)$$

$$= 2^2 + 2 \times 2 + 4$$

$$= 12.$$

Now, when $x = 2$, $y = 2^3 = 8$. Since the slope at $x = 2$ is 12, the normal line at the point $(2, 8)$ has slope $-\frac{1}{12}$. The point-slope equation of the normal line to the graph of $f(x)$ is

$$y - 8 = -\frac{1}{12}(x - 2) \quad \text{or} \quad y = -\frac{1}{12}x + \frac{49}{6}.$$

Example 3.1.7. Find $\frac{df}{dx}\big|_{x=0}$ for the function

$$f(x) = \begin{cases} x^2 \sin \frac{1}{x}, & \text{if } x \neq 0 \\ 0, & \text{if } x = 0. \end{cases}$$

Solution. We have

$$\frac{df}{dx}\bigg|_{x=0} = f'(0) = \lim_{x \to 0} \frac{f(x) - f(0)}{x - 0} = \lim_{x \to 0} \frac{x^2 \sin \frac{1}{x} - 0}{x - 0}$$

$$= \lim_{x \to 0} \left(x \sin \frac{1}{x} \right)$$

$$= 0.$$

The value of the last limit is zero because $0 \leqslant |x \sin \frac{1}{x}| \leqslant |x|$, so, by the squeeze theorem from Chapter 2, we have

$$0 \leqslant \lim_{x \to 0} \left| x \sin \frac{1}{x} \right| \leqslant \lim_{x \to 0} |x| = 0.$$

This means $\lim_{x \to 0} (x \sin \frac{1}{x}) = 0$.

3.1.2 One-sided derivatives

Since the derivative of a function $y = f(x)$ at $x = a$ is defined by the limit

$$f'(a) = \lim_{h \to 0} \frac{f(a+h) - f(a)}{h},$$

as we did in one-sided limits, the corresponding two one-sided limits are given special names.

Definition 3.1.2. The limits

$$f'_+(a) = \lim_{h \to 0^+} \frac{f(a+h) - f(a)}{h},$$

$$f'_-(a) = \lim_{h \to 0^-} \frac{f(a+h) - f(a)}{h}$$

are called the *one-sided derivatives of f at a*. The *right-hand derivative of f at a* (or the *right derivative*) is $f'_+(a)$ and the *left-hand derivative of f at a* (or the *left derivative*) is $f'_-(a)$.

In view of the results we have obtained from limits and one-sided limits, we conclude that $f'(a)$ exists if and only if the two one-sided limits $f'_+(a)$ and $f'_-(a)$ both exist and are equal. That is,

$$f'(a) = L \quad \text{if and only if } f'_+(a) = f'_-(a) = L.$$

Example 3.1.8. Investigate $f'(0)$ for $f(x) = |x|$.

Solution. Since

$$\lim_{h \to 0} \frac{f(0+h) - f(0)}{h} = \lim_{h \to 0} \frac{|h|}{h}$$

and

$$f'_+(0) = \lim_{h \to 0^+} \frac{|h|}{h} = 1 \quad \text{but } f'_-(0) = \lim_{h \to 0^-} \frac{|h|}{h} = -1,$$

$f'(0)$ does not exist.

As seen from the graph of $|x|$, when $x < 0$, the slope of the curve (and tangent line) is -1 and when $x > 0$, the slope of the curve (and tangent line) is 1. There is a sharp "corner" in the graph at $x = 0$, where the slope of the curve abruptly changes from -1 to 1. Whenever this happens, the function has no derivative at this point.

3.1.3 A function may fail to have a derivative at a point

We have seen that $y = |x|$ has no derivative at $x = 0$. Intuitively speaking, there is a corner at $x = 0$ where $f(x)$ changes abruptly. The analytical reason why there is no derivative is because the two one-sided derivatives do not match. From a geometrical point of view, the left derivative is the slope of the "left tangent line", which is the limiting secant line MN when the point M approaches N from the left. The right derivative is the slope of the "right tangent line", which is the limiting secant line PN when P approaches N from the right, as shown in Figure 3.1.5. If the slopes of the two tangents do not match, then at that point, there will be no unique tangent line. Figure 3.1.6 and Figure 3.1.7 show cases that a function fails to have a derivative. This lack of derivative can also be interpreted in terms of rates of change. At any point on the graph of a function $f(x)$, the instantaneous rate of change of $f(x)$, which is the slope of the curve at

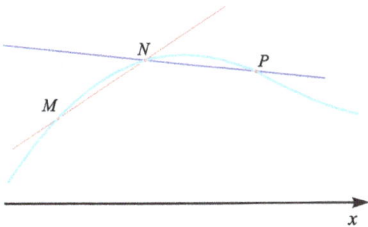

Figure 3.1.5: "Smooth" curves have tangents.

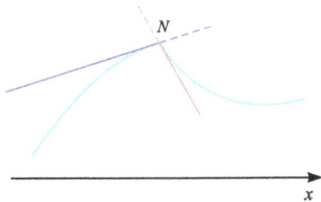

Figure 3.1.6: There is no derivative at N.

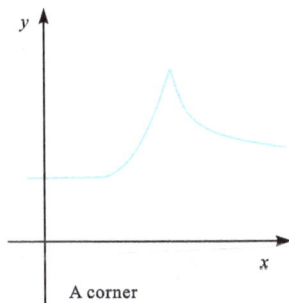

A corner

Figure 3.1.7: A function has no derivative at a corner.

that point, must be unique if it exists. You cannot have two different slopes/tangents at a single point.

Now let us think about another case where the derivative of a function does not exist: when $f(x)$ has a vertical tangent at $x = a$. The slope of a vertical line is undefined, or you could say the slope is $\pm\infty$. This happens when

$$\lim_{h \to 0} \frac{f(a+h) - f(a)}{h} = \pm\infty,$$

or, equivalently,

$$\lim_{x \to a} \frac{f(x) - f(a)}{x - a} = \pm\infty.$$

For example, if $y = \sqrt[3]{x}$, then, at $x = 0$,

$$f'(0) = \lim_{x \to 0} \frac{f(x) - f(0)}{x - 0} = \lim_{x \to 0} \frac{\sqrt[3]{x} - 0}{x - 0} = \lim_{x \to 0} \frac{1}{\sqrt[3]{x^2}} = \infty.$$

In this case, the graph of $f(x) = \sqrt[3]{x}$ has a vertical tangent at $x = 0$, as seen in Figure 3.1.9.

There is one other situation where a function $f(x)$ fails to have a derivative at $x = a$. If $f(x)$ is undefined at $x = a$, then it is easy to see that $f'(a)$ does not exist, since, in the limit

$$\lim_{h \to 0} \frac{f(a+h) - f(a)}{h},$$

$f(a)$ is undefined. If $f(a)$ is defined, as in Figure 3.1.8, but $f(x)$ has a removable or jump discontinuity at $x = a$, then we see from the graph that the "left tangent line" and the "right tangent line" either are not the same, or one (both) tangent line(s) is (are) vertical. In fact, analytically, we have the following theorem.

Theorem 3.1.1. *If the function $f(x)$ has a derivative at $x = a$, then f is continuous at $x = a$. That is, $\lim_{x \to a} f(x) = f(a)$.*

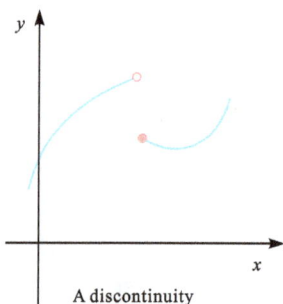

A discontinuity

Figure 3.1.8: A function has no derivative at any of its discontinuities.

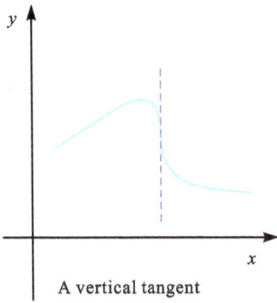

A vertical tangent

Figure 3.1.9: A function has no derivative where its tangent is vertical.

Proof. Take the limit

$$\lim_{x \to a}(f(x) - f(a)) = \lim_{x \to a}\frac{f(x) - f(a)}{x - a}(x - a)$$
$$= \lim_{x \to a}\frac{f(x) - f(a)}{x - a}\lim_{x \to a}(x - a)$$
$$= f'(a) \times 0 = 0.$$

This shows that $\lim_{x \to a}f(x) = f(a)$, which means that $f(x)$ is continuous at $x = a$. □

If $f(x)$ has a derivative at $x = a$, that is, if $f'(a)$ exists, then we say that $f(x)$ is *differentiable* at $x = a$. Otherwise, f is not differentiable at $x = a$. For example, $y = |x|$ is not differentiable at $x = 0$. To be differentiable, intuitively, a curve must be continuous and smooth (have a unique tangent line at all points). Theorem 3.1.1 says that *differentiability implies continuity* and thus we also know that $f(x)$ is not differentiable at any of its discontinuities.

NOTE. Continuity is a necessary but not sufficient condition for a function f to have a derivative at point $x = a$. For example, $y = |x|$ is continuous at $x = 0$, but not differentiable there.

Example 3.1.9. Find a and b such that the function

$$f(x) = \begin{cases} x^2 + 1, & \text{when } x \geqslant 1 \\ ax + b, & \text{when } x < 1 \end{cases}$$

is differentiable at $x = 1$.

Solution. First, $f(x)$ must be continuous at $x = 1$, so the left and right limits of f as x approaches 1 must be equal. Thus

$$\lim_{x \to 1^+} f(x) = \lim_{x \to 1^-} f(x) = f(1),$$

$$\lim_{x\to1^-}(ax+b) = a+b = 2 = f(1).$$

Since $f(x)$ needs to be differentiable at $x = 1$, the two one-sided derivatives must exist and must be equal to each other. This means we must have

$$f'_+(1) = f'_-(1).$$

Hence, we have

$$\lim_{x\to1^-}\frac{f(x)-f(1)}{x-1} = \lim_{x\to1^+}\frac{f(x)-f(1)}{x-1},$$

$$\lim_{x\to1^-}\frac{ax+b-2}{x-1} = \lim_{x\to1^+}\frac{x^2+1-2}{x-1}.$$

However, since $a+b=2$, $b-2=-a$, so

$$\lim_{x\to1^-}\frac{a(x-1)}{x-1} = \lim_{x\to1^+}\frac{x^2-1}{x-1},$$

$$a\lim_{x\to1^-}1 = \lim_{x\to1^+}x+1,$$

$$a = 2.$$

Therefore, $b = 2 - a = 0$. Moreover, we have $f'(1) = 2$. Figure 3.1.10 shows the graph of $f(x)$.

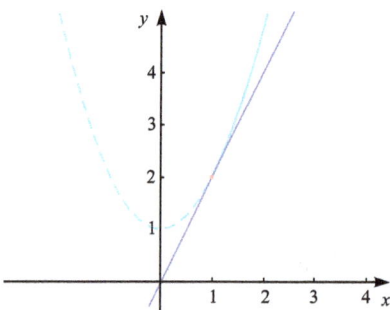

Figure 3.1.10: Graphs of $y = 2x$ and $y = f(x)$ in Example 3.1.9.

Notice that the line $y = 2x$ is the tangent to the curve $y = x^2 + 1$ at $x = 1$.

There are other ways in which the derivative fails to exist, but these are unusual. For example,

$$f(x) = \begin{cases} x\sin(\frac{1}{x}), & \text{if } x \neq 0 \\ 0, & \text{if } x = 0 \end{cases}$$

is continuous at $x = 0$ but it has no derivative there since

$$\lim_{x\to0}\frac{f(x)-f(0)}{x-0} = \lim_{x\to0}\frac{x\sin\frac{1}{x}-0}{x-0} = \lim_{x\to0}\sin\frac{1}{x} \quad \text{does not exist.}$$

Geometrically, this is because this curve oscillates up and down, with large negative and positive slopes, as $x \to 0$, see Figure 3.1.12. Notice that this is different from the case shown in Example 3.1.7, where

$$f(x) = \begin{cases} x^2 \sin(\frac{1}{x}), & \text{if } x \neq 0 \\ 0, & \text{if } x = 0, \end{cases}$$

where the factor x^2 "smooths out" the oscillations as $x \to 0$, as shown in Figure 3.1.11.

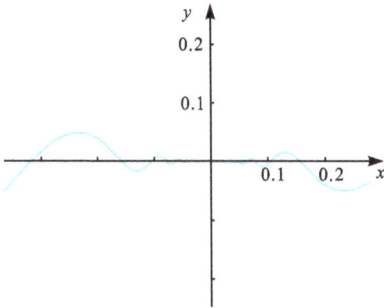

Figure 3.1.11: Graph of $f(x) = x^2 \sin \frac{1}{x}$.

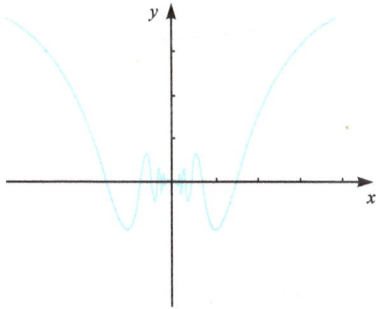

Figure 3.1.12: Graph of $f(x) = x \sin \frac{1}{x}$.

3.2 Derivative as a function

If x is equal to a specific number, say, $x = a$, then the derivative of f at a is the number denoted by $f'(a)$. If $f(x)$ is differentiable at any point x in its domain, the derivative $f'(x)$ is a variable depending on the value of x. Thus we can think of the values $f'(x)$ as defining a function f' defined by the equation

$$f'(x) = \lim_{h \to 0} \frac{f(x+h) - f(x)}{h}. \tag{3.1}$$

This function is defined for all x in the domain of f for which this limit exists. When $y = f(x)$, the derivative of f as a function can be written

$$f'(x), \quad \frac{dy}{dx}, \quad y', \quad \frac{df}{dx}, \quad (f(x))' \quad \text{or} \quad \frac{d}{dx}f(x).$$

To find the derivative of $f(x)$ at a point $x = a$, we can find the derivative of f first and then substitute x with a, that is,

$$f'(x)|_{x=a} = f'(a).$$

3.2.1 Graphing the derivative of a function

Notice that the derivative of a function is the slope of that function, so we could graph the derivative for some simple functions.

Example 3.2.1. Graph the derivative of

$$f(x) = \begin{cases} 2x + 1, & \text{when } x \geqslant 1 \\ -x + 2, & \text{when } x < 1. \end{cases}$$

Solution. The graph of the function $f(x)$ consists of two half lines. When $x > 1$, the half line is $y = 2x + 1$ and has slope 2. When $x < 1$, the half line is $y = -x + 2$ and the slope is -1, so we have

$$f'(x) = \begin{cases} 2, & \text{for } x > 1 \\ -1, & \text{for } x < 1. \end{cases}$$

The graph of $f(x)$ is shown in Figure 3.2.1. It is no surprise that $f(x)$ has no derivative at $x = 1$ since f is not continuous there, as shown in Figure 3.2.2.

NOTE. If the graph is increasing, then the slope must be positive. If the graph is decreasing, the slope is negative.

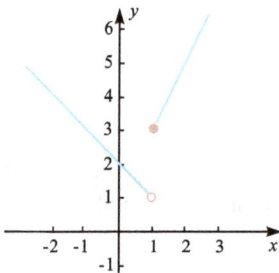

Figure 3.2.1: Graph of $f(x)$ in Example 3.2.1.

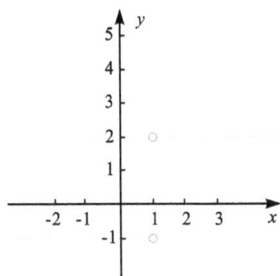

Figure 3.2.2: The derivative of $f(x)$ in Example 3.2.1.

3.2.2 Derivatives of some basic functions

Example 3.2.2. If $f(x) = C$, where C is a constant, find $f'(x)$.

Solution. By equation (3.1),

$$f'(x) = \lim_{h \to 0} \frac{f(x+h) - f(x)}{h} = \lim_{h \to 0} \frac{C - C}{h} = \lim_{h \to 0} 0 = 0.$$

This means

$$\frac{d}{dx}(C) = 0 \quad \text{or} \quad (C)' = 0.$$

This example proves that the derivative of a constant function is always zero. This is to be expected because there is no change in $f(x)$, so the rate of change is of course 0. Geometrically, the graph of a constant function is a horizontal line, whose tangent line always has slope 0.

Example 3.2.3. Find the derivative of $y = x^n$, where n is a positive integer.

Solution. We use the binomial theorem to expand $(x + h)^n$. We have

$$y' = \lim_{h \to 0} \frac{f(x+h) - f(x)}{h} = \lim_{h \to 0} \frac{(x+h)^n - x^n}{h}$$

$$= \lim_{h \to 0} \frac{(x^n + nx^{n-1}h + \frac{n(n-1)}{2}x^{n-2}h^2 + \cdots + h^n) - x^n}{h}$$

$$= \lim_{h \to 0} \left(\frac{nx^{n-1}h + \frac{n(n-1)}{2}x^{n-2}h^2 + \cdots + h^n}{h} \right)$$

$$= \lim_{h \to 0} \left(nx^{n-1} + \frac{n(n-1)}{2}x^{n-2}h + \cdots + h^{n-1} \right)$$

$$= nx^{n-1}.$$

Therefore,

$$\frac{d}{dx}(x^n) = nx^{n-1} \quad \text{or} \quad (x^n)' = nx^{n-1}.$$

NOTE. This is a special case of the power rule $(x^u)' = ux^{u-1}$, for $u \in \mathbf{R}$. We have proved that the rule is valid when u is a positive integer.

Example 3.2.4. Find the derivative of $y = \sin x$.

Solution. Use the trigonometric formula

$$\sin A - \sin B = 2\cos\left(\frac{A+B}{2}\right)\sin\left(\frac{A-B}{2}\right)$$

as follows:

$$y' = \lim_{h \to 0} \frac{\sin(x+h) - \sin x}{h} = \lim_{h \to 0} \frac{2\cos(\frac{2x+h}{2})\sin(\frac{h}{2})}{h}$$

$$= \lim_{h \to 0} \cos\left(x + \frac{h}{2}\right)\frac{\sin(\frac{h}{2})}{(\frac{h}{2})}$$

$$= \lim_{h \to 0} \cos\left(x + \frac{h}{2}\right) \times \lim_{\frac{h}{2} \to 0} \frac{\sin(\frac{h}{2})}{(\frac{h}{2})}$$

$$= \cos x,$$

so

$$\frac{d}{dx}(\sin x) = \cos x \quad \text{or} \quad (\sin x)' = \cos x.$$

NOTE. In a similar way, one can prove that

$$\frac{d}{dx}(\cos x) = -\sin x \quad \text{or} \quad (\cos x)' = -\sin x.$$

Example 3.2.5. Find equations for the line tangent to the graph of $f(x) = \sin x$ at the point $x = \frac{\pi}{3}$ and for the normal line at the same point.

Solution. Since

$$f'(x) = (\sin x)' = \cos x,$$

we find

$$f'\left(\frac{\pi}{3}\right) = \cos x|_{x=\frac{\pi}{3}} = \cos\frac{\pi}{3} = \frac{1}{2}.$$

This means the line tangent to the graph of $\sin x$ at $x = \frac{\pi}{3}$ has slope $\frac{1}{2}$, so the line normal to the graph of $\sin x$ at the same point has slope -2. When $x = \frac{\pi}{3}$, $y = f(\frac{\pi}{3}) = \sin\frac{\pi}{3} = \frac{\sqrt{3}}{2}$, so the point-slope form of the tangent at $x = \frac{\pi}{3}$ is

$$y - \frac{\sqrt{3}}{2} = \frac{1}{2}\left(x - \frac{\pi}{3}\right)$$

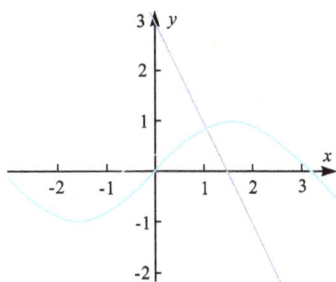

Figure 3.2.3: Graphs of $y = \sin x$, and its tangent and normal at $x = \frac{\pi}{3}$.

and the normal line at $x = \frac{\pi}{3}$ is

$$y - \frac{\sqrt{3}}{2} = -2\left(x - \frac{\pi}{3}\right).$$

Graphs are shown in Figure 3.2.3.

NOTE. If two coplanar lines are perpendicular to each other, then the product of their slopes, if both exist, is –1.

Example 3.2.6. Find the derivative of $y = a^x$, where $a > 0$, $a \neq 1$ is a constant.

Solution. We have

$$y' = \lim_{h \to 0} \frac{a^{x+h} - a^x}{h} = a^x \lim_{h \to 0} \frac{a^h - 1}{h}.$$

Let $a^h - 1 = t$. Then $h = \frac{\ln(1+t)}{\ln a}$ and $t \to 0$ as $h \to 0$, so the last limit becomes

$$\lim_{h \to 0} \frac{a^h - 1}{h} = \lim_{t \to 0} \frac{t}{\frac{\ln(1+t)}{\ln a}} = \lim_{t \to 0} \frac{t \ln a}{\ln(1 + t)}$$

$$= \lim_{t \to 0} \frac{\ln a}{\frac{1}{t} \ln(1 + t)} = \ln a \times \lim_{t \to 0} \frac{1}{\frac{1}{t} \ln(1 + t)}$$

$$= \ln a \times \lim_{t \to 0} \frac{1}{\ln(1 + t)^{\frac{1}{t}}}$$

$$= \ln a \frac{\lim_{t \to 0} 1}{\lim_{t \to 0} \ln(1 + t)^{\frac{1}{t}}} = \ln a \frac{\lim_{t \to 0} 1}{\ln \lim_{t \to 0} (1 + t)^{\frac{1}{t}}}$$

$$= \frac{\ln a}{\ln e}$$

$$= \ln a.$$

Hence, we conclude

$$\frac{d}{dx}(a^x) = a^x \ln a \quad \text{or} \quad (a^x)' = a^x \ln a.$$

NOTE. In particular, when $a = e$ we have

$$\frac{d}{dx}(e^x) = e^x \quad \text{or} \quad (e^x)' = e^x,$$

since $\ln e = 1$, so the derivative of e^x is itself at any point x. This function is a fixed point under the differential operator.

There is another way to define the number e. Since

$$(a^x)' = a^x \lim_{h\to 0} \frac{a^h - 1}{h},$$

as shown above, the limit

$$\lim_{h\to 0} \frac{a^h - 1}{h} = \lim_{h\to 0} \frac{a^{0+h} - a^0}{h}$$

is in fact the derivative of a^x at $x = 0$. That is, this limit is the slope of the tangent line to a^x at $x = 0$. Numerical and graphical views (see Figure 3.2.4 and Figure 3.2.5) show that the slope of 2^x is less than 1 and the slope of 3^x is greater than 1, so it is sensible

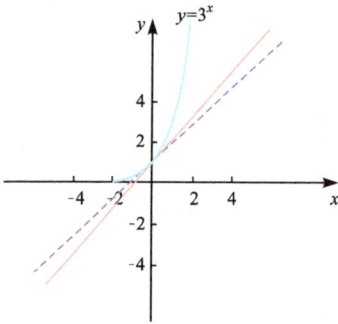

Figure 3.2.4: Graphs of $y = x + 1$, $y = 3^x$ and its tangent at $x = 0$.

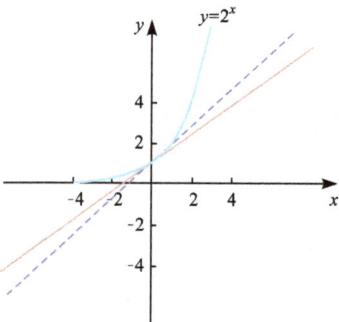

Figure 3.2.5: Graphs of $y = x + 1$, $y = 2^x$ and its tangent at $x = 0$.

to believe that there is a number such that the exponential function with this base has slope 1 at $x = 0$. We define this number by the letter e, so we obtain

$$\lim_{h \to 0} \frac{e^h - 1}{h} = 1.$$

In other words, e is defined as the number such that this limit is 1. This definition agrees with the previous one in Section 2.3 in Chapter 2. Observe that, when h is small, $\frac{e^h - 1}{h} \approx 1$, which implies that $e^h \approx 1 + h$, so $e \approx (1 + h)^{\frac{1}{h}}$ when h is small. If we take the limit as $h \to 0$, we will have $e = \lim_{h \to 0} (1 + h)^{\frac{1}{h}}$.

Example 3.2.7. Find the derivative of $y = \ln x$ for all $x > 0$.

Solution. Again, we will use $\lim_{t \to 0} (1 + t)^{\frac{1}{t}} = e$ as follows:

$$y' = (\ln x)' = \lim_{h \to 0} \frac{\ln(x + h) - \ln x}{h} = \lim_{h \to 0} \frac{1}{h} \ln\left(\frac{x + h}{x}\right)$$

$$= \lim_{h \to 0} \frac{1}{h} \ln\left(1 + \frac{h}{x}\right) = \lim_{h \to 0} \frac{1}{x} \times \frac{x}{h} \ln\left(1 + \frac{h}{x}\right)$$

$$= \frac{1}{x} \lim_{h \to 0} \ln\left(1 + \frac{h}{x}\right)^{\frac{x}{h}}, t = \frac{h}{x}, \quad t \to 0 \text{ when } h \to 0$$

$$= \frac{1}{x} \times \lim_{t \to 0} \ln(1 + t)^{\frac{1}{t}}$$

$$= \frac{1}{x} \times \ln e$$

$$= \frac{1}{x}.$$

Thus we have

$$\frac{d}{dx}(\ln x) = \frac{1}{x} \quad \text{or} \quad (\ln x)' = \frac{1}{x}.$$

3.3 Derivative laws

In the previous section, using the definition of the derivative, we found the derivative of some simple functions. Since the derivative is defined as a limit, we can use the limit laws to help us find derivatives of more complicated functions. In fact, we have the following theorem.

Theorem 3.3.1. *Suppose that $f(x)$ and $g(x)$ are two differentiable functions. Then:*
(1) $[f(x) \pm g(x)]' = f'(x) \pm g'(x)$ *(sum/difference rule);*
(2) $[cf(x)]' = cf'(x)$, $c \subset \mathbf{R}$ *(constant multiple rule);*
(3) $[f(x)g(x)]' = f'(x)g(x) + f(x)g'(x)$ *(product rule);*
(4) $[\frac{f(x)}{g(x)}]' = \frac{f'(x)g(x) - f(x)g'(x)}{g^2(x)}$, *if $g(x) \neq 0$ (quotient rule).*

Proof. (1) Let $F(x) = f(x) \pm g(x)$. Then

$$[f(x) \pm g(x)]' = F'(x)$$
$$= \lim_{h \to 0} \frac{F(x+h) - F(x)}{h}$$
$$= \lim_{h \to 0} \frac{[f(x+h) \pm g(x+h)] - [f(x) \pm g(x)]}{h}$$
$$= \lim_{h \to 0} \frac{[f(x+h) - f(x)] \pm [g(x+h) - g(x)]}{h}$$
$$= \lim_{h \to 0} \frac{f(x+h) - f(x)}{h} \pm \lim_{h \to 0} \frac{g(x+h) - g(x)}{h}$$
$$= f'(x) \pm g'(x).$$

(2) Let $F(x) = f(x)g(x)$. Then

$$F'(x) = \lim_{h \to 0} \frac{F(x+h) - F(x)}{h}$$
$$= \lim_{h \to 0} \frac{f(x+h)g(x+h) - f(x)g(x)}{h}$$
$$= \lim_{h \to 0} \frac{f(x+h)g(x+h) - f(x)g(x+h) + f(x)g(x+h) - f(x)g(x)}{h}$$
$$= \lim_{h \to 0} \left(\frac{f(x+h) - f(x)}{h} g(x+h) \right) + \lim_{h \to 0} \left(\frac{g(x+h) - g(x)}{h} f(x) \right)$$
$$= \lim_{h \to 0} \left(\frac{f(x+h) - f(x)}{h} \right) \lim_{h \to 0} g(x+h) + f(x) \lim_{h \to 0} \left(\frac{g(x+h) - g(x)}{h} \right)$$
$$= f'(x)g(x) + f(x)g'(x).$$

NOTE. We have $\lim_{h \to 0} g(x+h) = g(x)$. Since $g(x)$ is differentiable, it must be continuous.

The proofs of (3) and (4) are left to the reader. □

Example 3.3.1. Find the derivative of $f(x) = \sin x - 2\cos x + \frac{\pi}{2}$ and then find $f'(\pi)$.

Solution. Use the sum rule and the constant multiple rule to obtain

$$f'(x) = \left(\sin x - 2\cos x + \frac{\pi}{2} \right)'$$
$$= (\sin x)' - 2(\cos x)' + \left(\frac{\pi}{2} \right)'$$
$$= \cos x - 2(-\sin x) + 0$$
$$= \cos x + 2\sin x.$$

Then $f'(\pi) = (\cos x + 2\sin x)|_{x=\pi} = \cos \pi + 2\sin \pi = -1$.

Example 3.3.2. Find the derivative of $y = \log_a x$ where $x > 0$, $a > 0$, and $a \neq 1$.

Solution. Since $\log_a x = \frac{\ln x}{\ln a}$ and $(\ln x)' = \frac{1}{x}$, we have

$$(\log_a x)' = \left(\frac{\ln x}{\ln a}\right)' = \frac{1}{\ln a}(\ln x)' = \frac{1}{x \ln a}.$$

Example 3.3.3. Differentiate $y = x^2 \sin x$.

Solution. Using the product rule, we have

$$\frac{dy}{dx} = \frac{d}{dx}(x^2)\sin x + x^2 \frac{d}{dx}(\sin x)$$
$$= 2x \sin x + x^2 \cos x.$$

Example 3.3.4. Differentiate $y = xe^x \ln x$.

Solution. Use the product rule twice to obtain

$$y' = (xe^x \ln x)'$$
$$= (xe^x)' \ln x + (xe^x)(\ln x)'$$
$$= (x'e^x + x(e^x)') \ln x + xe^x \frac{1}{x}$$
$$= (e^x + xe^x) \ln x + e^x.$$

Example 3.3.5. Use the quotient rule to find $\frac{d}{dx}(\frac{1}{x^n})$, where n is a positive integer.

Solution. Applying the quotient rule with $f(x) = 1$, $g(x) = x^n$, we obtain

$$\frac{d}{dx}\left(\frac{1}{x^n}\right) = \frac{\frac{d}{dx}(1)x^n - 1 \times \frac{d}{dx}(x^n)}{(x^n)^2}$$
$$= \frac{0 - nx^{n-1}}{x^{2n}} = -\frac{n}{x^{n+1}}$$
$$= -nx^{-n-1}.$$

NOTE. This means $(x^{-n})' = (-n)x^{(-n)-1}$. We have now proved another special case of the power rule $(x^u)' = ux^{u-1}$ (for u is a negative integer). If $n = 0$, then $x^0 = 1$. Moreover, we have $(x^0)' = 0 = 0x^{-1}$. Together with Example 3.2.3, we have proved that $(x^n)' = nx^{n-1}$ is valid for all integers n (for $x \neq 0$). Some special cases of the rule include

$$(x)' = 1x^{1-1} = 1 \quad \text{and} \quad \left(\frac{1}{x}\right)' = (x^{-1})' = -x^{-1-1} = -\frac{1}{x^2}.$$

Example 3.3.6. If $y = \tan x$, then show that $y' = \sec^2 x$.

Solution. Since $\tan x = \frac{\sin x}{\cos x}$, we find, using the quotient rule,

$$
\begin{aligned}
y' = (\tan x)' &= \left(\frac{\sin x}{\cos x} \right)' \\
&= \frac{(\sin x)' \cos x - (\sin x)(\cos x)'}{\cos^2 x} \\
&= \frac{\cos^2 x + \sin^2 x}{\cos^2 x} \\
&= \frac{1}{\cos^2 x} \\
&= \sec^2 x.
\end{aligned}
$$

NOTE. Similarly, applying the quotient rule to $\cot x$, $\csc x$, and $\sec x$, we have

$$(\cot x)' = -\csc^2 x, (\sec x)' = \sec x \tan x, \quad \text{and} \quad (\csc x)' = -\csc x \cot x.$$

Example 3.3.7. Find the line tangent to the graph of the function

$$y = x + \frac{1}{x}$$

at the point $(1, 2)$.

Solution. We first find the derivative of y. We have

$$
\begin{aligned}
y' &= \left(x + \frac{1}{x} \right)' = (x)' + \left(\frac{1}{x} \right)' \\
&= 1 - \frac{1}{x^2}.
\end{aligned}
$$

Then observe that, when $x = 1$, $y'(1) = 1 - \frac{1}{1} = 0$, so the equation of the tangent line is

$$y - 2 = 0(x - 1), \quad \text{or} \quad y = 2.$$

Figure 3.3.1 shows these graphs.

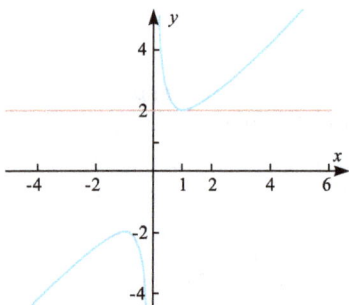

Figure 3.3.1: Graphs of $y = x + \frac{1}{x}$ and $y = 2$.

Example 3.3.8. Biologists use the equation

$$R = M^2\left(\frac{C}{2} - \frac{M}{3}\right)$$

to model the reaction of a change in temperature (measured in degrees) of a body to a dose of medicine, where C is a positive constant and M is the amount of medicine absorbed in the blood. The sensitivity is defined as the rate of change of R with respect to M, that is, $\frac{dR}{dM}$. Find the *sensitivity* of the body to the medicine.

Solution. We have

$$\frac{dR}{dM} = \left(M^2\left(\frac{C}{2} - \frac{M}{3}\right)\right)' = \left(\frac{CM^2}{2} - \frac{M^3}{3}\right)'$$

$$= \frac{C \cdot 2M}{2} - \frac{3M^2}{3}$$

$$= MC - M^2.$$

3.4 Derivative of an inverse function

Consider an inverse function $x = \phi(y)$ of a one-to-one function $y = f(x)$ defined on an interval **I**. If the original function is differentiable, then we would expect the function $x = \phi(y)$ to also be differentiable. As discussed previously, the inverse of a continuous function is also continuous. The two functions $x = \phi(y)$ and $y = f(x)$ have the same graph (the two equations are the same), although in the function $y = f(x)$ we regard y as the dependent variable while in the other function, $x = \phi(y)$, we regard x as the dependent variable. We first restate the definition of a strictly monotonic function.

Definition 3.4.1. A function $f(x)$ is *strictly increasing* on an interval **I** if $f(x_1) < f(x_2)$ whenever $x_1 < x_2$ in **I**. It is *strictly decreasing* if $f(x_1) > f(x_2)$ whenever $x_1 < x_2$ in **I**. The function $f(x)$ is said to be *strictly monotonic* on **I** if it is either strictly increasing on **I** or strictly decreasing on **I**.

A function that is strictly monotonic on an interval **I** clearly must be one-to-one, so it must have an inverse function (see Section 1.3.4 on inverse and composite functions in Chapter 1).

Theorem 3.4.1. *Suppose that $y = f(x)$ is differentiable and strictly monotonic on some interval I_y. Then the inverse function $x = \phi(y)$ is also differentiable on the interval $I_x = \{x | x = \phi(y) \text{ and } y \in I_y\}$. Furthermore,*

$$\frac{d}{dx}[f(x)] = \frac{1}{\frac{d}{dy}[\phi(y)]}, \quad \text{at any } x, y \text{ satisfying } y = f(x)$$

$$\text{or simply} \quad \frac{dy}{dx} = \frac{1}{\frac{dx}{dy}}, \quad \text{or} \quad \frac{dy}{dx} \times \frac{dx}{dy} = 1.$$

Proof. Since $y = f(x)$ is differentiable, the derivative is given by

$$\frac{d}{dx}[f(x)] = \lim_{\Delta x \to 0} \frac{\Delta y}{\Delta x}, \quad \text{where } \Delta y = f(x + \Delta x) - f(x).$$

Since $y = f(x)$ is one-to-one, we have $\Delta y \neq 0$ as $\Delta x \to 0$. Furthermore, since $y = f(x)$ is differentiable, it is continuous, which means that $\Delta y \to 0$ as $\Delta x \to 0$. Thus we have

$$\frac{d}{dx}[f(x)] = \lim_{\Delta x \to 0} \frac{\Delta y}{\Delta x} = \lim_{\Delta y \to 0} \frac{1}{\frac{\Delta x}{\Delta y}} = \frac{1}{\lim_{\Delta y \to 0} \frac{\Delta x}{\Delta y}}.$$

Since $\Delta y = f(x + \Delta x) - f(x)$ and $y = f(x)$, we have

$$\Delta y = f(x + \Delta x) - y, \quad \text{since } y = f(x),$$

so

$$y + \Delta y = f(x + \Delta x).$$

This means

$$\phi(y + \Delta y) = x + \Delta x.$$

Since $x = \phi(y)$, we have

$$\Delta x = \phi(y + \Delta y) - x = \phi(y + \Delta y) - \phi(y).$$

Thus we obtain

$$\lim_{\Delta y \to 0} \frac{\Delta x}{\Delta y} = \lim_{\Delta y \to 0} \frac{\phi(y + \Delta y) - \phi(y)}{\Delta y} = \frac{d}{dy}[\phi(y)].$$

Therefore,

$$\frac{d}{dx}[f(x)] = \frac{1}{\frac{d}{dy}[\phi(y)]}. \qquad \square$$

NOTE. This means that the derivative of the inverse function $x = \phi(y)$ with respect to y is the reciprocal of the derivative of the original function $y = f(x)$ with respect to x, where (x, y) is any point satisfying $y = f(x)$.

Example 3.4.1. If $y = 3x - 2$, find:
1. $\frac{dy}{dx}$;
2. the inverse function of y in the form $x = \phi(y)$;
3. $\frac{dx}{dy}$ and verify that $\frac{dy}{dx} \frac{dx}{dy} = 1$.

Solution. Since $y = 3x - 2$, we find

$$\frac{dy}{dx} = (3x - 2)' = (3x)' - 2' = 3(x)' - 0 = 3.$$

Solving the equation for x gives

$$x = \frac{y + 2}{3}.$$

This is the inverse of $y = 3x - 2$, while

$$\frac{dx}{dy} = \left(\frac{y + 2}{3}\right)' = \left(\frac{y}{3}\right) + \left(\frac{2}{3}\right)' = \frac{1}{3}.$$

Of course, we have

$$\frac{dy}{dx} \times \frac{dx}{dy} = 3 \times \frac{1}{3} = 1.$$

The geometric interpretation of this fact is shown in Figure 3.4.1.

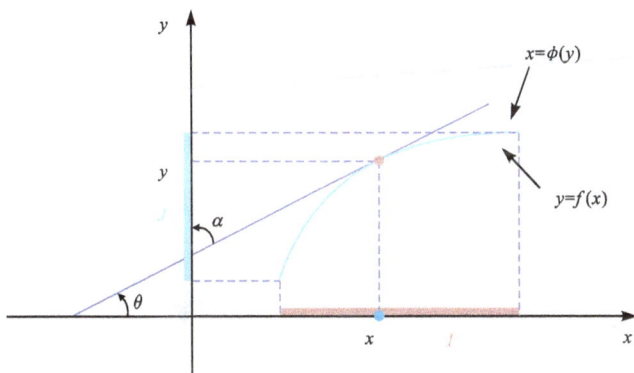

Figure 3.4.1: Geometric interpretation of $\frac{dy}{dx}$ and $\frac{dx}{dy}$.

Because $y = f(x)$ and $x = \phi(y)$ are the same equation, the graphs and the tangent lines at (x, y) are the same. However, the slope/derivative is interpreted differently. The slope of $y = f(x)$ is the steepness of the tangent line with respect to the positive x-axis, or $\tan \theta$, as shown in Figure 3.4.1, while the slope of $x = \phi(y)$ is the steepness of the tangent line with respect to the positive y-axis (in this case y is the independent variable), or $\tan \alpha$, as shown in Figure 3.4.1. In this case we have

$$\alpha + \theta = \frac{\pi}{2}, \quad \text{so } \tan \alpha = \cot \theta.$$

This means

$$\frac{dx}{dy} = \tan\alpha = \cot\theta = \frac{1}{\tan\theta} = \frac{1}{\frac{dy}{dx}}.$$

Rewritten in the inverse notation, we have $y = f(x)$ and $x = f^{-1}(y)$, so

$$\frac{dy}{dx} = \frac{1}{\frac{dx}{dy}} \quad \text{or} \quad [f(x)]' = \frac{1}{[f^{-1}(y)]'}.$$

Example 3.4.2. Find the derivative of the function $y = \arcsin x$, $-1 \leqslant x \leqslant 1$.

Solution. Since $x = \sin y$ is the inverse function for $-\frac{\pi}{2} \leqslant y \leqslant \frac{\pi}{2}$, $-1 \leqslant x \leqslant 1$, and the function $x = \sin y$ is differentiable and monotonic on the interval $(-\frac{\pi}{2}, \frac{\pi}{2})$, we have

$$y' = (\arcsin x)' = \frac{1}{[\sin y]'_y} = \frac{1}{\cos y}.$$

We know that, for any angle y, $\sin^2 y + \cos^2 y = 1$, so $\cos y = \sqrt{1 - \sin^2 y}$ ($\cos y \geqslant 0$, since $-\frac{\pi}{2} \leqslant y \leqslant \frac{\pi}{2}$), but $\sin y = x$, so we have $\cos y = \sqrt{1 - x^2}$. Hence,

$$y' = (\arcsin x)' = \frac{1}{\cos y} = \frac{1}{\sqrt{1 - x^2}}.$$

NOTE. Using the derivative of the inverse function, we show in a similar way that

$$(\arccos x)' = -\frac{1}{\sqrt{1 - x^2}},$$
$$(\arctan x)' = \frac{1}{1 + x^2},$$
$$(\operatorname{arccot} x)' = -\frac{1}{1 + x^2}.$$

Now we use another way, different from our previous example, to show that $(\ln x)' = \frac{1}{x}$.

Example 3.4.3. Find the derivative of the function $y = \ln x$, where $x > 0$.

Solution. The inverse function $x = e^y$ is differentiable and monotonic on the interval $(-\infty, \infty)$. Hence, using the derivative of an exponential function from Example 3.2.6, we have

$$y' = (\ln x)' = \frac{1}{[e^y]'_y} = \frac{1}{e^y} = \frac{1}{x}.$$

In the case that x and y are swapped in the inverse function, they play their usual roles. This means $y = f(x)$ and $y = f^{-1}(x)$, so the graphs of the two functions are reflections in the line $y = x$. This theorem then becomes the following.

Theorem 3.4.2. *The derivative of $y = f(x)$ at (x_0, y_0) is the reciprocal of the derivative of $y = f^{-1}(x)$ at (y_0, x_0).*

Example 3.4.4. For the function $f(x) = e^x$, find:

1. $\frac{df}{dx}$ and $\frac{df}{dx}\big|_{(1,e)}$;
2. the inverse function for f and find $\frac{df^{-1}}{dx}$, $\frac{df^{-1}}{dx}\big|_{(e,1)}$ for this function.

Solution. If $f(x) = e^x$, then $\frac{df}{dx} = e^x$ and $\frac{df}{dx}\big|_{(1,e)} = e^x\big|_{x=1} = e$.

Let $y = e^x$. Solving for x, we have $x = \ln y$. Swapping x and y, we obtain $y = \ln x$. In other words, $f^{-1}(x) = \ln x$. For this function,

$$\frac{df^{-1}}{dx} = \frac{1}{x} \quad \text{and} \quad \frac{df^{-1}}{dx}\bigg|_{x=e} = \frac{1}{e}.$$

Thus, we see that the derivative of $f(x) = e^x$ at $(1, e)$ is the reciprocal of the derivative of $f^{-1}(x) = \ln x$ at $(e, 1)$.

Recall from Chapter 1 that, geometrically, the graph of a function and its inverse are mirror images of each other through the line $y = x$. As an example, we present the graphs of $y = x$, $y = e^x$, and $y = \ln x$ together with a tangent drawn at two mirror image points on each curve, as shown in Figure 3.4.2.

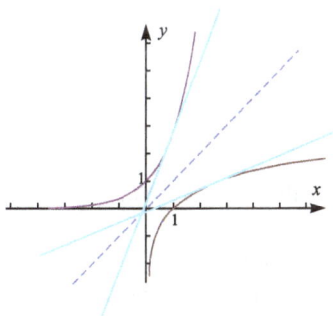

Figure 3.4.2: Graphs of $y = x$, $y = e^x$ and $y = \ln x$ and tangents.

3.5 Differentiating a composite function – the chain rule

So far, we have obtained the derivatives of basic functions such as power functions (when the exponent is an integer), exponential functions, logarithms, trigonometric functions, and inverse trigonometric functions. What about a more complicated function formed by composing those basic functions? For example, how would we differentiate the function $y = \sin(x^2)$? Notice that $y = \sin(x^2)$ is a composition of the two functions $y = \sin u$ and $u = x^2$, and we already know how to find the derivatives of both of these functions. We now explore the connection between the two derivatives. Let us first consider a simple case.

Example 3.5.1. Find the derivative of $y = (x^2 + 1)^2$.

Solution. *Method 1.* We have

$$y = (x^2 + 1)^2 = x^4 + 2x^2 + 1,$$

so

$$\frac{dy}{dx} = (x^4 + 2x^2 + 1)' = 4x^3 + 4x.$$

Method 2. Let $u = x^2 + 1$ and $y = u^2$. Then

$$\frac{dy}{du} = 2u \quad \text{and} \quad \frac{du}{dx} = 2x,$$

so

$$\frac{dy}{du} \times \frac{du}{dx} = 2u \times 2x$$
$$= 2(x^2 + 1) \times 2x = 4x^3 + 4x.$$

Therefore, when $y = f(u(x))$ we have

$$\frac{dy}{dx} = \frac{dy}{du} \times \frac{du}{dx}.$$

This pattern is called the *chain rule* and is stated in the following theorem.

Theorem 3.5.1 (Chain rule). *Suppose the function $y = f(u)$ is differentiable on an interval I_u and the function $u = g(x)$ is differentiable on some interval I_x, such that, for any $x \in I_x$, $g(x) = u \in I_u$. The composition $y = f(g(x))$ is also differentiable on I_x and*

$$\frac{dy}{dx} = \frac{dy}{du} \cdot \frac{du}{dx},$$

or equivalently $\quad \dfrac{d}{dx} f(g(x)) = f'(g(x))g'(x).$

Proof. If Δx is an increment in x, then let Δu and Δy be the corresponding increments in u and y so that

$$\Delta u = g(x + \Delta x) - g(x) = g(x + \Delta x) - u,$$
$$\Delta y = f(u + \Delta u) - f(u) = f(g(x + \Delta x)) - f(g(x)).$$

Note that $u = g(x)$ is differentiable at x and $y = f(u)$ is differentiable at $u = g(x)$, so

$$\lim_{\Delta x \to 0} \frac{\Delta u}{\Delta x} = g'(x) \quad \text{and} \quad \lim_{\Delta u \to 0} \frac{\Delta y}{\Delta u} = f'(u).$$

By Corollary 2.5.2 from Chapter 2, we write

$$\Delta u = g'(x)\Delta x + \alpha \cdot \Delta x, \quad \text{for some } \alpha \text{ which approaches 0 as } \Delta x \to 0,$$
$$\Delta y = f'(u)\Delta u + \beta \cdot \Delta u, \quad \text{for some } \beta \text{ which approaches 0 as } \Delta u \to 0.$$

If we substitute the expression for Δu, we get

$$\Delta y = [f'(u) + \beta] \cdot [g'(x)\Delta x + \alpha \cdot \Delta x]$$
$$\implies \frac{\Delta y}{\Delta x} = [f'(u) + \beta] \cdot [g'(x) + \alpha].$$

If $\Delta x \to 0$, then $\Delta u \to 0$ because $u(x)$ is continuous, so both $\alpha \to 0$ and $\beta \to 0$ as $\Delta x \to 0$. Therefore,

$$\begin{aligned}
\frac{dy}{dx} &= \lim_{\Delta x \to 0} \frac{\Delta y}{\Delta x} \\
&= \lim_{\Delta x \to 0} [f'(u) + \beta][g'(x) + \alpha] \\
&= \lim_{\Delta x \to 0} [f'(u) + \beta] \lim_{\Delta x \to 0} [g'(x) + \alpha] \\
&= f'(u)g'(x) \\
&= f'(g(x))g'(x).
\end{aligned}$$

This proves the chain rule. □

NOTE. $f'(g(x))$ means: calculate $f'(u) = \frac{df}{du}$ and then replace u with $g(x)$.

Example 3.5.2. Find the derivative of the function $y = \sin(x^2)$.

Solution. Let $u = x^2$. Then both $y = \sin u$ and $u = x^2$ are defined and differentiable for all $x \in \mathbf{R}$. By the chain rule, we have

$$\frac{dy}{dx} = \frac{dy}{du}\frac{du}{dx} = \cos u \times 2x = 2x \cos u = 2x \cos x^2.$$

Example 3.5.3. Find the derivative of the function $y = \cos(x^2 + x)$.

Solution. Let $u = x^2 + x$. Then $y = \cos u$ and $u = x^2 + x$ are defined and differentiable for all $x \in \mathbf{R}$. By the chain rule, we have

$$y' = \frac{dy}{dx} = \frac{dy}{du}\frac{du}{dx} = -\sin u \cdot (2x + 1) = -(2x + 1)\sin(x^2 + x).$$

There is another way to interpret the chain rule:

$$f'(g(x)) = f'(g(x))g'(x).$$

It says that the derivative of a composition is the derivative of the outside function with respect to the inside function $g(x)$, multiplied by the derivative of the inside function with respect to x.

Example 3.5.4. Find the derivative of $y = (x^2 + 1)^{10}$.

Solution. The outside is a power function u^{10} and the inside is a polynomial function $x^2 + 1$, so

$$\frac{dy}{dx} = \underbrace{10(x^2 + 1)^{10-1}}_{\text{derivative of the outside function}} \times \underbrace{\frac{d}{dx}(x^2 + 1)}_{\text{derivative of the inside function}}$$

$$= 10(x^2 + 1)^9 \times (2x + 0)$$

$$= 20x(x^2 + 1)^9.$$

Example 3.5.5. Prove $(x^u)' = ux^{u-1}$, where $u \in \mathbf{R}$.

Proof. Since $x^u = e^{u \ln x}$ and the outside function is e^x, the inside function is $u \ln x$. We have

$$\frac{dx^u}{dx} = \frac{de^{u \ln x}}{dx} = e^{u \ln x} \frac{d(u \ln x)}{dx} = x^u \frac{u}{x} = ux^{u-1}. \qquad \square$$

NOTE. From this result, we now have the generalized *power rule* of differentiation:

$$(x^n)' = nx^{n-1} \quad \text{is valid for all } n \in \mathbf{R} \text{ at each } x \text{ where } x^n \text{ is defined.}$$

A special case of the power rule is

$$(\sqrt{x})' = (x^{\frac{1}{2}})' = \frac{1}{2}x^{\frac{1}{2}-1} = \frac{1}{2}x^{-\frac{1}{2}} = \frac{1}{2\sqrt{x}} \quad \text{for } x > 0.$$

Example 3.5.6. Find $f'(x)$ if $f(x) = \frac{1}{\sqrt[3]{x^2+2x+3}}$.

Solution. First rewrite the function f as

$$f(x) = (x^2 + 2x + 3)^{-1/3} = u^{-1/3}$$

and then use the power rule for derivatives ($\frac{d}{du}u^{-1/3} = -\frac{1}{3}u^{-4/3}$) and the chain rule as follows:

$$f'(x) = -\frac{1}{3}(x^2 + 2x + 3)^{-4/3}\frac{d}{dx}(x^2 + 2x + 3)$$

$$= -\frac{1}{3}(x^2 + 2x + 3)^{-4/3}(2x + 2).$$

Example 3.5.7. Let $y = \ln|x|$, $x \neq 0$. Find $\frac{dy}{dx}$.

Solution. When $x > 0$, then $y = \ln|x| = \ln x$ and

$$\frac{dy}{dx} = (\ln x)' = \frac{1}{x}.$$

When $x < 0$, then $y = \ln|x| = \ln(-x)$, so

$$\frac{dy}{dx} = (\ln(-x))' = \frac{1}{-x}(-x)' = \frac{1}{-x} \times (-1) = \frac{1}{x}.$$

Therefore,

$$\frac{d}{dx}\ln|x| = \frac{1}{x} \quad \text{for all } x \in \mathbf{R}, x \neq 0.$$

NOTES. 1. In the case that u is a function of x, using the chain rule, we obtain

$$\frac{d}{dx}(\ln|u|) = \frac{1}{u}\frac{du}{dx}.$$

2. The chain rule can be extended to compositions of three or more functions. For example, if $y = f(u)$, $u = g(v)$ and $v = h(x)$, so that $y = f(u) = f(g(h)) = f(g(h(x)))$, then we use the chain rule twice to obtain

$$\frac{dy}{dx} = \frac{dy}{du}\frac{du}{dx} = \frac{dy}{du}\frac{du}{dv}\frac{dv}{dx},$$

or, in another notation,

$$\frac{d}{dx}f(g(h(x))) = f'(g(h(x)))g'(h(x))h'(x).$$

Example 3.5.8. Find the derivative of the function $y = e^{\sin\frac{2}{x}}$.

Solution. We write this as $y = e^u$, where $u = \sin v$ and $v = \frac{2}{x}$. Then we use the chain rule as follows:

$$\frac{dy}{dx} = \frac{dy}{du}\frac{du}{dv}\frac{dv}{dx}$$

$$= e^u \cdot \cos v \cdot \left(-\frac{2}{x^2}\right)$$

$$= -e^{\sin\frac{2}{x}} \cdot \cos\frac{2}{x} \cdot \frac{2}{x^2}.$$

Example 3.5.9. Find $f'(x)$ when $f(x) = e^{\cos(\cot x)}$.

Solution. We write this as $f(x) = e^u$, $u = \cos(v)$, $v = \cot(x)$. Then

$$\frac{dy}{dx} = \frac{dy}{du}\frac{du}{dv}\frac{dv}{dx}$$
$$= e^u(-\sin(v))(-\csc^2(x))$$
$$= e^{\cos v}\sin(\cot(x))\csc^2(x)$$
$$= e^{\cos(\cot x)}\sin(\cot(x))\csc^2(x).$$

Alternatively, we apply the chain rule directly two times as follows:

$$f'(x) = e^{\cos(\cot x)} \cdot \frac{d}{dx}\cos(\cot x)$$
$$= e^{\cos(\cot x)} \cdot -\sin(\cot x)\frac{d}{dx}\cot x$$
$$= e^{\cos(\cot x)} \cdot -\sin(\cot x)\cdot -\csc^2 x$$
$$= e^{\cos(\cot x)}\sin(\cot(x))\csc^2(x).$$

3.6 Derivatives of higher orders

In mechanics, the first order derivative of the displacement function with respect to time is the velocity function. If we differentiate the velocity function with respect to time, we obtain the acceleration function. In fact, this acceleration function is obtained by differentiating the displacement function twice with respect to time.

In mathematics, if we differentiate the function $y' = f'(x)$, then we write this as $y'' = [f'(x)]' = f''(x)$. We call this the *second order derivative* of $y = f(x)$ or simply the *second derivative* of $y = f(x)$. In other words, we differentiate $f(x)$ twice to obtain f''. Similarly, y''' denotes the third order derivative of $y = f(x)$ (differentiate three times) and $y^{(n)}$ is the nth order derivative of $y = f(x)$ (differentiate n times). Possible notations for the nth order derivative of the function $y = f(x)$ are

$$\frac{d^n y}{dx^n}, \quad f^{(n)}(x), \quad \text{or} \quad y^{(n)}.$$

Example 3.6.1. If $f(x) = x\sin x$, then find $f''(x)$.

Solution. Using the product rule, we have

$$f'(x) = \frac{d}{dx}(x)\sin x + x\frac{d}{dx}(\sin x)$$
$$= \sin x + x\cos x.$$

To find $f''(x)$ we differentiate $f'(x)$ and obtain

$$f''(x) = \frac{d}{dx}(\sin x + x\cos x)$$

$$= \frac{d}{dx}(\sin x) + \frac{d}{dx}(x)\cos x + x\frac{d}{dx}(\cos x)$$
$$= \cos x + \cos x - x\sin x$$
$$= 2\cos x - x\sin x.$$

For reference, the graphs of f, f', and f'' are shown in Figure 3.6.1. Be aware that you can interpret $f''(x)$ as the slope of the curve $y = f'(x)$ at the point $(x, f'(x))$ or $y = f''(x)$ as the rate of change of a rate of change.

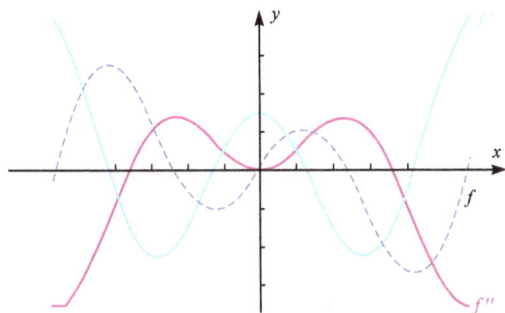

Figure 3.6.1: Graphs of f, f' and f'' in Example 3.6.1.

Example 3.6.2. If $y = \sin x$, find $y^{(n)}$.

Solution. Compute the first few derivatives:

$$y' = \cos x = \sin\left(x + \frac{\pi}{2}\right),$$
$$y'' = (\cos x)' = -\sin x = \sin\left(x + \frac{2\pi}{2}\right),$$
$$y''' = (-\sin x)' = -\cos x = \sin\left(x + \frac{3\pi}{2}\right),$$
$$y^{(4)} = (-\cos x)' = \sin x = \sin\left(x + \frac{4\pi}{2}\right).$$

This suggests the following result, which can be proved by induction:

$$y^{(n)} = \sin^{(n)} x = \sin\left(x + \frac{n\pi}{2}\right).$$

NOTE. y', y'', y''' and $y^{(4)}$ are $\cos x$, $-\sin x$, $-\cos x$, $\sin x = y$, respectively. Hence, all future derivatives just repeat this sequence of four derivatives.

Example 3.6.3. If $y = x^a$, find $y^{(n)}$.

Solution. We have

$$y' = ax^{a-1},$$
$$y'' = a(a-1)x^{a-2},$$
$$y''' = a(a-1)(a-2)x^{a-3},$$
$$\vdots$$
$$y^{(n)} = a(a-1)(a-2)\cdots(a-n+1)x^{a-n}.$$

NOTE. If a is a positive integer n, then $\frac{d^n}{dx^n}(x^n) = n!$ and all higher order derivatives are zero.

Example 3.6.4. If $y = \ln(1+x)$, find $y^{(n)}$.

Solution. We have

$$y' = \frac{1}{1+x},$$
$$y'' = \left(\frac{1}{1+x}\right)' = -\frac{1}{(1+x)^2},$$
$$y''' = (-1)^2 1 \cdot 2 \cdot \frac{1}{(1+x)^3},$$
$$y^{(4)} = (-1)^3 1 \cdot 2 \cdot 3 \cdot \frac{1}{(1+x)^4},$$
$$\vdots$$
$$y^{(n)} = (-1)^{n-1}(n-1)! \frac{1}{(1+x)^n}.$$

3.7 Implicit differentiation

In Chapter 1, we remarked that, when an equation is used to define a function, we often use our knowledge to analyze the formula and supply a natural domain and range for the function. However, for functions defined implicitly by an equation, the domain and range of the function are not always obvious.

Consider the equation

$$x^2 + y^2 = 1.$$

This equation defines two functions

$$y = \sqrt{1-x^2} \quad \text{and} \quad y = -\sqrt{1-x^2}, \quad \text{both with domain } [-1,1].$$

Notice that the graph of the entire equation is the unit circle and the graphs of the above two functions are semi-circles – one above the x-axis, the other below.

However, if we consider the *folium of Descartes*, which is given by

$$x^3 + y^3 = 6xy.$$

The graph is shown in Figure 3.7.1. It is hard to express y explicitly as a function of x. However, we can still view this equation as defining y in terms of x, because, if we put an x-value into the equation, we can solve it to find one or more corresponding y-values.

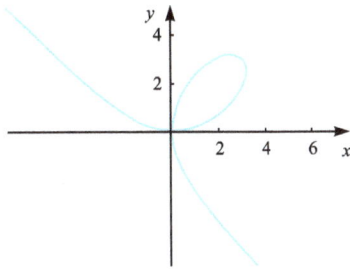

Figure 3.7.1: Folium of Descartes.

For example, when $x = 3$ we find that there are three points of intersection between the line $x = 3$ and the graph of the equation $x^3 + y^3 = 6xy$. Solving the equation when $x = 3$, we obtain

$$27 + y^3 = 18y,$$

which gives $y = 3$, $y = -\frac{3}{2} + \frac{3}{2}\sqrt{5}$, and $y = -\frac{3}{2} - \frac{3}{2}\sqrt{5}$. A function defined by this equation might map the domain element 3 to the range element 3 or $-\frac{3}{2} + \frac{3}{2}\sqrt{5}$ or $-\frac{3}{2} - \frac{3}{2}\sqrt{5}$, but this would be a choice that we make (and we would have to make a choice in order to have a function).

Notice that the tangents at these three points

$$(3,3), \quad \left(3, -\frac{3}{2} + \frac{3\sqrt{5}}{2}\right), \quad \text{and} \quad \left(3, -\frac{3}{2} - \frac{3\sqrt{5}}{2}\right)$$

are not vertical, so the restricted part of the graph which is near any of these points is the graph of a function. Therefore, we are able to find the derivative of that function at each point.

The process of finding the derivative of a function defined implicitly by an equation is called *implicit differentiation* and is illustrated in the following examples.

Example 3.7.1. If $x^2 + y^2 = 25$, then find $\frac{dy}{dx}$ and $\frac{dy}{dx}\big|_{(4,3)}$.

Solution. We differentiate each term in the equation with respect to x. We have

$$\frac{d}{dx}(x^2) + \frac{d}{dx}(y^2) = \frac{d}{dx}(25), \quad \text{so}$$

$$2x + \frac{d}{dx}(y^2) = 0.$$

Since y^2 is a composition of a power function u^2 and the function $u = y$, we must use the chain rule to find its derivative. We have

$$\frac{d}{dx}(y^2) = \frac{d}{dy}(y^2) \times \frac{dy}{dx} = 2y\frac{dy}{dx},$$

so

$$2x + 2y\frac{dy}{dx} = 0$$

and

$$\frac{dy}{dx} = -\frac{x}{y}.$$

When $x = 4$ and $y = 3$, $\frac{dy}{dx}\big|_{(4,3)} = -\frac{4}{3}$. This is the slope of the line tangent to the circle at $(4, 3)$.

In this example, we could have solved for y to obtain

$$y = \pm\sqrt{25 - x^2}.$$

Because we want the derivative at $(4, 3)$, we only consider the positive y-values.

$$\frac{dy}{dx} = \frac{1}{2\sqrt{25 - x^2}} \times (-2x)\Big|_{x=4} = \frac{-x}{\sqrt{25 - x^2}}\Big|_{x=4}$$

$$= \frac{-4}{\sqrt{25 - 4^2}}$$

$$= -\frac{4}{3}.$$

Example 3.7.2. Find the tangent line and the normal line to the folium of Descartes

$$x^3 + y^3 = 6xy$$

at the point $(3, 3)$.

Solution. We differentiate each term with respect to x and use implicit differentiation to obtain

$$\frac{d}{dx}x^3 + \frac{d}{dx}y^3 = \frac{d}{dx}(6xy),$$

$$3x^2 + \frac{d}{dy}y^3 \times \frac{dy}{dx} = 6\left(\frac{dx}{dx}\right)y + 6x \times \frac{dy}{dx},$$

$$3x^2 + 3y^2\frac{dy}{dx} = 6y + 6x \times \frac{dy}{dx}.$$

Solving $\frac{dy}{dx}$ gives

$$\frac{dy}{dx} = \frac{3x^2 - 6y}{6x - 3y^2} = \frac{x^2 - 2y}{2x - y^2}.$$

At $(3,3)$, $\frac{dy}{dx}\big|_{x=3,y=3} = \frac{x^2-2y}{2x-y^2}\big|_{x=3,y=3} = \frac{9-6}{6-9} = -1$, so the slope of the tangent line at $(3,3)$ is -1. The slope of the normal line at $(3,3)$ is 1. Therefore, an equation of the tangent line at $(3,3)$ is

$$y - 3 = -(x - 3) \quad \text{or} \quad y = -x + 6,$$

so an equation of the normal line is

$$y - 3 = x - 3 \quad \text{or} \quad y = x.$$

Example 3.7.3. Given the equation

$$x^2 + xy + y^2 = 3,$$

find:

1. y';
2. the equation of the tangent to $x^2 + xy + y^2 = 3$ at the point $(1,1)$;
3. the points on the curve where the tangent line is horizontal.

Solution. 1. Differentiate both sides of $x^2 + xy + y^2 = 3$ with respect to x. Regarding y as a function of x and applying the chain rule to the y^2 term and the product rule to the xy term, we get

$$2x + y + xy' + 2yy' = 0.$$

We now solve for y' to obtain

$$y' = \frac{-2x - y}{x + 2y}.$$

2. When $x = y = 1$, $y' = -1$. A glance at Figure 3.7.2 confirms that this is a reasonable value for the slope at $(1,1)$, so an equation of the tangent to the curve at $(1,1)$ is

$$y - 1 = -(x - 1) \quad \text{or} \quad x + y = 2.$$

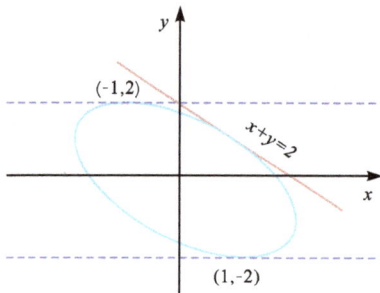

Figure 3.7.2: Graphs of $y = 2$, $y = -2$, $y + x = 2$ and $x^2 + xy + y^2 = 3$.

3. The tangent line is horizontal if $y' = 0$. Using the expression for y' from part 1, we see that $y' = 0$ when $2x + y = 0$. However, (x, y) must also satisfy the equation of the curve, so substituting $y = -2x$ in this equation, we get

$$x^2 - 2x^2 + 4x^2 = 3,$$

which simplifies to $x = \pm 1$. When $x = 1$, $y = -2x = -2$, so one point with horizontal tangent is $(1, -2)$. When $x = -1$, $y = -2x = 2$, so the only other point is $(-1, 2)$. Therefore, the tangent is horizontal at the points $(1, -2)$ and $(-1, 2)$, as can be observed in Figure 3.7.2.

Example 3.7.4. Find $\frac{d^2y}{dx^2}$ when the function $y = y(x)$ is implicitly defined by the equation

$$x - y + \frac{\sin y}{2} = 0.$$

Solution. Differentiate both sides with respect to x to get

$$1 - \frac{dy}{dx} + \frac{1}{2}\cos y \cdot \frac{dy}{dx} = 0,$$

$$\frac{dy}{dx} = \frac{2}{2 - \cos y}.$$

Differentiate again with respect to x, using the chain rule twice and the power rule as follows:

$$\frac{d^2y}{dx^2} = \frac{d}{dx}\left(\frac{2}{2 - \cos y}\right) = \frac{d}{dy}\left(\frac{2}{2 - \cos y}\right)\frac{dy}{dx}$$

$$= 2\frac{d}{dy}\left((2 - \cos y)^{-1}\right)\frac{dy}{dx} = -2(2 - \cos y)^{-2}(\sin y)\frac{dy}{dx}$$

$$= \frac{-2\sin y}{(2 - \cos y)^2} \cdot \frac{2}{2 - \cos y} = -\frac{4\sin y}{(2 - \cos y)^3}.$$

Logarithmic differentiation

In some equations involving exponents and/or products/quotients, it is advantageous to take logarithms of both sides of an equation before differentiating (often using implicit differentiation).

Example 3.7.5. The function $y = y(x)$ is defined implicitly by $x^y = y^x$ for $x > 0$ and $y > 0$. Find $\frac{dy}{dx}$.

Solution. Take the natural logarithm of both sides to get

$$y \ln x = x \ln y.$$

Using implicit differentiation and the product rule, we obtain

$$y' \ln x + \frac{y}{x} = \ln y + \frac{x}{y} y',$$

$$y' = \frac{\ln y - \frac{y}{x}}{\ln x - \frac{x}{y}}.$$

Example 3.7.6. Suppose $y = \sqrt{\frac{(x-1)(x-2)}{(x-3)(x-4)}}$ for $x \in (-\infty, 1) \cup (2, 3) \cup (4, \infty)$. Find y'.

Solution. Take the natural logarithm of both sides to obtain

$$\ln y = \ln \sqrt{\frac{(x-1)(x-2)}{(x-3)(x-4)}} = \frac{1}{2} \ln \left| \frac{(x-1)(x-2)}{(x-3)(x-4)} \right|$$

$$= \frac{1}{2} (\ln |x-1| + \ln |x-2| - \ln |x-3| - \ln |x-4|).$$

Differentiate implicitly to obtain

$$\frac{1}{y} y' = \frac{1}{2} \left[\frac{1}{x-1} + \frac{1}{x-2} - \frac{1}{x-3} - \frac{1}{x-4} \right],$$

so

$$y' = \frac{1}{2} \sqrt{\frac{(x-1)(x-2)}{(x-3)(x-4)}} \left[\frac{1}{x-1} + \frac{1}{x-2} - \frac{1}{x-3} - \frac{1}{x-4} \right].$$

3.8 Functions defined by parametric and polar equations

3.8.1 Functions defined by parametric equations

In physics, you probably have already encountered projectile motion. A ball in space is given an initial speed of u m/sec, at angle α radians to the horizontal. The subsequent motion of the ball follows a free fall motion if the air resistance is negligible.

The motion can be resolved in two directions. Horizontally, it moves with a constant speed, and its horizontal distance is a function of time. Vertically, it is a motion under gravity, so we have

$$\begin{cases} x = u \cos \alpha \cdot t \\ y = u \sin \alpha \cdot t - \frac{1}{2}gt^2. \end{cases}$$

At any time t, the ordered pair of numbers (x,y) gives the coordinates of the ball at that instant. This is an example of a function defined by parametric equations.

In mathematics, the *parameterization* of a curve is a representation of this curve through equations expressing the coordinates of the points of the curve as functions of a variable called a parameter. For example:

$$(1) \begin{cases} x = t \\ y = t^2; \end{cases} \qquad (2) \begin{cases} x = R\cos\theta \\ y = R\sin\theta; \end{cases}$$

$$(3) \begin{cases} x = 2\sin 2t \\ y = 3\cos 2t; \end{cases} \qquad (4) \begin{cases} x = 7\cos u + 3\sin(8u - 1) \\ y = 7\sin u - 3\cos(8u - 1). \end{cases}$$

If x and y are defined by parametric equations

$$\begin{cases} x = \phi(t) \\ y = \psi(t) \end{cases} \qquad \text{where } t \in [\alpha, \beta],$$

then this may or may not define y as a function of x. However, if ϕ and ψ are differentiable at a particular t and $\phi'(t) \neq 0$, then the parametric equations may define a function y of x in a neighborhood of that point. We consider the question of how to find $\frac{dy}{dx}$ in this case.

Suppose that $\phi(t)$ and $\psi(t)$ are differentiable and $\phi'(t) \neq 0$ at a particular t. In some neighborhood of this t, if there exists an inverse function $t = h(x)$ of $x = \phi(t)$,

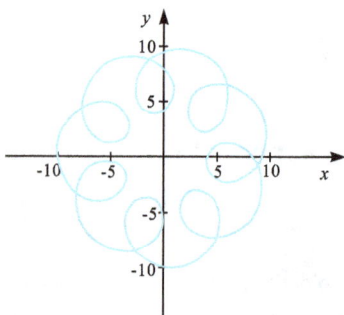

Figure 3.8.1: Graph of a curve defined by parametric equations.

then apply the chain rule to the composite function $y = \psi(h(x))$ to get

$$\frac{dy}{dx} = \frac{dy}{dt}\frac{dt}{dx}.$$

Recall from Theorem 3.4.2 that the derivative of the inverse function satisfies

$$\frac{dt}{dx} = \frac{1}{\left(\frac{dx}{dt}\right)}.$$

Thus $\frac{dy}{dx}$ is given by the following expression:

$$\frac{dy}{dx} = \frac{\frac{dy}{dt}}{\frac{dx}{dt}} \quad \text{or} \quad \frac{dy}{dx} = \frac{y'(t)}{x'(t)} = \frac{\psi'(t)}{\phi'(t)}. \tag{3.2}$$

The second derivative of $y(x)$ with respect to x is

$$\frac{d^2y}{dx^2} = \frac{d(y')}{dx} = \frac{d(y')}{dt}\frac{dt}{dx} = \frac{\left(\frac{d(y')}{dt}\right)}{\left(\frac{dx}{dt}\right)} \quad \text{or}$$

$$\frac{d^2y}{dx^2} = \frac{\left(\frac{y'(t)}{x'(t)}\right)'}{x'(t)} = \frac{y''(t)x'(t) - y'(t)x''(t)}{(x'(t))^3} \tag{3.3}$$

$$= \frac{\psi''(t)\phi'(t) - \psi'(t)\phi''(t)}{(\phi'(t))^3}.$$

NOTE. In equation (3.3), all the primes (derivatives) are with respect to t.

Example 3.8.1. If $y = y(x)$ is defined by the parametric equations

$$\begin{cases} x = t \\ y = 1 + t^2, \end{cases}$$

then find $\frac{dy}{dx}$, $\frac{dy}{dx}|_{t=2}$, and the line tangent to the graph of y at the point $(2,5)$.

Solution. We have

$$\frac{dy}{dx} = \frac{\frac{dy}{dt}}{\frac{dx}{dt}} = \frac{(1+t^2)'}{(t)'} = \frac{2t}{1} = 2t.$$

When $t = 2$, $\frac{dy}{dx}|_{t=2} = 2t|_{t=2} = 4$.
The point $(2,5)$ corresponds to the value of parameter $t = 2$, so

$$\left.\frac{dy}{dx}\right|_{(2,5)} = \left.\frac{dy}{dx}\right|_{t=2} = 2t|_{t=2} = 4.$$

Therefore, the point-slope form of the tangent line at $(2, 5)$ is

$$y - 5 = 4(x - 2).$$

We could check this by eliminating the parameter t to get an equation connecting x and y. We have

$$y = 1 + x^2$$

and

$$\frac{dy}{dx} = 2x \quad \text{and} \quad \frac{dy}{dx}\bigg|_{t=2} = \frac{dy}{dx}\bigg|_{x=2} = 2 \times 2 = 4.$$
$$\underbrace{\phantom{\frac{dy}{dx}\bigg|_{x=2}}}_{\text{Note: } x=2 \text{ when } t=2.}$$

This agrees with the previous calculation.

Example 3.8.2. Find $\frac{dy}{dx}$ and $\frac{d^2y}{dx^2}$ when the function $y = y(x)$ is defined by the parametric equations

$$\begin{cases} x = \ln(1 + t^2) \\ y = t - \arctan t. \end{cases}$$

Solution. Recall from Example 3.4.2 that $\frac{d}{dt}(\arctan t) = \frac{1}{1+t^2}$. Hence, we have

$$\frac{dy}{dx} = \frac{\frac{dy}{dt}}{\frac{dx}{dt}} = \frac{1 - \frac{1}{1+t^2}}{(\frac{2t}{1+t^2})} = \frac{t}{2}.$$

For the second derivative $\frac{d^2y}{dx^2}$, one needs to be aware that we are going to differentiate $\frac{dy}{dx}$ with respect to x, not t. We have

$$\frac{d^2y}{dx^2} = \frac{d}{dx}\left(\frac{dy}{dx}\right) = \frac{d}{dx}\left(\frac{t}{2}\right).$$

We cannot say the derivative $\frac{d}{dx}(\frac{t}{2})$ is $\frac{1}{2}$, since the derivative is with respect to x, not t! We have to use the chain rule and we obtain

$$\frac{d}{dx}\left(\frac{t}{2}\right) = \frac{d}{dt}\left(\frac{t}{2}\right) \times \frac{dt}{dx} = \frac{1}{2}\frac{dt}{dx}$$
$$= \frac{1}{2}\frac{1}{\frac{dx}{dt}} = \frac{1}{2} \cdot \frac{1}{\frac{2t}{1+t^2}}$$
$$= \frac{1 + t^2}{4t}.$$

We could also have used equation (3.3) to find

$$x'(t) = (\ln(1 + t^2))' = \frac{2t}{1 + t^2} \quad \text{and}$$

$$y'(t) = (t - \arctan t)' = 1 - \frac{1}{1+t^2} = \frac{t^2}{1+t^2}$$

and

$$x''(t) = \left(\frac{2t}{1+t^2}\right)' = \frac{(2t)'(1+t^2) - 2t \times (1+t^2)'}{(1+t^2)^2}$$
$$= \frac{2(1+t^2) - 4t^2}{(1+t^2)^2} = \frac{2 - 2t^2}{(1+t^2)^2},$$
$$y''(t) = \left(\frac{t^2}{1+t^2}\right)' = \frac{(t^2)'(1+t^2) - t^2 \times (1+t^2)'}{(1+t^2)^2}$$
$$= \frac{2t(1+t^2) - t^2(2t)}{(1+t^2)^2} = \frac{2t}{(1+t^2)^2}.$$

Therefore,

$$\frac{d^2y}{dx^2} = \frac{y''(t)x'(t) - y'(t)x''(t)}{(x'(t))^3}$$
$$= \frac{\frac{2t}{(1+t^2)^2}\frac{2t}{1+t^2} - \frac{t^2}{1+t^2}\frac{2-2t^2}{(1+t^2)^2}}{(\frac{2t}{1+t^2})^3}$$
$$= \frac{1+t^2}{4t}.$$

3.8.2 Polar curves

The position of point P in a plane can also be given in terms of its directed distance r from a fixed point O, called the pole, and the counterclockwise angle θ which OP makes with a fixed half line, called the initial line (the positive x-axis). The angle θ is normally measured in radians and its principal value is taken to be in the range $[-\pi, \pi]$. A negative value of this angle means that the angle is measured in a clockwise direction from the x-axis. In Figure 3.8.2, the Cartesian coordinates of the point P are (x, y) and its polar coordinates are (r, θ).

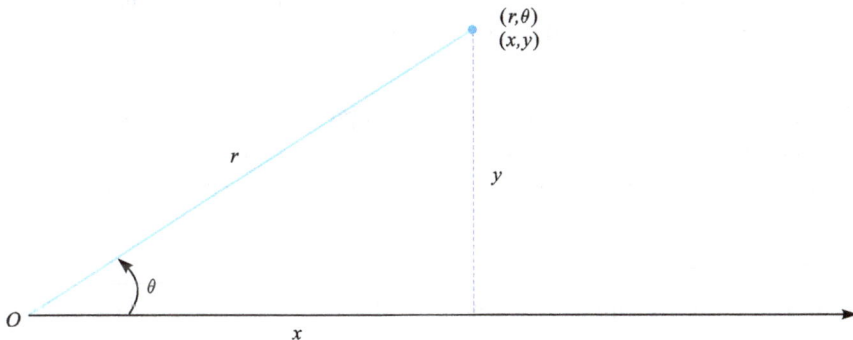

Figure 3.8.2: Polar coordinates.

Example 3.8.3. Plot the point P with polar coordinates $(2, \frac{\pi}{3})$ and the point Q with polar coordinates $(3, -\frac{\pi}{4})$.

Solution. Draw the line OP that makes an angle of $\frac{\pi}{3}$ radians with the x-axis and make the length $OP = 2$ units. Then P is the point identified. For Q, the negative value of the angle means that $\frac{\pi}{4}$ is measured in a clockwise direction from the x-axis, so we draw the line OQ at angle $-\frac{\pi}{4}$ radians with the x-axis and make $OQ = 3$ units. Then Q is the point identified.

NOTE. The point P also has polar coordinates $(-2, \frac{4\pi}{3})$, $(2, -\frac{5\pi}{3})$, and $(-2, -\frac{2\pi}{3})$.

We see from Figure 3.8.1, for the point P with coordinates (x, y) and (r, θ),

$$x = r\cos\theta, \quad y = r\sin\theta \quad \text{and} \quad r^2 = x^2 + y^2, \quad \tan\theta = \frac{y}{x}$$

These are *conversion formulas*. We could use them to convert the equation of a curve from its Cartesian form to its polar form, or vice versa.

Example 3.8.4. Use the conversion formulas to find the Cartesian equation of each of the curves

$$(1)\ r = 2; \quad (2)\ r = -2; \quad (3)\ r = 2\sin\theta; \quad \text{and} \quad (4)\ \theta = \frac{\pi}{2}.$$

Solution. (1) Since $r^2 = 4$, its Cartesian equation is $x^2 + y^2 = 2^2$. This curve is a circle centered at the origin with radius 2.
    (2) This is the same as (1).
    (3) Since $r^2 = 2r\sin\theta$, its Cartesian equation is $x^2 + y^2 = 2y$ or $x^2 + (y-1)^2 = 1$. This is a circle centered at $(0, 1)$ with radius 1.
    (4) $x = r\cos\frac{\pi}{2} = 0$, but note $\theta = \frac{\pi}{2}$ is not the entire y-axis. Since θ is measured in counterclockwise direction, it is the positive y-axis only.

NOTE. In general, if $\theta = \alpha$ and $\alpha > 0$, then the graph is a half line which makes an angle α measured in counterclockwise direction with the positive x-axis. If $\alpha < 0$, then the graph is a half line which makes an angle $-\alpha$ measured in clockwise direction with the positive x-axis.

Example 3.8.5. Use the conversion formulas to find the polar equation of each of the following curves:
(1) $x^2 + y^2 = 9$; (2) $xy = 25$; (3) $x^2 + y^2 = 4x$; (4) $(x^2 + y^2)^2 = x^2 - y^2$.

Solution. We have

(1) $r^2 = 9$, so $r = 3$ or $r = -3$;

(2) $x = r\cos\theta$, $y = r\sin\theta$, so $r\cos\theta \cdot r\sin\theta = 25$, so $r^2 = \frac{25}{\cos\theta\sin\theta}$, or $r^2 = 50/\sin 2\theta$;

(3) $r^2 = 4r\cos\theta$, or $r = 4\cos\theta$;

(4) $(r^2)^2 = r^2\cos^2\theta - r^2\sin^2\theta$, so $r^2 = \cos^2\theta - \sin^2\theta$, that is, $r^2 = \cos 2\theta$.

Example 3.8.6. Suppose a function $y = y(x)$ is described by the equation $r = 1 + \cos\theta$ in polar coordinates. Find the equation of the tangent line to the function at $\theta = \frac{\pi}{2}$.

Solution. Recall the relationship between the polar coordinates r, θ and the rectangular coordinates x, y is $x = r\cos\theta$, $y = r\sin\theta$ (with the same origin and x-axis corresponding to $\theta = 0$). Hence,

$$\begin{cases} x = r(\theta)\cos\theta = (1 + \cos\theta)\cos\theta = \cos^2\theta + \cos\theta, \\ y = r(\theta)\sin\theta = (1 + \cos\theta)\sin\theta = \sin\theta + \cos\theta\sin\theta. \end{cases}$$

These are now in the form of parametric equations, so differentiating (using the product rule and the chain rule) we obtain

$$\left.\frac{dy}{dx}\right|_{\theta=\frac{\pi}{2}} = \left.\frac{\frac{dy}{d\theta}}{\frac{dx}{d\theta}}\right|_{\theta=\frac{\pi}{2}} = \left.\frac{\cos\theta - \sin^2\theta + \cos^2\theta}{-\sin\theta - 2\sin\theta\cos\theta}\right|_{\theta=\frac{\pi}{2}} = 1.$$

When $\theta = \frac{\pi}{2}$, $r = 1 + \cos\frac{\pi}{2} = 1$, in which case the point (x,y) in rectangular coordinates is

$$(x,y) = (r\cos\theta,\, r\sin\theta) = (0,1),$$

so the equation of the desired tangent line is

$$y - 1 = x - 0, \quad \text{or} \quad y = x + 1.$$

Graphs are shown in Figure 3.8.3.

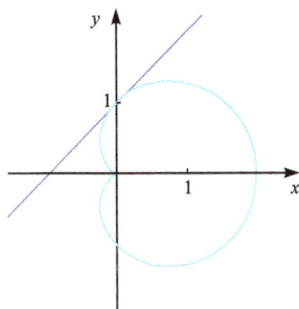

Figure 3.8.3: Graphs of $r = 1 + \cos\theta$ and $y = x + 1$.

3.9 Related rates of change

A radar on a post is detecting a speeding car as shown in Figure 3.9.1. If the car is moving at a rate of change of 100 km/h, then the rate of change of the car detected by the radar is not 100 km/h, but we can find the equation connecting the two rates of change. They are called related rates of change.

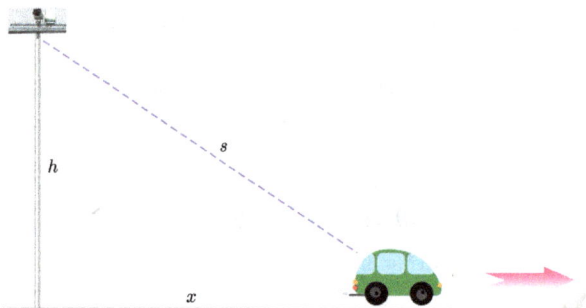

Figure 3.9.1: Radar detects speeding cars.

The equation connecting the variables is

$$x^2 + h^2 = s^2,$$

(3.4)

where x and s are both functions of the time t. The rate of change of the speeding car is $\frac{dx}{dt}$. $\frac{ds}{dt}$ is the rate of change of s with respect to the time t, the radar detects exactly this rate of change.

Now we differentiate each term of the equation with respect to t to find

$$\frac{d}{dt}(x^2) + \frac{d}{dt}(h^2) = \frac{d}{dt}(s^2).$$

Using the chain rule, we find

$$2x\frac{dx}{dt} + 0 = 2s\frac{ds}{dt},$$

so

$$\frac{ds}{dt} = \frac{x}{s}\frac{dx}{dt}.$$

This shows that $\frac{ds}{dt}$ is not equal to $\frac{dx}{dt}$. For instance, when $x = 50$ m, $s = \sqrt{50^2 + 10^2} \approx 50.99$ m, and at this instant

$$\frac{ds}{dt} = \frac{50}{50.99} \times 100 \approx 98.06 \text{ km/h}.$$

In general, we may need to calculate the rate at which one quantity is changing when we know the rate at which a related quantity is changing. To solve the "related rates" problem, we need to:

1. find an equation linking the variables and then use the chain rule to differentiate both sides with respect to time;
2. solve for the unknown rate of change using the known rates of change.

Example 3.9.1. Air is being pumped into a spherical balloon so that its volume increases at a constant rate of $100\,\text{mm}^3/\text{sec}$. How fast is the radius of the balloon increasing when the radius is exactly $10\,\text{mm}$?

Solution. There are two variables involved: the volume V and the radius r of the balloon. An equation connecting them is

$$V = \frac{4}{3}\pi r^3.$$

Both V and r are functions of time t and $\frac{dV}{dt} = 100\,\text{mm}^3/\text{sec}$ is known.

Now, differentiating each term of the equation with respect to t and using the chain rule, we obtain

$$\frac{dV}{dt} = \frac{d}{dt}\left(\frac{4}{3}\pi r^3\right) = \frac{4\pi}{3}\frac{d}{dr}(r^3)\frac{dr}{dt}$$
$$= \frac{4\pi}{3} \times 3r^2 \times \frac{dr}{dt},$$

so

$$\frac{dr}{dt} = \frac{1}{4\pi r^2}\frac{dV}{dt}$$
$$= \frac{1}{4\pi r^2} \times 100$$
$$= \frac{25}{\pi r^2}.$$

When $r = 10\,\text{mm}$,

$$\frac{dr}{dt} = \frac{25}{\pi(10)^2} = \frac{1}{4\pi} \approx 0.007\,96\,\text{mm/sec}.$$

This means that, at the instant when $r = 10\,\text{mm}$, the radius r of the balloon is changing at an approximate rate of $0.007\,96\,\text{mm/sec}$.

3.10 The tangent line approximation and the differential

3.10.1 Linearization

Sometimes we need to estimate the change in one variable by a small change in another variable. For example, if a circular lamina with radius $r = 10\,\text{cm}$ is heated and the radius expands by 0.02 cm, what is the change in its area? This is an easy question since we have

$$A(r) = \pi r^2, \quad \text{so}$$
$$\Delta A = A(r + \Delta r) - A(r) = \pi(r + \Delta r)^2 - \pi r^2$$
$$= 2\pi r \Delta r + \pi \Delta r^2$$
$$= 2\pi \times 10 \times 0.02 + \pi \times 0.02^2$$
$$\approx 1.258 \text{ cm}^2.$$

We are lucky here since we can find a nice connection between the change in r and the change in area: we write the change in area as a function of the change in radius. Both $A(r + \Delta r)$ and $A(r)$ are easily evaluated. However, it may be easy to calculate one particular value $f(a)$ of a function, but difficult or even impossible to compute nearby values of f. For example with $f(x) = \sqrt{x}$, it is easy to evaluate $f(1) = \sqrt{1} = 1$, but it is hard to find $f(1.02) = \sqrt{1.02}$ without a calculator.

Let us go back to the graph of a function to see if it helps us in our dilemma. If you observe the line tangent to a curve at a point as seen in Figure 3.10.1, you will see that the curve is very close to its tangent line near the point of contact. The nearer we are to the point of contact, the closer the tangent line is to the curve. This provides us a method of finding approximate values of functions.

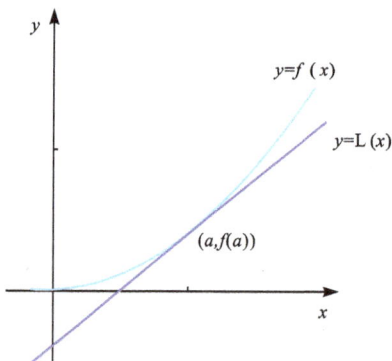

Figure 3.10.1: Graphs of a function and its tangent at $x = a$.

That is, we use the tangent line at the point $(a, f(a))$ as an approximation to the curve $y = f(x)$ when x is near a. An equation of this tangent line at the point $(a, f(a))$ is

$$y = f(a) + f'(a)(x - a)$$

and the approximation of the curve values $f(x)$ is

$$f(x) \approx f(a) + f'(a)(x - a). \tag{3.5}$$

This is called the *linear approximation* or *tangent line approximation of f at a*. The linear tangent line function $L(x) = f(a) + f'(a)(x - a)$ is called the *linearization of f at a*.

Example 3.10.1. Estimate $\sqrt{1.02}$ and $e^{0.012}$ by using linearizations of suitable functions at $x = 1$ and 0, respectively.

Solution. Let $f(x) = \sqrt{x}$. Then $f'(x) = \frac{1}{2\sqrt{x}}$ and the linear approximation of f at $x = 1$ is given by

$$f(x) \approx f(1) + f'(1)(x - 1),$$
$$\sqrt{x} \approx \sqrt{1} + \frac{1}{2\sqrt{1}}(x - 1),$$
$$\sqrt{1.02} \approx 1 + \frac{1}{2}(1.02 - 1) = 1.01.$$

Let $g(x) = e^x$. Then $g'(x) = e^x$. The linear approximation of g at $x = 0$ is given by

$$g(x) \approx g(0) + g'(0)(x - 0),$$
$$e^x \approx 1 + e^0 x \approx 1 + x,$$

so $e^{0.02} \approx 1 + 0.02 = 1.02$. A calculator gives $\sqrt{1.02} = 1.0099505$ and $e^{0.02} = 1.02020134$, so the tangent line approximation gives a good estimate with an error less than 0.02%.

Example 3.10.2. Find the linearization of the function $f(x) = \sqrt{x + 8}$ at $x = -4$ and use it to approximate the numbers $\sqrt{3.98}$ and $\sqrt{4.05}$. Are these approximations overestimates or underestimates?

Solution. The derivative of $f(x) = \sqrt{x + 8}$ is

$$f'(x) = \frac{1}{2}(x + 8)^{-1/2} = \frac{1}{2\sqrt{x + 8}},$$

so we have $f(-4) = 2$ and $f'(-4) = \frac{1}{4}$. Putting these values into the equation of its tangent line at $x = -4$, we see that the linearization is

$$L(x) = f(-4) + f'(-4)(x + 4)$$
$$= 2 + \frac{1}{4}(x + 4) = 3 + \frac{x}{4}.$$

That is, the linear approximation of $f(x)$ is

$$\sqrt{x + 8} \approx 3 + \frac{x}{4} \quad \text{when } x \text{ is near } -4.$$

In particular, we have

$$\sqrt{3.98} = \sqrt{-4.02 + 8} \approx 3 - \frac{4.02}{4} = 1.995$$

and

$$\sqrt{4.05} = \sqrt{-3.95 + 8} \approx 3 - \frac{3.95}{4} = 2.0125.$$

The linear approximation is illustrated in Figure 3.10.2. We see that the tangent line approximation is a good approximation to the given function when x is near -4. We also see that our approximations are overestimates because the tangent line lies above the curve.

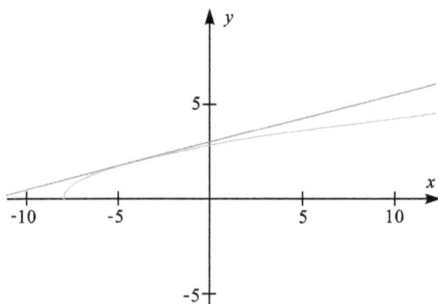

Figure 3.10.2: Graphs of $y = \sqrt{x + 8}$ and $y = 3 + x/4$.

NOTE. Of course, a calculator could give approximations for $\sqrt{3.98}$ and $\sqrt{4.05}$, but the linear approximation gives an approximation over an interval of x-values.

3.10.2 Differentials

The differential dy and the differential dx

Linear approximations can be formulated using the idea and notation of *differentials*. The Leibniz notation of the derivative $\frac{dy}{dx}$ looks like a quotient of two variables dy and dx (it is indeed a limit of a quotient in which both the numerator and the denominator tend to zero). This makes it tricky to define dy and dx as separate entities so that $\frac{dy}{dx}$ behaves like a quotient whether it was one or not, as we have seen in the chain rule.

If $y = f(x)$, where f is a differentiable function, then the *differential dx* is an independent variable that can be given the value of any real number. The *differential dy* is then defined in terms of the differential dx by

$$dy = f'(x) \, dx, \tag{3.6}$$

so dy is a dependent variable; it is determined by the values of x and dx. If $dx \neq 0$, we divide both sides of $dy = f'(x) \, dx$ by dx to obtain

$$\frac{dy}{dx} = f'(x).$$

We have seen similar equations before, but now the left side can genuinely be interpreted as the ratio of two differentials.

Example 3.10.3. If $y = xe^{-x}$, then find the differential dy in terms of the differential dx.

Solution. Since

$$y' = (xe^{-x})' = (x)'e^{-x} + x(e^{-x})' = e^{-x} - xe^{-x},$$

$dy = (e^{-x} - xe^{-x})\,dx.$

Example 3.10.4. Find a function y such that $dy = x^2\,dx$.

Solution. Let $y = x^3/3$. Then

$$dy = d\left(\frac{x^3}{3}\right) = \left(\frac{x^3}{3}\right)' dx = x^2\,dx.$$

NOTE. In fact, this is not the only choice for the function y. We could also use $y = x^3/3 + 5$ or $y = x^3/3 - 7$. In fact, $y = x^3/3 + C$ will work for any constant C.

Theorem 3.10.1. *For any differentiable function $u(x)$ and $v(x)$ and any constant k and l, we have:*

(1) $d(C) = 0$;   (2) $d(ku(x) \pm lv(x)) = kd(u(x)) \pm ld(v(x))$;

(3) $d(uv) = udv + vdu$;   (4) $d(\frac{u}{v}) = \frac{vdu - udv}{v^2}$, $v \neq 0$.

Proof. The proofs are not hard and we only give the proof of (2). Since $\frac{d(uv)}{dx} = u\frac{dv}{dx} + v\frac{du}{dx}$, by the product rule of derivatives, multiplying the differential dx on both sides gives $d(uv) = udv + vdu$. $\qquad\square$

Example 3.10.5. If $y = e^{2x}\sin 3x$, then find dy.

Solution. We have

$$dy = d(e^{2x}\sin 3x) = e^{2x}\,d\sin(3x) + \sin(3x)\,de^{2x}$$
$$= e^{2x} \cdot 3\cos 3x\,dx + \sin 3x \cdot 2e^{2x}\,dx = (3e^{2x}\cos 3x + 2e^{2x}\sin 3x)\,dx.$$

Example 3.10.6. If y is implicitly defined by $y + xe^y = 1$, find dy.

Solution. We have

$$d(y + xe^y) = d1 \quad\Longrightarrow\quad dy + d(xe^y) = 0.$$

Since $d(xe^y) = xde^y + e^y\,dx$ and $de^y = e^y\,dy$, we have

$$dy + xe^y\,dy + e^y\,dx = 0 \quad\Longrightarrow\quad dy = \frac{-e^y}{1 + xe^y}\,dx.$$

Linearization and differentiability

Usually, if the function $y = f(x)$ is not linear, the change in y arising from a small change in x will be nonlinear. As seen previously, if a circular disk is heated, the radius of the disk has a change Δr. The area A of the disk therefore has a change $\Delta A = \pi(r + \Delta r)^2 - \pi r^2 = 2\pi r \Delta r + \pi(\Delta r)^2$. Notice that ΔA is equal to a linear part in Δr plus a negligible term $\pi(\Delta r)^2 = o(\Delta r)$ as $\Delta r \to 0$.

Definition 3.10.1. Let $y = f(x)$. *The change in y can be linearized at $x = a$ if there is a number A depending on a such that*

$$\Delta y = A\Delta x + o(\Delta x),$$

where $o(\Delta x)$ is negligible with respect to Δx as $\Delta x \to 0$.

If $y = f(x)$ can be linearized at $x = a$, then there is a number A such that $\Delta y = A\Delta x + o(\Delta x)$. This implies

$$\frac{\Delta y}{\Delta x} = A + \frac{o(\Delta x)}{\Delta x}.$$

Taking the limits as $\Delta x \to 0$ on both sides gives $A = \frac{dy}{dx} = f'(a)$. Therefore, $f(x)$ has a derivative at $x = a$. From the arguments, we also know that, if the linearization of $y = f(x)$ at $x = a$ exists, the choice of A can only be $f'(a)$.

On the other hand, if the function $f(x)$ has a derivative at $x = a$, i.e., $\lim_{\Delta x \to 0} \frac{\Delta y}{\Delta x} = f'(a)$ exists, then this means that there is a function $\alpha(x)$ (by Corollary 2.5.2 from Chapter 2) such that

$$\frac{\Delta y}{\Delta x} = f'(a) + \alpha(x), \quad \text{where } \alpha(x) \to 0 \text{ as } \Delta x \to 0.$$

Therefore, $\Delta y = f'(a)\Delta x + \alpha\Delta x$. Notice that $\alpha\Delta x = o(\Delta x)$ as $\Delta x \to 0$, so

$$\Delta y = f'(a)\Delta x + o(\Delta x). \tag{3.7}$$

This implies that Δy can be linearized at $x = a$. Therefore, we have the following theorem.

Theorem 3.10.2. *A function $y = f(x)$ can be linearized at $x = a$ if and only if $f(x)$ has a derivative at $x = a$.*

NOTE. Nowadays, there are some books that define the term "differentiable" (at $x = a$) to mean that Δy can be linearized at $x = a$ and define $dy = A\Delta x$, that is, $dy = f'(a)\Delta x$. This is the same as the previous definition of dy when $dx = \Delta x$.

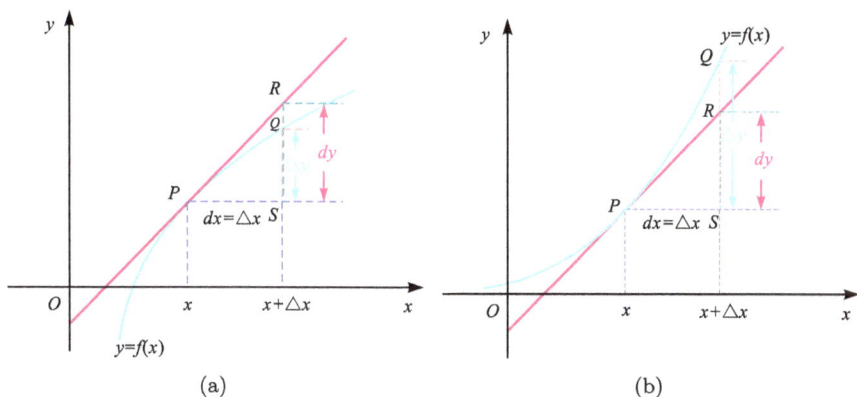

Figure 3.10.3: Geometric interpretation of dy.

In fact, when $dx = \Delta x$, then there is a nice geometric meaning of the differential dy, as shown in Figure 3.10.3.

That is, dx and dy are the changes in x and y along the tangent line. The slope of the tangent line at P is the derivative $f'(x)$. Thus, $f'(x) = \frac{SR}{PS} = \frac{dy}{dx}$.

NOTE. The tangent line approximation is also called the differential approximation, since $\Delta y \approx dy$ when Δx is small.

Example 3.10.7. Find an approximation of the value $\sin(0.05)$ using the differential approximation.

Solution. Let $f(x) = \sin x$, with $f'(x) = \cos x$. Use the differential approximations

$$\Delta y \approx dy = f'(x)\Delta x = \cos x \Delta x$$

and

$$\sin 0.05 - \sin 0 \approx \cos 0 \times (0.05 - 0).$$

Then

$$\sin 0.05 \approx \sin 0 + \cos 0 \times (0.05 - 0)$$
$$\approx 0 + 1(0.05)$$
$$\approx 0.05.$$

Note that we have chosen $a = 0$, because 0 is close to 0.05 and both $f(0)$ and $f'(0)$ are easily evaluated.

Example 3.10.8. A ball of radius $r = 3\,\text{cm}$ is heated such that the radius increases by $0.03\,\text{cm}$. Estimate the change in its volume by using a differential approximation.

Solution. Since $V = \frac{4\pi r^3}{3}$, we have

$$\Delta V \approx dV = V'(r)\Delta r = 4\pi r^2 \Delta r.$$

Now that $\Delta r = 0.03$ cm and $r = 3$ cm, $\Delta V \approx 4\pi 3^2 (0.03) \approx 3.3929$ cm^3.

Some useful differential approximations

Some useful differential approximations computed at $x = 0$ are listed below (assuming $|x|$ is small). Check these for yourself by using the approximation formula $f(x) \approx f(a) + f'(a)(x - a)$ with $a = 0$:

1. $\sqrt[n]{1 + x} \approx 1 + \frac{1}{n}x$;
2. $\sin x \approx x$;
3. $\tan x \approx x$;
4. $e^x \approx 1 + x$;
5. $\ln(1 + x) \approx x$.

In fact, these are pairs of asymptotic functions when $x \to 0$.

3.11 Derivative rules – summary

In this chapter, we have proved the following rules of differentiation (provided all the derivatives involved exist):

(1) $\frac{d(C)}{dx} = 0$;   (2) $\frac{d(x^n)}{dx} = nx^{n-1}$;   (3) $\frac{d(\sin x)}{dx} = \cos x$;

(4) $\frac{d(\cos x)}{dx} = -\sin x$;   (5) $\frac{d(a^x)}{dx} = a^x \ln a$;   (6) $\frac{d(\ln |x|)}{dx} = \frac{1}{x}$;

(7) $\frac{d(\tan x)}{dx} = \sec^2 x$;   (8) $\frac{d(\cot x)}{dx} = -\csc^2 x$;   (9) $\frac{d(\sec x)}{dx} = \sec x \tan x$;

(10) $\frac{d(\csc x)}{dx} = -\csc x \cot x$;   (11) $\frac{d(\arctan x)}{dx} = \frac{1}{1+x^2}$;

(12) $\frac{d(\text{arccot}\, x)}{dx} = -\frac{1}{1+x^2}$;   (13) $\frac{d(\arcsin x)}{dx} = \frac{1}{\sqrt{1-x^2}}$;

(14) $\frac{d(\arccos x)}{dx} = -\frac{1}{\sqrt{1-x^2}}$;

(15) $\frac{d}{dx}(f(x) \pm g(x)) = \frac{d}{dx}f(x) \pm \frac{d}{dx}g(x)$;

(16) $\frac{d}{dx}(cg(x)) = c(\frac{d}{dx}g(x))$, where c is a constant;

(17) $\frac{d}{dx}\left(\frac{f(x)}{g(x)}\right) = \frac{(\frac{d}{dx}f(x))g(x) - f(x)(\frac{d}{dx}g(x))}{(g(x))^2}$;

(18) $\frac{d}{dx}(f(x)g(x)) = f(x)\frac{d}{dx}g(x) + g(x)\frac{d}{dx}f(x)$;

(19) $\frac{d}{dx}f(g(x)) = f'(g(x))g'(x)$ or $\frac{dy}{dx} = \frac{dy}{du}\frac{du}{dx}$ (the chain rule);

(20) $\frac{df(y)}{dx} = \frac{df(y)}{dy}\frac{dy}{dx}$ (implicit differentiation);

(21) $\frac{dy}{dx} = \frac{dy/dt}{dx/dt}$ (parametric differentiation).

3.12 Exercises

1. The graph of a function over a closed interval **I** is given below. Determine the domain points at which the graph appears to be (a) differentiable; (b) continuous but not differentiable; (c) neither continuous nor differentiable.

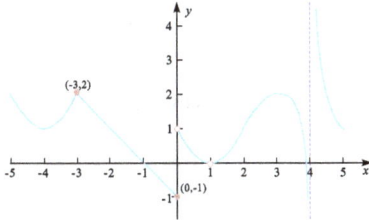 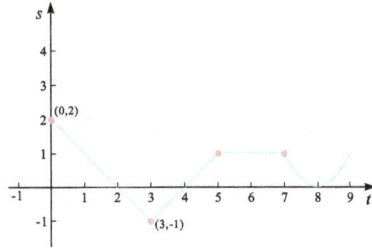

| Question 1 | Question 3 |

2. Values of a function $y = f(x)$ at selected values of x are given in the following table.

x	0	0.1	0.2	0.3	0.4	0.5	0.6	0.7	0.8	0.9
y	0.5	0.7	1.3	1.5	1.9	2.3	1.5	1.7	2.2	2.5

Estimate $y'(0.15)$ using a symmetric difference quotient.

3. **(Particle motion)** A particle P moves on the number line. The graph above shows the position of the particle as a function of time t.

   (a) During what time periods is P moving to the left? Moving to the right? And standing still?

   (b) Graph the particle's velocity and speed (where defined).

4. Use the definition of the derivative $f'(a) = \lim_{\Delta x \to 0} \frac{f(a+\Delta x)-f(a)}{\Delta x}$ to find the derivative of each of the following functions at the given point a and then find the tangent line and the normal line to the graph of the functions at $x = a$:

   (a) $f(x) = \frac{1}{x}$, $a = 2$; (b) $f(x) = x^2 + 4$, $a = 1$.

5. Use the limit definition $\lim_{x \to a} \frac{f(x)-f(a)}{x-a}$ to evaluate the derivative of each of the following functions at the given point $x = a$:

   (a) $f(x) = \sqrt{x+1}$, $a = 3$; (b) $f(x) = \begin{cases} x^2 \cos\frac{1}{x}, & x \neq 0 \\ 0, & x = 0, \end{cases}$ $a = 0$.

6. Assume $f'(x_0)$ exists. Evaluate each of the following limits:

   (a) $\lim_{\Delta x \to 0} \frac{f(x_0-\Delta x)-f(x_0)}{\Delta x}$; (b) $\lim_{h \to 0} \frac{f(x_0+2h)-f(x_0)}{h}$;

   (c) $\lim_{t \to 0} \frac{f(x_0)-f(x_0-t)}{t}$; (d) $\lim_{h \to 0} \frac{f(x_0+ah)-f(x_0+bh)}{h}$.

7. Let $f(x) = \begin{cases} -x, & x < 0 \\ x^2, & x \geq 0. \end{cases}$ Find $f'_+(0)$ and $f'_-(0)$. Does $f'(0)$ exist?

8. If $f(x) = 2|x-1| + (x-1)^2$, find $f'(1)$.

9. Find the constants a and b such that $f(x) = \begin{cases} ax+b, & x>1 \\ ax^3+x+2b, & x\leqslant 1 \end{cases}$ is differentiable every-where.

10. Find the constants a and b such that $f(x) = \begin{cases} xe^x, & x\leqslant 0 \\ ax+b, & x>0 \end{cases}$ is differentiable every-where.

11. If $f(x) = \begin{cases} \sin x, & x\leqslant 0 \\ x^2, & x>0, \end{cases}$ then find $f'(x)$.

12. If $f(0) = 0$ and $f'(0) = 2$, then find $\lim_{x\to 0} \frac{f(\sin x)}{x}$.

13. Assume u and v are two differentiable functions of x, $u(2) = 1$, $u'(2) = -3$, $v(2) = -1$, and $v'(2) = 4$. Find the values of:

   (a) $\frac{d}{dx}(uv)|_{x=2}$; (b) $\frac{d}{dx}(\frac{u}{v} + 2u - v)|_{x=2}$; (c) $(\frac{2u}{v} + 7u - 2uv)'|_{x=2}$.

14. Show that:

   (a) $\frac{d}{dx}(\cos x) = -\sin x$; (b) $\frac{d}{dx}(\cot x) = -\csc^2 x$;

   (c) $\frac{d}{dx}(\sec x) = \sec x \tan x$; (d) $\frac{d}{dx}(\csc x) = -\csc x \cot x$.

15. Find the derivative of each of the following functions:

   (a) $y = \ln x + \cos x - 2\sin x$; (b) $f(x) = e^x + x \ln x$; (c) $s = \frac{e^t}{t} - 2\sec t + \pi^2$;

   (d) $V = \frac{4}{3}\pi r^3$; (e) $y = \frac{3x-2}{2x+5}$; (f) $y = \frac{x^2+5x-2}{x^2}$;

   (g) $y = \frac{x^2}{1-x^3}$; (h) $y = \frac{\ln\theta}{\theta}$; (i) $g(u) = ue^u$.

16. (**Witch of Maria Agnesi**) The witch of Maria Agnesi has the Cartesian equation $y = \frac{8a^3}{x^2+4a^2}$. Find the line tangent to the curve at $(2,1)$ for $a = 1$. Also find the normal line at the same point. http://en.wikipedia.org/wiki/Witch_of_Agnesi

17. (**Newton's serpentine**) This curve, named and studied by Newton in 1701, is defined by $y = \frac{abx}{x^2+a^2}$. Find the tangents to the curve at the origin and the point $(1,2)$ for the case $a = 1$ and $b = 4$. http://mathworld.wolfram.com/SerpentineCurve.html

18. Assume $y = \arctan x$. Show that $\frac{dy}{dx} = \frac{1}{1+x^2}$.

19. Assume $f(1) = 2$, $f(2) = 3$, $f'(1) = 3$, and $f'(2) = -1$. If $g(x) = f^{-1}(x)$, then find $g'(2)$.

20. Find the derivative of each of the following functions:

   (a) $\rho = \sqrt{3r - r^2}$; (b) $s = \sin(\frac{3\pi t}{2}) + \cos(\frac{3\pi t}{2})$;

   (c) $r = (\csc\theta + \cot\theta)^{-1}$; (d) $h = \theta^2 \sec\frac{1}{\theta}$;

   (e) $y = x^2 \sin^2 x + x\cos 2x$; (f) $y = \frac{1}{7}(3x-2)^7 + (1-\frac{1}{2x})^{-1}$;

   (g) $h = \theta^2 \sec\frac{1}{\theta}$; (h) $f = (\frac{\cos t}{1+\sin t})^2$;

   (i) $y = t\tan^2\sqrt{2t} + 6$; (j) $\theta = \tan t^2 + \tan^2 t$;

   (k) $r = \cos(\tan\theta)$; (l) $y = \sin(\cos x)$;

   (m) $y = (\sin\sqrt{x^2+1})^\pi$; (n) $y = \sqrt{x + \sqrt{x}}$;

   (o) $y = e^{\tan^{-1}2x}$; (p) $y = \ln(\sin 2x)$;

   (q) $y = \sin(\cos(2x))$.

21. Given that f is differentiable, find $\frac{dy}{dx}$ for each of the following functions:

   (a) $y = f(\sqrt{x})$; (b) $y = \sqrt{f(x)}$; (c) $y = f(\frac{1}{\ln x})$;

   (d) $y = \sin(f(\sin x))$; (e) $y = f(f(e^{2x}))$.

22. Determine each of the following higher order derivatives:
    (a) y'' for $y = x^3 - 2x^2 - 1$; (b) $\frac{d^3y}{dx^3}$ for $y = e^{2x}$;
    (c) $\frac{d^4y}{dx^4}$ for $y = x^2 \ln x$; (d) $\frac{d^5y}{dx^5}$ for $y = \frac{1}{1+x}$.

23. Show that $\frac{d^n}{dx^n}(x^{n-1}e^{\frac{1}{x}}) = \frac{(-1)^n}{x^{n+1}}e^{\frac{1}{x}}$.

24. Find $\frac{d^ny}{dx^n}$ for each of the following functions:
    (a) $y = e^x$; (b) $y = \cos x$; (c) $y = \sin^2 x$; (d) $y = x^n \ln x$;
    (e) $y = x^2 \cos x$; (f) $y = e^{-3x}$.

25. If $y = f(x)$ and $\frac{dy}{dx} = y'$, show that:
    (a) $\frac{d^2x}{dy^2} = -\frac{y''}{(y')^3}$; (b) $\frac{d^3x}{dy^3} = \frac{3(y'')^2 - y'y'''}{(y')^5}$.

26. Use implicit differentiation to find $\frac{dy}{dx}$ for:
    (a) $x^2 - xy + y^3 = 6$; (b) $y^5 + 3x^2y^2 - 2x^4 = 2$;
    (c) $\sqrt{x+y} + \sqrt{xy} = 4$; (d) $e^{xy} = x + y$.

27. Given $y + x = \cos(xy)$, find $\frac{dy}{dx}$ and the tangent line at $(0,1)$.

28. Find $\frac{dy}{dx}$ for each of the following functions using logarithmic differentiation:
    (a) $y = x^x$; (b) $y = \frac{\sqrt{x+1}\sin x}{(x^2+1)(x-2)}$;
    (c) $y = (\cos x)^{\sqrt{x}}$; (d) $y = (x-a)^a(x-b)^b(x-c)^c$.

29. If the function y is implicitly defined by $y^3 + (\frac{dy}{dx})^3 = x^4 + 6$ and $y(1) = -1$, find $\frac{d^2y}{dx^2}\big|_{(1,-1)}$.

30. Let (x_0, y_0) be a point on the curve defined by $x = \sqrt{a^2 - y^2} + a\ln\frac{a-\sqrt{a^2-y^2}}{y}$, $0 < y < a$. Show that the distance between the point (x_0, y_0) and the point where the tangent line to the curve at (x_0, y_0) meets the x-axis is constant.

31. Write the following Cartesian equations in polar form:
    (a) $x^2 + y^2 = 2xy$; (b) $(x^2 + y^2)^2 = 2xy$; (c) $y = 2x$;
    (d) $x = 2$; (e) $y = -3$.

32. **(Lissajous curve)** The parametric equations $x = A\sin(at + \delta)$ and $y = B\sin(bt)$ describe *complex harmonic motion*. This family of curves was investigated by Nathaniel Bowditch in 1815 and later in more detail by Jules Antoine Lissajous in 1857. Graph the curve when $A = B = 1$, $a = 2$, $\delta = 0$, and $b = 3$. Then find the horizontal tangent lines for $0 \leqslant t \leqslant \frac{\pi}{2}$.

33. If $x = t - \sin t$ and $y = 1 - \cos t$, for $0 < t < 2\pi$, find $\frac{d^2y}{dx^2}$.

34. Show that the length of the portion of any tangent line to the astroid $x = a\cos^3 t$, $y = a\sin^3 t$ cut off by the coordinate axes is constant.

35. Show that the distance from the origin to any normal line to the curve $x = a(\cos t + t\sin t)$, $y = a(\sin t - t\cos t)$ is constant.

36. Sketch the following curves, given in polar form:
    (a) $r = 2\cos\theta$; (b) $r = 2(1 + \cos\theta)$; (c) $r = a + a\sin\theta$;
    (d) $r = a\sin 2\theta$; (e) $r^2 = 4\cos 2\theta$.

37. (**Cardioid**) Find the points on the polar curve $r = 2(1 + \cos\theta)$ where the tangent line is horizontal.

38. Use a local linearization to estimate the following numbers:
    (a) $\sqrt{8.99}$; (b) $e^{0.01}$; (c) $\sin(0.03)$; (d) $\ln(1.02)$; (e) $\cos 0.03$.

39. Find the differential dy for each of the following functions:
    (a) $y = \sin 2x$; (b) $y = e^{-x}\cos x$; (c) $y = \frac{x}{1+x^2}$; (d) $y = \tan^{-1}(e^{-2x})$.

40. (**Sensitivity to change**) The equation $dy = f'(x)\,dx$ tells us how sensitive the output of f is to a change in input at various values of x. The larger the value of f' at x, the greater the effect of a given change dx. Now consider the following two problems.
    (a) Suppose we can determine the radius of a sphere within 1% of its true value. What effect would the tolerance of 1% have on our estimate of the surface area of the sphere $(S = 4\pi R^2)$?
    (b) How accurately should we measure the radius R of a sphere approximately so that, when we calculate the surface area, the error is within 0.5% of its true value?

41. (*Group activity) Given $f(x) = \begin{cases} x^2\sin\frac{1}{x}, & x \neq 0 \\ 0, & x = 0, \end{cases}$ is $f'(x)$ continuous?

42. If $f(xy) = f(x) + f(y)$ for all $x, y > 0$ and $f'(1) = 2$, show that $f'(x) = \frac{2}{x}$.

43. Show that, if $f(0) = 0$, $f'(0)$ exists if and only if there is a function $g(x)$ which is continuous at $x = 0$ and $f(x) = xg(x)$.

44. Let f be the function defined by $f(x) = \begin{cases} g(x)\cos\frac{1}{x}, & \text{if } x \neq 0 \\ 0, & \text{if } x = 0. \end{cases}$ If $g(x)$ is differentiable and $g'(0) = g(0) = 0$, then find $f'(0)$.

45. Assume $f(x) = |x - a|g(x)$ and $g(x)$ is differentiable everywhere. Prove:
    (a) if $g(a) \neq 0$, then $f'(a)$ does not exist;
    (b) if $g(a) = 0$, then $f'(a)$ exists.

46. For what value of k does the equation $e^{2x} = k\sqrt{x}$ have exactly one solution?

47. Assume $f(x) = a_1\sin x + a_2\sin 2x + a_3\sin 3x + \cdots + a_n\sin nx$, where $a_1, a_2, \ldots, a_n$ are real numbers and n is a positive integer. If $|f(x)| \leq |\sin x|$ for all x, show that $|a_1 + 2a_2 + \cdots + na_n| \leq 1$.

48. (**Hyperbolic function**) The hyperbolic functions $\sinh x$, $\cosh x$, $\tanh x$, and $\coth x$ are defined in Chapter 1. Show that:
    (a) $(\sinh x)' = \cosh x$; (b) $(\cosh x)' = \sinh x$;
    (c) $(\tanh x)' = \frac{1}{\cosh x}$; (d) $(\coth x)' = \frac{-1}{\sinh x}$.

49. (**Related rates of change**) James is flying a kite at a height of 60 feet above his head. If the kite moves horizontally at a constant speed of 4 ft/sec, at what rate is the string being let out when the kite is 80 feet away from him?

50. (**Pendulum**) When a pendulum with length l performs a simple harmonic motion, the period of the motion is given by $T = 2\pi\sqrt{\frac{l}{g}}$. In winter, if the length of the pendulum is shortened by 0.01 cm, what effect will this have on the period?

51. **(Debye's law)** In physical chemistry, Debye's law about the orientation polarization P of a gas satisfies

$$P = \frac{4}{3}\pi N\left(\frac{\mu^2}{3kT}\right),$$

where N, μ, and k are constants and T is the temperature of the gas.
Find the rate of change of P with respect to the temperature T.

52. **(Marginal cost)** Assume the cost (in dollars) of producing q units of a certain commodity is $C(q) = 2000 + 2q + 3q^2$. Find:
    (a) the average rate of change of the cost when the production level is changed from $q = 100$ to $q = 140$;
    (b) the marginal cost (the instantaneous rate of change of C) with respect to q when $q = 50$.

53. **(Wage pay plan)** A company is willing to pay a graduate in computer science a starting salary of \$56 000 and the employee will get a raise of \$2 500 each year.
    (a) At what percentage rate will the employee's salary be increasing after 1 year?
    (b) What will happen to the percentage rate of change of this employee's salary in the long run?

54. **(Elasticity)** In economics, elasticity is the measurement of how responsive an economic variable is to a change in another. The **price elasticity of demand** is given by

$$E(p) = \frac{p}{q} \times \frac{dq}{dp},$$

where $q = D(p)$ is the amount of a commodity that is demanded by the market at a unit price p.
Interpret $E(p)$ in terms of the percentage rate of change in demand q and the percentage rate of change in price p.

4 Applications of the derivative

In this chapter, you will learn about:
- *Fermat's theorem and the closed interval test;*
- *the mean value theorem and the first derivative test;*
- *the extended mean value theorem and L'Hôpital's rule;*
- *concavity and the second derivative test;*
- *Taylor's theorem and Taylor's polynomial approximation;*
- *how to sketch curves using derivatives;*
- *numerical solutions to equations;*
- *curvatures.*

Derivatives have many applications in solving a large variety of problems arising from mathematical or nonmathematical areas of study. In this chapter, we will use derivatives to determine the extreme values for a function, analyze monotonic functions, find intervals of increase or decrease, prove inequalities, and find limits of some indeterminate forms using L'Hôpital's rule. We will also show that derivatives can make significant contributions to incredibly diverse parts of mathematics, including curve sketching, approximating functions by polynomials, numerical approximation of solutions of equations, and curvature of plane curves.

4.1 Extreme values and the candidate theorem

In Chapter 2, we introduced the absolute/global maximum and absolute/global minimum of a function defined on an interval **I**. We know that a continuous function defined on a closed interval must obtain its global extreme values somewhere within the closed interval. In addition to the global extrema, there are some other extreme values that may be of interest. For example, you might be the tallest one in your class, even though you are not the tallest in your school. Your height is still an extreme value when we consider a small group around you. Returning to mathematics, we have the following definition for relative/local extrema.

Definition 4.1.1. If f is defined on D, then:
1. f has a *local minimum* (also called *relative minimum*) at $x = c \in D$, if there is a neighborhood U of c, such that $f(x) \geqslant f(c)$ for each $x \in U \subset D$.
2. f has a *local maximum* (also called *relative maximum*) at $x = c \in D$, if there is neighborhood U of c, such that $f(x) \leqslant f(c)$ for each $x \in U \subset D$.

NOTE. If there is a local maximum of f at $x = c$, then the definition states that there is an interval $[a, b]$ in D with $a < c < b$ such that $f(x) \leqslant f(c)$ for all $x \in (a, b)$. Similarly, if

https://doi.org/10.1515/9783110527780-004

f has a local minimum at $x = c$, then there is an interval (a, b) containing c such that $f(x) \geqslant f(c)$ for all $x \in (a, b)$. Hence, a local maximum or minimum cannot occur at a boundary point of the domain of f (there are some books which recognize that $f(x)$ can have its local extrema at boundary points of its domain, but we will not in this text).

Figure 4.1.1 shows a graph of a function f with several extrema.

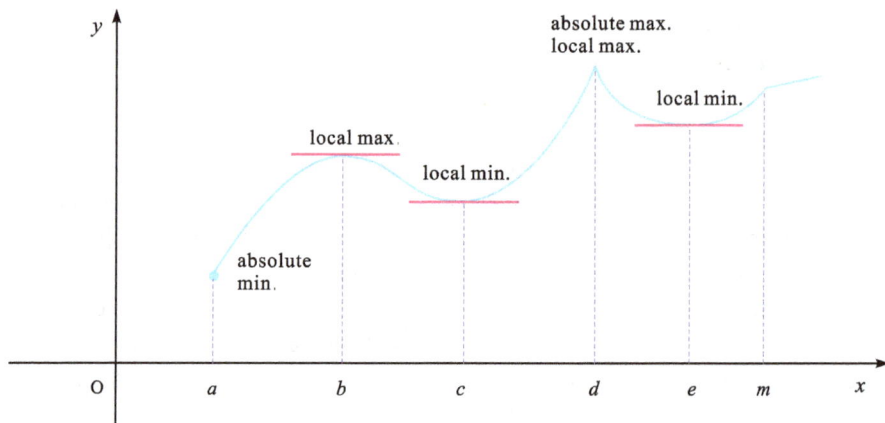

Figure 4.1.1: Local and global extrema.

Notice in Figure 4.1.1 the situations where f could have a local extremum. You should see that, at any local extremum, the tangent line is either horizontal or does not exist. For example, at points where $x = b$, c, and e, the tangent line is horizontal, and the derivative at these points is zero. However, at the point where $x = d$, there is no tangent line, and the derivative does not exist. This gives some clues for finding local extrema. Now we confirm this analytically. The following theorem is credited to a French mathematician, Pierre de Fermat.

Pierre de Fermat (1601 or 1607–1665) was a French lawyer at the Parliament of Toulouse, France, and a mathematician who is given credit for early developments that led to infinitesimal calculus. In particular, he is recognized for his discovery of an original method to find the greatest and the smallest ordinates of curved lines, which is analogous to that of the differential calculus, then unknown, and his research into number theory. He made notable contributions to analytic geometry, probability, and optics. He is best known for his last theorem, which he described in a note at the margin of a copy of Diophantus' *Arithmetica*. http://en.wikipedia.org/wiki/Fermat

Theorem 4.1.1 (Fermat's theorem). *Suppose $f(x)$ is defined on some open interval **I** containing the point $x = c$ and $f'(c)$ exists. If f has a local maximum or a local minimum at $x = c$, then $f'(c) = 0$.*

Proof. We only prove the result when $f(c)$ is a local minimum, since the proof for the local maximum is analogous. For all sufficiently small positive or negative increments $\Delta x = x - c$ of x, we have $x = c + \Delta x \in I$ and

$$f(x) - f(c) \geqslant 0.$$

Hence, if $x - c$ is positive, then we have $\frac{f(x)-f(c)}{x-c} \leqslant 0$ and if $x - c$ is negative, we have $\frac{f(x)-f(c)}{x-c} \geqslant 0$. Thus,

$$f'_+(c) = \lim_{x \to c^+} \frac{f(x) - f(c)}{x - c} \geqslant 0 \quad \text{and}$$
$$f'_-(c) = \lim_{x \to c^-} \frac{f(x) - f(c)}{x - c} \leqslant 0,$$

by the definition of one-sided derivatives and the properties of limits.

Since $f'(c)$ exists, the two one-sided derivatives $f'_+(c)$ and $f'_-(c)$ must exist and $f'_+(c) = f'_-(c)$. However, we have shown that $f'_+(c) \leqslant 0$ and $f'_-(c) \geqslant 0$. This only happens when

$$f'_+(c) = f'_-(c) = 0,$$

so $f'(c) = 0.$ □

We give a name of those points at which the derivative of a function f is 0.

Definition 4.1.2. If $f'(c) = 0$, then $x = c$ is called a *stationary point* of the function f.

Example 4.1.1. Find all the stationary points of $y = 2x^3 - 3x^2 + 1$.

Solution. Compute

$$\frac{dy}{dx} = 6x^2 - 6x.$$

Then solving $\frac{dy}{dx} = 0$ gives

$$6x^2 - 6x = 0, \quad \text{so} \quad 6x(x - 1) = 0.$$

Therefore, there are two stationary points: $x = 1$ and $x = 0$, as shown in Figure 4.1.2.

Fermat's theorem says that, if $f(x)$ is differentiable, then its local extrema must occur at a stationary point. If $f(x)$ is not differentiable at a point, then $f(x)$ may or may not also have a local extreme value at that point, as shown in Figure 4.1.1 (when $x = d$ and $x = m$). The next example shows that a function f can have a local minimum (or maximum) at a point where the derivative $f'(x)$ does not exist.

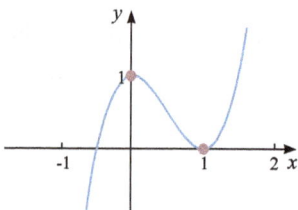

Figure 4.1.2: Graph of $y = 2x^3 - 3x + 1$.

Example 4.1.2. The function $f(x) = |x|$ has no derivative at $x = 0$, as we saw in Chapter 3. However, it has a local minimum value at $x = 0$, because $f(x) = |x| \geqslant 0$ for all x and $f(0) = 0$.

We give a special name for those points where the derivative of f is 0 or the derivative of f does not exist, since they are candidates for extreme values.

Definition 4.1.3. A *critical number/point* of a function f is an interior point c in the domain of f such that either $f'(c) = 0$ or $f'(c)$ does not exist.

Theorem 4.1.2 (Candidate theorem). *A local maximum or local minimum of a function f can only occur at the critical points of f.*

Theorem 4.1.2 shows that a local maximum or minimum of a function f on an open interval **I** must occur at a critical point of f. However, these points are only candidates for the extreme values and they may actually not be an extremum. For example, when $f(x) = x^3$, with domain R (all real numbers), the derivative always exists. Hence, the only candidates for local extrema are stationary points. Since $f'(x) = 3x^2$, solving $f'(x) = 0$ gives only one stationary point, $x = 0$. However, $f(x)$ has neither a local maximum nor a local minimum at the point $x = 0$, as seen in Figure 4.1.3 (it has a point of inflection at $x = 0$, as we will see in Section 4.6).

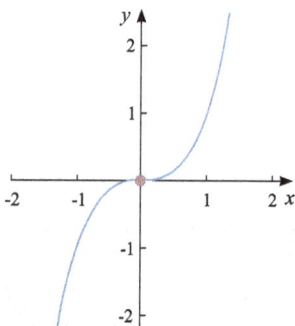

Figure 4.1.3: Graph of $y = x^3$.

Nevertheless, the candidate theorem gives a list of candidates for the local extrema of a function. It does not give any idea whether a candidate is a local minimum or maximum. Luckily, this is enough for us to determine the global extreme values of a continuous function over a closed interval.

As we know from the extreme value theorem, if a continuous function f is defined on a closed interval, then it must take its global maximum and global minimum values somewhere in that interval. If $f(x)$ takes its global extrema at an interior point in $[a, b]$, then the global extremum is also a local extremum. However, the function may take its global extreme values at one or both of the endpoints of that interval. A graphical illustration is shown in Figure 4.1.1. Another easy example is $f(x) = x$, for $0 \leqslant x \leqslant 2$, which takes its global maximum at $x = 2$ and its global minimum at $x = 0$. Both of the two global extrema are at the endpoints of the interval as shown in Figure 4.1.4.

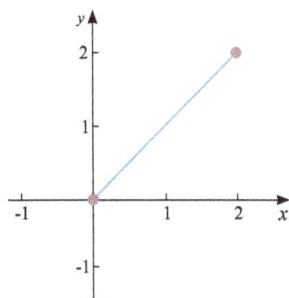

Figure 4.1.4: A function may take its global extrema at endpoints.

In light of the above discussion, we have the following test for global extrema of a continuous function defined on a closed interval.

Closed interval test for global extrema of a continuous function
If a continuous function f is defined on a closed interval $[a, b]$, then it takes its global extreme values either at its critical points or at the endpoints of the interval. To identify its global extreme values on $[a, b]$, we take the following steps.

Step 1 Find the derivative of f and all its critical points.

Step 2 Evaluate f at these critical points and at the endpoints.

Step 3 Compare all function values found in Step 2. The largest function value is the global maximum value of f on $[a, b]$ and the smallest value is the global minimum value of f on $[a, b]$.

Example 4.1.3. Find the global extreme values for the function

$$f(x) = x^3 - 3x^2 - 9x - 2 \quad \text{when } -2 \leqslant x \leqslant 6.$$

Solution. First we find the derivative as follows:

$$f'(x) = 3x^2 - 6x - 9 = 3(x^2 - 2x - 3)$$
$$= 3(x - 3)(x + 1).$$

Therefore, the critical points are $x = 3$ and $x = -1$. Now we evaluate f at these two points and at the two endpoints as follows:

$$f(3) = -29, \quad f(-1) = 3, \quad f(-2) = -4, \quad f(6) = 52.$$

We conclude f attains its global maximum value 52 at $x = 6$ and its global minimum value -29 at $x = 3$. The graph is shown in Figure 4.1.5.

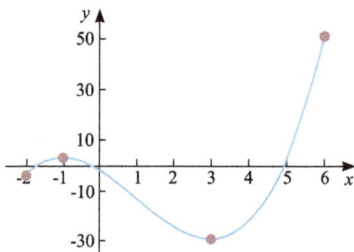

Figure 4.1.5: Graph of $f(x) = x^3 - 3x^2 - 9x - 2$ for $-2 \leqslant x \leqslant 6$.

Example 4.1.4. Find the global extreme values of $f(x) = x^{\frac{2}{3}}(x - 1)$ when $-1 \leqslant x \leqslant 1$.

Solution. The product rule gives the derivative

$$f'(x) = x^{2/3} + (x - 1)\frac{2}{3}x^{-1/3} = \frac{3x + 2(x - 1)}{3x^{1/3}}$$
$$= \frac{5x - 2}{3x^{1/3}}.$$

Therefore, $f'(x) = 0$ only when $x = \frac{2}{5}$, while $f'(x)$ does not exist when $x = 0$. Thus, the critical numbers of f are $\frac{2}{5}$ and 0. The candidates for global extreme values of f are $x = -1$, $x = 0$, $x = \frac{2}{5}$, and $x = 1$. Evaluating f at these points gives

$$f(-1) = -2, \quad f(0) = 0, \quad f\left(\frac{2}{5}\right) \approx -0.326, \quad f(1) = 0.$$

Therefore, over the interval $[-1, 1]$, the function f reaches its global maximum value 0 at $x = 0$ and $x = 1$. The function f attains its global minimum value -2 at $x = -1$, as shown in Figure 4.1.6.

NOTE. A function $f(x)$ may obtain its global extrema at many points in its domain.

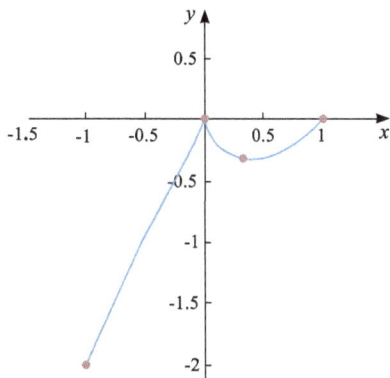

Figure 4.1.6: Graph of $f(x) = x^{\frac{2}{3}}(x-1)$ for $-1 \leqslant x \leqslant 1$.

Example 4.1.5. The function $f(x) = \sin x$ takes on its local and absolute maximum value of 1 infinitely many times, since $\sin(2n + 1/2)\pi = 1$ for any integer n. Likewise, it takes the local and absolute minimum value of -1 infinitely many times, since $\sin(2n - 1/2)\pi = -1$ for any integer n.

Example 4.1.6. Find the global extrema for

$$f(x) = \ln \frac{x}{1+x^2} \quad \text{when } 1 \leqslant x \leqslant 2.$$

Solution. The derivative of $f(x)$ is

$$f'(x) = \frac{1+x^2}{x}\left(\frac{x}{1+x^2}\right)' = \frac{(1+x^2)}{x}\left(\frac{x'(1+x^2) - x(1+x^2)'}{(1+x^2)^2}\right)$$

$$= \frac{(1+x^2)}{x} \times \frac{(1+x^2) - x \times 2x}{(1+x^2)^2}$$

$$= \frac{1-x^2}{x(1+x^2)}.$$

We see that, at $x = 1$, $x = -1$, or $x = 0$, $f'(x) = 0$ or $f'(x)$ does not exist, but -1 and 0 are not in the interval of interest $[1, 2]$, so the only candidates for extrema are $x = 1$ (which is both a critical point and an endpoint) and $x = 2$. Since $f(1) = \ln\frac{1}{1+1^2} = -\ln 2 \approx -0.693$ and $f(2) = \ln\frac{2}{1+2^2} \approx -0.916$, the global maximum value of $f(x)$ is -0.693 and the global minimum value is -0.916. The graph of $f(x)$ is shown in Figure 4.1.7.

We have seen how to find candidates for local extreme values and how to find the global extreme values for functions defined over a closed interval. What happens if $f(x)$ is defined on an open interval or half open interval? How do we determine whether or not a local extremum candidate is indeed a local maximum or local minimum? We can sketch some graphs in order to develop our intuitive ideas. For example, if f keeps

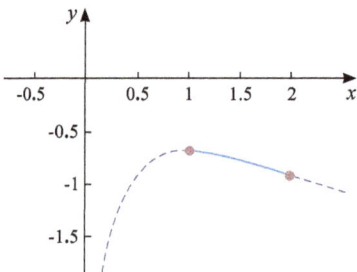

Figure 4.1.7: Graph of $f(x) = \ln\frac{1}{1+x^2}$ for $1 \leqslant x \leqslant 2$.

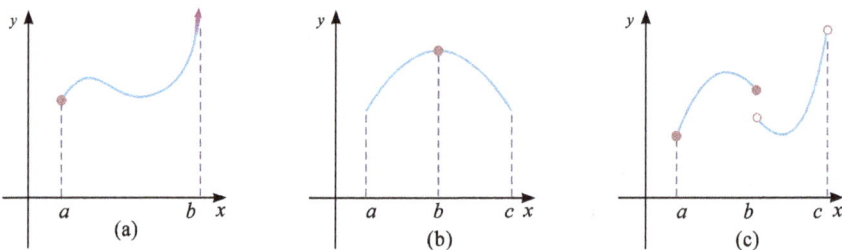

Figure 4.1.8: Cases that functions are not continuous or not defined on a closed interval.

increasing towards the right endpoint, then f may or may not have a global extreme value; if f is continuous and changes from increasing to decreasing at $x = a$, then f takes a local maximum value at $x = a$. Several cases are shown in Figure 4.1.8.

To further study extreme values, we need to investigate the monotonic behavior of f over some interval. We thus need the mean value theorem, which plays a central role in differential calculus and has many applications as well, including determining the monotonicity of functions, proving inequalities, and developing theories.

4.2 The mean value theorem

We start the mean value theorem by considering a mechanics problem. Assume a car moves along the x-axis back and forth, starting at a point A. It will initially move toward the right but after a while it reverses direction and moves back to the starting point. Then there should be a time such that the velocity at that instant is zero. In particular, the velocity at the instant when it makes the turn (reverses direction) must be zero. You also can see that this is true when a ball is thrown vertically upwards and eventually falls to the ground. There must be an instant in time when the ball's velocity is zero. If a function $f(x)$ is continuous and differentiable on some interval $[a, b]$ and $f(a) = f(b)$ (the car returns to its starting point), then there must be a point c in (a, b) such that $f'(c) = 0$, as seen in Figure 4.2.1 (note again, the velocity is the derivative of

(a)

(b)

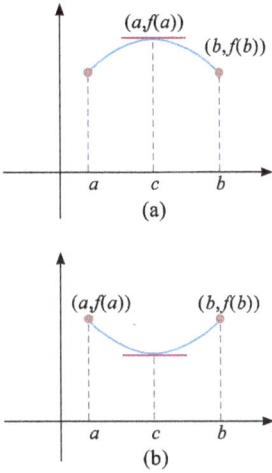

Figure 4.2.1: On a smooth curve, there is a horizontal tangent if $f(a) = f(b)$.

the displacement function). This can be confirmed algebraically by Rolle's theorem, which is proved using Fermat's theorem.

NOTE. The first known formal proof was offered by Michel Rolle in 1691 and used the methods of differential calculus. The name "Rolle's theorem" was first used by Moritz Wilhelm Drobisch (Germany) in 1834 and by Giusto Bellavitis (Italy) in 1846. http://en.wikipedia.org/wiki/Rolle%27s_theorem

Theorem 4.2.1 (Rolle). *Let f be a function that satisfies the following three conditions:*
1. *f is continuous on the closed interval $[a, b]$;*
2. *f is differentiable on the open interval (a, b);*
3. *$f(a) = f(b)$.*

Then there is a number $c \in (a, b)$ such that $f'(c) = 0$.

Proof. We consider three separate cases:

1. Case I: $f(x)$ is constant for all $x \in [a, b]$. Then $f'(x) = 0$ for every $x \in (a, b)$, so the number c can be taken to be any number in (a, b).

2. Case II: $f(x) > f(a)$ for some x in (a, b). By the extreme value theorem, f attains a maximum value for some $x \in [a, b]$. However, the maximum cannot be at $x = a$ or $x = b$ because, by assumption, there is some $f(x) > f(a) = f(b)$, so f must attain its maximum value at a number c in the open interval (a, b). Then f also has a local maximum at c and f is differentiable at c, so $f'(c) = 0$ by Fermat's theorem.

3. Case III: $f(x) < f(a)$ for some x in (a, b). By the extreme value theorem, f has an absolute minimum value for some $x \in [a, b]$. However, the minimum cannot be at $x = a$ or $x = b$, because, by assumption, there is some $f(x) < f(a) = f(b)$, so f must attain

its minimum value at a number c in the open interval (a, b). Then f also has a local minimum at c and f is differentiable at c, so $f'(c) = 0$ by Fermat's theorem.

If any of these three hypotheses fails, then there may not be a number c such that $f'(c) = 0$. This is illustrated in Figure 4.2.2.

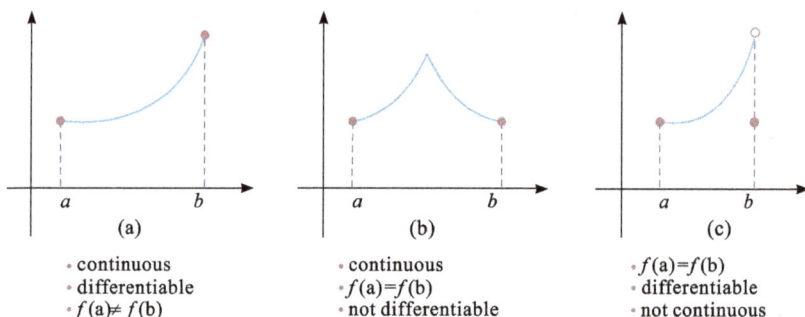

(a)	(b)	(c)
• continuous	• continuous	• $f(a) = f(b)$
• differentiable	• $f(a) = f(b)$	• differentiable
• $f(a) \neq f(b)$	• not differentiable	• not continuous

Figure 4.2.2: Cases that fail to satisfy conditions in Rolle's theorem.

However, if f fails to satisfy any one of the three hypotheses, its derivative may still be zero somewhere in the interval as seen in Figure 4.1.8 (c). There may also be more than one point where the derivative is 0, as seen in Figure 4.1.8 (c). □

Example 4.2.1. Check that the function $f(x) = x^2 - 3x - 4$ for $-1 \leqslant x \leqslant 4$ satisfies all the hypotheses required by Rolle's theorem. Find a number c such that $f'(c) = 0$.

Solution. Since $f(x)$ is a polynomial, it is continuous on $[-1, 4]$ and differentiable at every point in $(-1, 4)$. Then

$$f(-1) = (-1)^2 - 3(-1) - 4 = 0 \quad \text{and} \quad f(4) = 4^2 - 3(4) - 4 = 0.$$

Therefore, f satisfies all the hypotheses required by Rolle's theorem. Since

$$f'(x) = 2x - 3,$$

$x = \frac{3}{2}$ is a point in $(-1, 4)$ such that $f'(\frac{3}{2}) = 0$, as shown in Figure 4.2.3.

Example 4.2.2. Use Rolle's theorem to show that, for any real number k, the equation $x^5 - 6x + k = 0$ has at most one root in the interval $[-1, 1]$.

Solution. Let $f(x) = x^5 - 6x + k$. Suppose, on the contrary, that there are two roots, say, a and b, in $[-1, 1]$. Then we must have $f(a) = f(b) = 0$. By Rolle's theorem, there is a number c between a and b (so $c \in [-1, 1]$) such that $f'(c) = 0$. That is, $f'(c) = 5c^4 - 6 = 0$. Solving for c, we obtain $c = \sqrt[4]{\frac{6}{5}} > 1$, which is a contradiction to $c \in [-1, 1]$. Thus $f(x)$ has at most one root in the interval $[-1, 1]$.

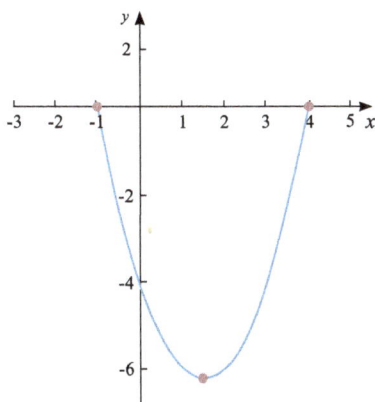

Figure 4.2.3: Graph of $f(x) = x^2 - 3x - 4$ for $-1 \leqslant x \leqslant 4$.

Rolle's theorem requires that $f(a) = f(b)$. What conclusions can we draw about the function if the third hypothesis ($f(a) = f(b)$) is not met? Consider the moving car with displacement function $s(t)$ and velocity $v(t)$. The average velocity during the time interval $[t_1, t_2]$ is given by

$$\frac{s(t_2) - s(t_1)}{t_2 - t_1}.$$

Intuitively speaking, since the car moves continuously, there should be a time such that its velocity at that instant is the same as the average velocity of the car during the time interval (because the average velocity is between the minimum velocity and the maximum velocity). That is, there is an instant $t = t_0$ such that

$$v(t_0) = \frac{s(t_2) - s(t_1)}{t_2 - t_1}.$$

Does this idea that there is a moment when the average rate of change is equal to the instantaneous rate of change hold for all differentiable functions f? That is, is there always a number $c \in (a, b)$ such that

$$f'(c) = \frac{f(b) - f(a)}{b - a}? \tag{4.1}$$

Before proving the statement, we can see that it is reasonable by interpreting it geometrically. The two graphs in Figure 4.2.4 show the points $A(a, f(a))$ and $B(b, f(b))$ on the graphs of two differentiable functions. In both graphs, the average rate of change over $[a, b]$ is equal to the instantaneous rate of change at one or more points in the interval. The slope of the secant line AB is

$$k_{AB} = \frac{f(b) - f(a)}{b - a}$$

and this is the same expression as on the right side of equation (4.1). Since $f'(c)$ is the slope of the tangent line at the point $(c, f(c))$, equation (4.1) says that there is at least

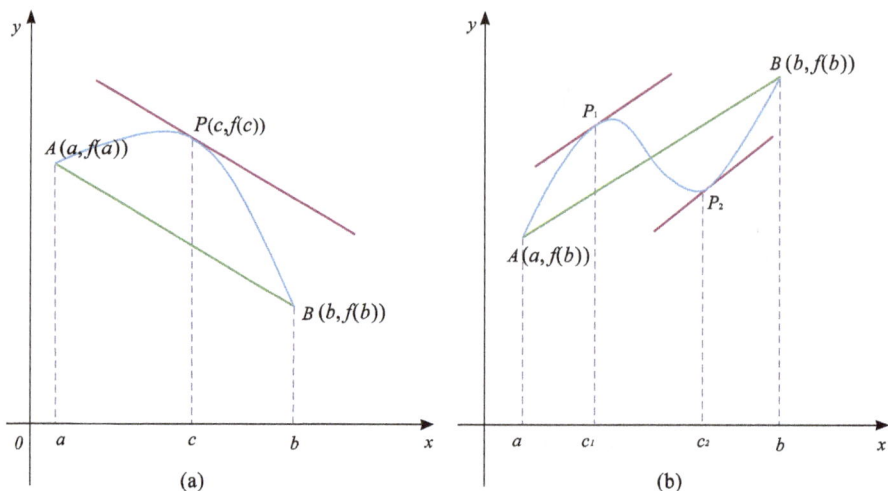

Figure 4.2.4: Illustrations for the mean value theorem.

one point $P(c, f(c))$ on the graph where the slope of the tangent line is the same as the slope of the secant line AB. Thus, the tangent line to the graph of $f(x)$ at $x = c$ is parallel to the secant line through A and B.

   We now establish the mean value theorem, due to the French mathematician Joseph-Louis Lagrange.

Joseph-Louis Lagrange
(1736–1813 in Paris) was an Italian mathematician and astronomer. He made significant contributions to the fields of analysis, number theory, and both classical and celestial mechanics. http://en. wikipedia.org/wiki/Joseph-Louis_Lagrange

Theorem 4.2.2 (Mean value theorem). *Let f be a function that satisfies the following hypotheses:*
1. *f is continuous on the closed interval $[a, b]$;*
2. *f is differentiable on the open interval (a, b).*

Then there is a number $c \in (a, b)$ such that

$$f'(c) = \frac{f(b) - f(a)}{b - a}, \quad \text{or equivalently}$$
$$f(b) - f(a) = f'(c)(b - a).$$

Proof. The equation of the secant line AB can be written as

$$y - f(a) = \frac{f(b) - f(a)}{b - a}(x - a) \quad \text{or}$$

$$y = f(a) + \frac{f(b) - f(a)}{b - a}(x - a).$$

We apply Rolle's theorem to a new function h, defined as the difference between f and the function giving the secant line AB. Let

$$h(x) = f(x) - f(a) - \frac{f(b) - f(a)}{b - a}(x - a).$$

Then

$$h'(x) = f'(x) - \frac{f(b) - f(a)}{b - a}.$$

It is easy to check that $h(x)$ is continuous on $[a, b]$ and differentiable on (a, b). Furthermore, $h(a) = h(b) = 0$. Therefore, by Rolle's theorem, there is a number c in (a, b) such that $h'(c) = 0$. That is,

$$h'(c) = f'(c) - \frac{f(b) - f(a)}{b - a} = 0, \quad \text{so}$$

$$f'(c) = \frac{f(b) - f(a)}{b - a}.$$

Example 4.2.3. Illustrate the mean value theorem with the function $f(x) = x^2 - x$, $a = 0$, $b = 3$.

Solution. Since f is a polynomial, it is continuous and differentiable for all values of x. By the mean value theorem, there is a $c \in (0, 3)$ such that

$$\frac{f(3) - f(0)}{3 - 0} = f'(c).$$

The derivative $f'(x) = 2x - 1$, so $f'(c) = 2c - 1$ and

$$\frac{f(3) - f(0)}{3 - 0} = \frac{3^2 - 3}{3 - 0} = 2,$$

so $2c - 1 = 2$, which gives $c = \frac{3}{2}$. We show this graphically in Figure 4.2.5.

Using the mean value theorem, we can also prove many inequalities, such as those in the next two examples.

Example 4.2.4. Use the mean value theorem to show that

$$|\sin x| \leqslant |x|.$$

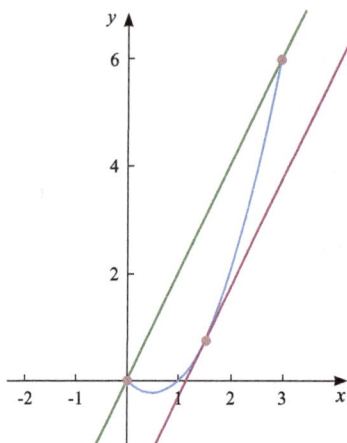

Figure 4.2.5: At $x = \frac{3}{2}$, the tangent is parallel to the secant.

Solution. Let $y = \sin t$. Then y is continuous and differentiable on $[0, x]$ if $x \geqslant 0$ and on $[x, 0]$ if $x < 0$. Since $(\sin t)' = \cos t$, by the mean value theorem, we get

$$\frac{\sin x - \sin 0}{x - 0} = \cos c, \quad \text{where } c \text{ is some point in } (x, 0) \text{ or } (0, x).$$

Taking the absolute value on both sides, we obtain

$$|\sin x| = |\cos c||x| \leqslant |x|.$$

Example 4.2.5. Use the mean value theorem to show that

$$\frac{x}{1+x} < \ln(1+x) < x \quad \text{for all } x > 0.$$

Solution. Let $f(t) = \ln(1+t)$. Then the function $f(t)$ satisfies all the conditions required by the mean value theorem on the interval $[0, x]$ and $f'(t) = \frac{1}{1+t}$. Applying the theorem to the function $f(t)$ on $[0, x]$, we must have some c satisfying $0 < c < x$ and

$$\ln(1+x) - \ln(1+0) = \frac{1}{1+c}(x - 0),$$

$$\ln(1+x) = \frac{x}{1+c}.$$

Since

$$\frac{1}{1+x} < \frac{1}{1+c} < 1,$$

we have

$$\frac{x}{1+x} < \frac{x}{1+c} < x.$$

Then

$$\frac{x}{1+x} < \ln(1+x) < x.$$

This completes the proof. □

The mean value theorem can be used to establish many of the basic results of differential calculus, such as the following theorem.

Theorem 4.2.3. *If $f'(x) = 0$ for all x in an interval (a, b), then f is constant on (a, b).*

Proof. Let x_1 and x_2 be any two numbers in (a, b) with $x_1 < x_2$. Since f is differentiable on (a, b), it must be differentiable on (x_1, x_2) and be continuous on $[x_1, x_2]$. Applying the mean value theorem to f on the interval $[x_1, x_2]$, there is a number c such that $x_1 < c < x_2$ and

$$f(x_2) - f(x_1) = f'(c)(x_2 - x_1).$$

Since $f'(x) = 0$ for all x, we have $f'(c) = 0$, so the above equation becomes

$$f(x_2) - f(x_1) = 0 \quad \text{or} \quad f(x_2) = f(x_1).$$

Therefore, f has the same value at any two numbers x_1 and x_2 in (a, b), so f is constant on (a, b). □

Theorem 4.2.3 can be used to prove some identities as well.

Example 4.2.6. Prove the identity $\arctan x + \operatorname{arccot} x = \pi/2$.

Solution. Although calculus is not needed to prove this identity, the proof using calculus is quite simple. Let

$$f(x) = \arctan x + \operatorname{arccot} x.$$

Then

$$f'(x) = \frac{1}{1+x^2} - \frac{1}{1+x^2} = 0$$

for all values of x. Therefore, $f(x) = C$, a constant. To determine the value of C, we substitute $x = 1$ (because we can evaluate $f(1)$ exactly) and we have

$$C = f(1) = \arctan 1 + \operatorname{arccot} 1 = \frac{\pi}{2}.$$

Thus $\arctan x + \operatorname{arccot} x = \pi/2$.

Definition 4.2.1. If $F'(x) = f(x)$ for all x in an interval (a, b), then $F(x)$ is called an *antiderivative* of $f(x)$ on (a, b).

For example, $\sin x$ is an antiderivative of $\cos x$, since $(\sin x)' = \cos x$; $\sin x + 1$ is also an antiderivative of $\cos x$ since $(\sin x + 1)' = \cos x$.

How many antiderivatives can a function have? What is the relationship between them? The following theorem gives the answer.

Theorem 4.2.4. *If $F(x)$ is an antiderivative of $f(x)$, so that $F'(x) = f(x)$, then all antiderivatives of $f(x)$ have the form $F(x) + C$, where C is an arbitrary constant.*

Proof. Clearly, for any constant C, the derivative of $F(x) + C$ is $f(x)$, so $F(x) + C$ is an antiderivative of $f(x)$. If $G(x)$ is any other antiderivative of $f(x)$, then $G'(x) = f(x)$. This means that

$$F'(x) = G'(x),$$

so

$$[G(x) - F(x)]' = 0.$$

Then, by Theorem 4.2.3, we have

$$G(x) - F(x) = C, \quad \text{for some constant } C.$$

Therefore,

$$G(x) = F(x) + C. \qquad \square$$

Example 4.2.7. Find all antiderivatives for each of the following functions:

(a) $f(x) = \cos x$; (b) $f(x) = e^{2x}$; (c) $f(x) = \frac{1}{x} + 3x$; (d) $f(x) = \frac{1}{1+x^2}$.

Solution. Since $(\sin x)' = \cos x$, $(\frac{1}{2}e^{2x})' = e^{2x}$, $(\ln|x| + \frac{3}{2}x^2)' = \frac{1}{x} + 3x$, and $(\arctan x)' = \frac{1}{1+x^2}$, the antiderivatives of $\cos x$, e^{2x}, $\frac{1}{x}$, and $\frac{1}{1+x^2}$ are $\sin x + C$, $\frac{1}{2}e^{2x} + C$, $\ln|x| + \frac{3}{2}x^2 + C$, and $\arctan x + C$, respectively, where C is an arbitrary constant.

4.3 Monotonic functions and the first derivative test

4.3.1 Monotonic functions

A function f is *increasing* if the points (x, y) on the graph of $y = f(x)$ rise as x increases. It is *decreasing* if the points (x, y) fall as x increases. A formal definition was given in Section 1.2.2 and is repeated here for convenience.

Definition 4.3.1. A function f, defined on a set which includes an interval **I**, is:
1. *increasing* on **I** if, for any x_1, $x_2 \in I$, $f(x_1) \leqslant f(x_2)$ whenever $x_1 < x_2$;
2. *strictly increasing* on **I** if $f(x_1) < f(x_2)$ whenever $x_1 < x_2$;
3. *decreasing* on **I** if, for any $x_1, x_2 \in I$, $f(x_1) \geqslant f(x_2)$ whenever $x_1 < x_2$;
4. *strictly decreasing* on **I** if $f(x_1) > f(x_2)$ whenever $x_1 < x_2$;
5. *monotonic* (or monotone) on **I** if it is either increasing or decreasing on **I**.

You probably have already noticed that, if the graph rises, its slope is positive and if it falls, its slope is negative, as seen in Figure 4.3.1. Our first goal in this section is to determine intervals on which a function is either increasing or decreasing by using derivatives. The next theorem shows that the sign of the derivative can indeed be used to determine intervals where the function is increasing or decreasing. The proof of this theorem is based on the mean value theorem.

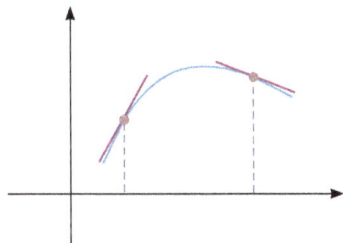

Figure 4.3.1: Positive slope implies increasing and negative slope implies decreasing.

Theorem 4.3.1 (Increasing/decreasing test). *Suppose that f is a function defined on an interval (a, b) and f' is defined on (a, b). Then:*

if $f'(x) > 0$ for all $x \in (a, b)$, then f is strictly increasing on (a, b);
if $f'(x) < 0$ for all $x \in (a, b)$, then f is strictly decreasing on (a, b).

Proof. Suppose $f'(x) > 0$ for all $x \in (a, b)$ and let x_1, $x_2 \in (a, b)$ with $x_1 < x_2$. By the mean value theorem, we have

$$f(x_2) - f(x_1) = f'(c)(x_2 - x_1),$$

where $x_1 < c < x_2$. Of course, $c \in (a, b)$, so $f'(c) > 0$ and it follows that

$$f(x_2) - f(x_1) = f'(c)(x_2 - x_1) > 0.$$

Hence, for any two numbers $x_1 < x_2$ in (a, b),

$$f(x_2) > f(x_1) \quad \text{or equivalently} \quad f(x_1) < f(x_2).$$

This shows that $f(x)$ is strictly increasing on (a, b).

A similar argument shows that, if $f'(x) < 0$ for $x \in (a, b)$, $f(x)$ is strictly decreasing on (a, b). □

NOTE. It is also easy to see that, if $f'(x) \geqslant 0$ for all $x \in (a, b)$, f is increasing on (a, b) and if $f'(x) \leqslant 0$ for all $x \in I$, then f is decreasing on (a, b). If $f'(x) \geqslant 0$ for all $x \in (a, b)$ and $f'(x) = 0$ at only a few points in the interval, then $f(x)$ is also strictly increasing. For example, if $f(x) = x^3$, then $f'(x) \geqslant 0$ and $f'(x) = 0$ only at the point $x = 0$. This cubic function is strictly increasing, as seen in Figure 4.3.2.

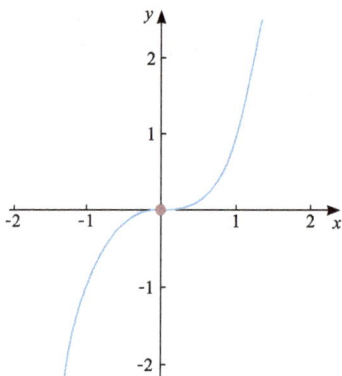

Figure 4.3.2: Graph of $y = x^3$.

Example 4.3.1. Use the increasing/decreasing test to determine the intervals on which f is increasing or decreasing, given

$$f(x) = 4x^3 - 18x^2 + 15x + 10, \quad -\infty < x < \infty.$$

Solution. The derivative of f is

$$f'(x) = 12x^2 - 36x + 15 = 3(2x - 1)(2x - 5).$$

From this factorization, we see that f' has two zeros, $x = \frac{1}{2}$ and $x = \frac{5}{2}$. These zeros determine the three intervals, $(-\infty, \frac{1}{2})$, $(\frac{1}{2}, \frac{5}{2})$, $(\frac{5}{2}, \infty)$, on each of which $f'(x)$ must have a constant sign.

A "sign table" is used to determine the sign of f' on each interval.

x	$(-\infty, \frac{1}{2})$	$\frac{1}{2}$	$(\frac{1}{2}, \frac{5}{2})$	$\frac{5}{2}$	$(\frac{5}{2}, \infty)$
$(2x - 1)$	$-$	0	$+$	0	$+$
$(2x - 5)$	$-$	0	$-$	0	$+$
$f'(x) = 3(2x - 1)(2x - 5)$	$+$	0	$-$	0	$+$
behavior of $f(x)$	↗ increasing	$\frac{27}{2}$	decreasing	$-\frac{5}{2}$	↗ increasing

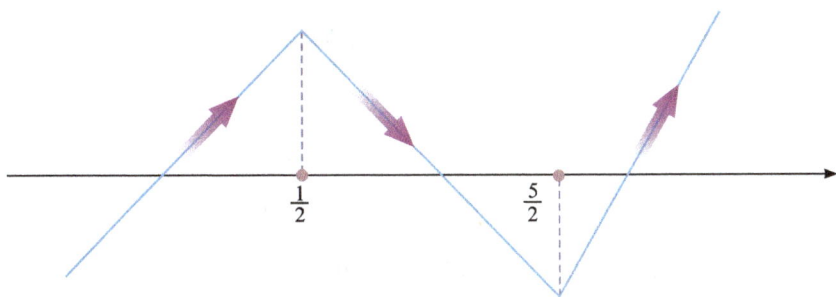

Figure 4.3.3: Rough graph for $f(x)$ in Example 4.3.1.

Thus, for any $x \in (-\infty, \frac{1}{2}) \cup (\frac{5}{2}, \infty)$, $f'(x) > 0$ and $f(x)$ is strictly increasing, while $f(x)$ is strictly decreasing on the interval $(\frac{1}{2}, \frac{5}{2})$, since $f'(x) < 0$ on the interval $(\frac{1}{2}, \frac{5}{2})$. We can sketch a rough graph of $f(x)$ as shown in Figure 4.3.3.

NOTE. The derivative $f'(x)$ must be always positive or always negative for all x-values in any interval between critical points. The reason for this is that, if x_1 and x_2 are both in one such interval and $f'(x_1)$ and $f'(x_2)$ have opposite signs, there must be a c between x_1 and x_2 such that $f'(c) = 0$ (by the intermediate value theorem applied to the derivative). This would contradict the definition of the intervals in that the critical points of f' are the endpoints of the intervals. Hence, we can determine the sign of f' on one of the intervals simply by testing a single x-value in that interval.

4.3.2 The first derivative test

As we saw from the previous example, once we know the intervals on which the function is decreasing or increasing, we can sketch a rough graph of the function. From the graph, we easily see that, if $f(x)$ changes from increasing to decreasing at a point, $f(x)$ must have a local maximum at this point; if $f(x)$ changes from decreasing to increasing at a point, then $f(x)$ must have a local minimum at that point. If the function is differentiable, in terms of the sign of the derivative, we could say that, if $f'(x)$ changes from positive to negative at $x = a$, $f(x)$ reaches a local maximum at $x = a$; if $f'(x)$ changes from negative to positive at $x = a$, then $f(x)$ has a local minimum at $x = a$. In fact, we have deduced the first derivative test.

Theorem 4.3.2 (First derivative test). *If $f(x)$ is continuous at $x = a$ and is differentiable on some interval (b, d) containing $x = a$ (perhaps not at $x = a$), while $x = a$ is a critical point (that is, $f'(a) = 0$ or $f'(a)$ does not exist), then:*

(a) *if $f'(x)$ changes from positive to negative at $x = a$, then $f(x)$ has a local maximum at $x = a$;*

(b) *if f'(x) changes from negative to positive at x = a, then f(x) has a local minimum at x = a.*

Proof. By the mean value theorem, we have

$$f(x) = f(a) + f'(c)(x - a) \quad \text{for some point } c \text{ between } x \text{ and } a.$$

(a) When $b < x < a$ and $f'(x) > 0$, then $f'(c)(x - a) < 0$, so $f(x) > f(a)$ for $x \in (b, a)$. When $a < x < d$ and $f'(x) < 0$, then $f'(c)(x - a) < 0$, so $f(x) > f(a)$ for $x \in (a, c)$. Therefore, if $f'(x)$ changes from positive to negative at $x = a$, then $f(x)$ takes a local maximum at $x = a$.

A similar argument applies to (b). □

Example 4.3.2. If

$$f(x) = \frac{3}{8}x^{\frac{8}{3}} - \frac{3}{2}x^{\frac{2}{3}}, \quad -\infty < x < +\infty,$$

find the intervals of increase and intervals of decrease for $f(x)$. Determine all local extrema of f.

Solution. First, using the power rule, we find the derivative of $f(x)$ as follows:

$$f'(x) = \frac{3}{8} \times \frac{8}{3} \times x^{\frac{8}{3}-1} - \frac{3}{2} \times \frac{2}{3}x^{\frac{2}{3}-1}$$

$$= x^{\frac{5}{3}} - x^{-\frac{1}{3}} = x^{\frac{5}{3}} - \frac{1}{\sqrt[3]{x}}$$

$$= \frac{x^{\frac{5}{3}}\sqrt[3]{x} - 1}{\sqrt[3]{x}}$$

$$= \frac{x^2 - 1}{\sqrt[3]{x}}.$$

There are three critical numbers, $x = 1$, $x = 0$, and $x = -1$. Since $f'(-2) < 0, f'(-0.5) > 0$, $f'(0.5) < 0$, and $f'(2) > 0$, we know that $f'(x) < 0$ for $x \in (-\infty, 1) \cup (0, 1)$ and $f'(x) > 0$ for $x \in (-1, 0) \cup (1, +\infty)$. We construct a sign and behavior diagram in Figure 4.3.4.

We see that the intervals of decrease are $(-\infty, -1) \cup (0, 1)$ and the intervals of increase are $(-1, 0) \cup (1, +\infty)$. By the first derivative test, the function has local minima

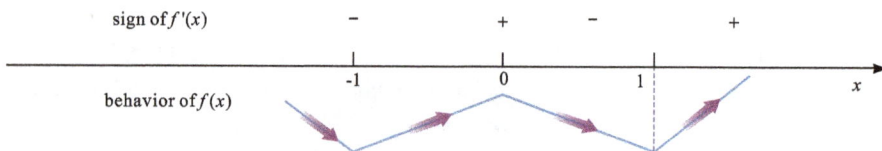

Figure 4.3.4: Signs of $f'(x)$ and behaviors of $f(x)$.

at $x = 1$ and $x = -1$ and a local maximum at $x = 0$. The local maximum value is $f(0) = 0$. The local minimum values are the same. We have

$$f(1) = \frac{3}{8} - \frac{3}{2} = -\frac{9}{8} = f(-1).$$

Example 4.3.3. Show that $e^x > 1 + x$ whenever $x \neq 0$.

Proof. Let $f(t) = e^t - t - 1$. Then $f'(t) = e^t - 1$, which means that

$$f'(t) \begin{cases} < 0, & \text{when } t < 0 \\ > 0, & \text{when } t > 0. \end{cases}$$

Case 1: When $t < 0$, $f(t)$ is strictly decreasing on the interval $[x, 0]$. Hence, $f(x) > f(0)$ and this implies $e^x > 1 + x$.

Case 2: When $t > 0$, $f(t)$ is strictly increasing on the interval $[0, x]$. This means that $f(0) < f(x)$, so $e^x > 1 + x$.

Hence, whenever $x \neq 0$, we have $e^x > 1 + x$. ☐

4.4 Extended mean value theorem and the L'Hôpital rules

4.4.1 Extended mean value theorem

There is an extension of the mean value theorem, which is sometimes called Cauchy's extended mean value theorem. This theorem deals with two differentiable functions $f(x)$ and $g(x)$. If we set $g(x) = x$, then Cauchy's theorem reduces to the mean value theorem.

Theorem 4.4.1 (Cauchy). *If f and g are continuous on $[a, b]$ and differentiable on (a, b) and $g'(x) \neq 0$, when $x \in (a, b)$, then there is at least one number $c \in (a, b)$ for which*

$$\frac{f(b) - f(a)}{g(b) - g(a)} = \frac{f'(c)}{g'(c)}, \quad \text{where } c \in (a, b).$$

Proof. Let

$$h(x) = (f(b) - f(a))g(x) - (g(b) - g(a))f(x).$$

Then

$$\begin{aligned} h(a) &= (f(b) - f(a))g(a) - (g(b) - g(a))f(a) \\ &= f(b)g(a) - f(a)g(h), \\ h(b) &= (f(b) - f(a))g(b) - (g(b) - g(a))f(b) \\ &= -f(a)g(b) + g(a)f(b), \end{aligned}$$

so $h(a) = h(b)$. Since $h(x)$ is also continuous on $[a,b]$ and differentiable on (a,b), by Rolle's theorem, there is at least one point $c \in (a,b)$ such that $h'(c) = 0$. This means

$$h'(x)|_{x=c} = (f(b) - f(a))g'(x) - (g(b) - g(a))f'(x)|_{x=c}$$
$$= (f(b) - f(a))g'(c) - (g(b) - g(a))f'(c) = 0.$$

Therefore,

$$(f(b) - f(a))g'(c) = (g(b) - g(a))f'(c),$$

so

$$\frac{f(b) - f(a)}{g(b) - g(a)} = \frac{f'(c)}{g'(c)}, \quad \text{where } c \in (a,b). \qquad \square$$

Example 4.4.1. Let $b > a > 0$. The function $f(x)$ is continuous on $[a,b]$ and differentiable on (a,b). Prove that there exists a number $c \in (a,b)$ such that $2c(f(b) - f(a)) = (b^2 - a^2)f'(c)$.

Proof. The desired result is equivalent to

$$\frac{f(b) - f(a)}{b^2 - a^2} = \frac{f'(c)}{2c}.$$

This gives hints to use the extended mean value theorem on $[a,b]$ by introducing another function $g(x) = x^2$. By this theorem, there is a number c such that

$$\frac{f(b) - f(a)}{g(b) - g(a)} = \frac{f'(c)}{g'(c)}.$$

Notice that $g(b) = b^2$, $g(a) = a^2$, and $g'(c) = 2c$. This completes the proof. $\qquad \square$

Example 4.4.2. Assume a function $f(x)$ is twice differentiable on an interval I containing a. Let

$$R(x) = f(x) - f(a) - f'(a)(x - a)$$

for all $x \in I$. Show that there is a number ξ, between a and x, such that

$$R(x) = \frac{f''(\xi)}{2}(x - a)^2.$$

Proof. Let $g(t) = (t - a)^2$. We first notice that $R(a) = 0$, $g(a) = 0$. We now apply Cauchy's theorem with two functions $R(t)$ and $g(t)$ on the interval $[a,x]$ (assume $x > a$ without loss of generality). Then there is a number (this time we denote this number by ξ_1 instead of c) such that

$$\frac{R(x)}{g(x)} = \frac{R(x) - R(a)}{g(x) - g(a)} = \frac{R'(\xi_1)}{g'(\xi_1)}.$$

Since $R'(a) = 0$ and $g'(a) = 0$, we apply Cauchy's theorem again, this time with functions $R'(t)$ and $g'(t)$, on the interval $[a, \xi_1]$. We will have a number ξ between ξ_1 and a (of course, this number ξ is also between x and a) such that

$$\frac{R'(\xi_1)}{g'(\xi_1)} = \frac{R'(\xi_1) - R'(a)}{g'(\xi_1) - g'(a)} = \frac{R''(\xi)}{g''(\xi)}.$$

Therefore, the above equations imply

$$\frac{R(x)}{g(x)} = \frac{R''(\xi)}{g''(\xi)} \quad \text{for some } \xi \text{ between } a \text{ and } x.$$

Notice that $g''(x) = 2$ and $R''(x) = f''(x)$. Then we obtain

$$\frac{R(x)}{g(x)} = \frac{f''(\xi)}{2} \quad \text{for some } \xi \text{ between } a \text{ and } x.$$

Because $g(x) = (x - a)^2$, we have $R(x) = \frac{f''(\xi)}{2}(x - a)^2$ for some number ξ between a and x. This completes the proof. $\qquad\qquad\square$

NOTE. This example gives the error when using linear approximation.

We will use Cauchy's theorem to prove the famous L'Hôpital rule for finding limits of indeterminate forms. We first introduce the indeterminate forms.

4.4.2 The indeterminate forms $\frac{0}{0}$, $\infty - \infty$, $\frac{\infty}{\infty}$, and $0 \times \infty$

4.4.2.1 The $\frac{0}{0}$ indeterminate form
The function

$$f(x) = \frac{e^x - \cos x}{x}$$

is undefined at $x = 0$, but we would still like to know how f behaves near 0 and, in particular, the value of the limit

$$\lim_{x \to 0} \frac{e^x - \cos x}{x}.$$

In computing this limit, we cannot apply the law that "the limit of a quotient is the quotient of the limits", because the limit of the denominator is 0. In fact, the limit exists but its value is not obvious because both the numerator and the denominator approach 0 and $\frac{0}{0}$ is not defined. This is an example of a $\frac{0}{0}$ indeterminate form. In Chapter 2 we used a geometric argument to show that $\lim_{x \to 0} \frac{\sin x}{x} = 1$, but the methods used in that proof cannot be adapted to a limit of this kind.

Definition 4.4.1. If $f(x) \to 0$ and $g(x) \to 0$ as $x \to a$, then $\frac{f(x)}{g(x)}$ is a $\frac{0}{0}$ indeterminate form as $x \to a$.

In general, if $\frac{f(x)}{g(x)}$ is a $\frac{0}{0}$ indeterminate form as $x \to a$, then $\lim_{x \to a} \frac{f(x)}{g(x)}$ may or may not exist. In the theorem below, we introduce a systematic method, known as L'Hôpital's rule, that in some cases enables us to find the value of a limit of this kind.

Theorem 4.4.2 (L'Hôpital's rule $\frac{0}{0}$). *Suppose that f and g are differentiable on some interval containing a except possibly at $x = a$, $g'(x) \neq 0$ on that interval, and $\frac{f(x)}{g(x)}$ is a $\frac{0}{0}$ indeterminate form as $x \to a$. If the limit of the ratio of the derivatives satisfies*

$$\lim_{x \to a} \frac{f'(x)}{g'(x)} = L, \quad \text{where L is finite or } \pm \infty,$$

then

$$\lim_{x \to a} \frac{f(x)}{g(x)} = \lim_{x \to a} \frac{f'(x)}{g'(x)} = L.$$

Proof. Since neither the value of $f(x)$ nor the value of $g(x)$ at the point $x = a$ affects the value of $\lim_{x \to a} \frac{f(x)}{g(x)}$, we can define $f(a) = g(a) = 0$, without changing the value of $\lim_{x \to a} \frac{f(x)}{g(x)}$. By Cauchy's theorem,

$$\lim_{x \to a} \frac{f(x)}{g(x)} = \lim_{x \to a} \frac{f(x) - f(a)}{g(x) - g(a)} = \lim_{x \to a} \frac{f'(c)}{g'(c)},$$

where c is a point, dependent on x, somewhere between a and x. As $x \to a$, c must approach a, so we have

$$\lim_{x \to a} \frac{f(x)}{g(x)} = \lim_{x \to a} \frac{f(x) - f(a)}{g(x) - g(a)} = \lim_{x \to a} \frac{f'(c)}{g'(c)} = \lim_{c \to a} \frac{f'(c)}{g'(c)} = \lim_{x \to a} \frac{f'(x)}{g'(x)}. \qquad \square$$

NOTE. The rule is named after the seventeenth-century French mathematician Guillaume de l'Hôpital (also written l'Hôspital), who published the rule in his 1696 book *Analyse des Infiniment Petits pour l'Intelligence des Lignes Courbes* (Analysis of the Infinitely Small for the Understanding of Curved Lines), the first textbook on differential calculus. However, it is believed that the rule was discovered by the Swiss mathematician Johann Bernoulli. http://en.wikipedia.org/wiki/L%27Hôpital%27s_rule

NOTES. 1. This theorem could also be extended to the case where $a = \pm\infty$.
2. If $\lim \frac{f'(x)}{g'(x)}$ does not exist, the rule will fail. A counterexample is $\lim_{x \to 0} \frac{x^2 \sin \frac{1}{x}}{e^x - 1}$.

Example 4.4.3. Find $\lim_{x \to 0} \frac{\ln(1+x)}{x}$.

Solution. Since $\lim_{x\to 0}\ln(1+x) = \ln 1 = 0$ and $\lim_{x\to 0} x = 0$, we can apply L'Hôpital's rule to obtain

$$\lim_{x\to 0}\frac{\ln(1+x)}{x} = \lim_{x\to 0}\frac{(\ln(1+x))'}{(x)'} = \lim_{x\to 0}\frac{1/(1+x)}{1} = \lim_{x\to 0}\frac{1}{1+x} = 1.$$

Example 4.4.4. Use L'Hôpital's rule to show that $\lim_{x\to 0}\frac{\sin x}{x} = 1$.

Solution. Since $\sin x/x$ is a $\frac{0}{0}$ indeterminate form as $x \to 0$, we apply L'Hôpital's rule to obtain

$$\lim_{x\to 0}\frac{\sin x}{x} = \lim_{x\to 0}\frac{(\sin x)'}{(x)'} = \lim_{x\to 0}\frac{\cos x}{1} = \lim_{x\to 0}\cos x = 1.$$

Example 4.4.5. Prove

$$\lim_{x\to 0}\frac{x\sin x}{1 - e^{-x^2}} = 1.$$

Solution. The function $\frac{x\sin x}{1-e^{-x^2}}$ is a $\frac{0}{0}$ indeterminate form as $x \to 0$, so we may apply L'Hôpital's rule to evaluate this limit as follows:

$$\lim_{x\to 0}\frac{x\sin x}{1 - e^{-x^2}} = \lim_{x\to 0}\frac{(x\sin x)'}{(1 - e^{-x^2})'} = \lim_{x\to 0}\frac{\sin x + x\cos x}{2xe^{-x^2}}.$$

However,

$$\frac{\sin x + x\cos x}{2xe^{-x^2}}$$

is still a $\frac{0}{0}$ indeterminate form as $x \to 0$, so we must apply L'Hôpital's rule again to obtain

$$\lim_{x\to 0}\frac{x\sin x}{1 - e^{-x^2}} = \lim_{x\to 0}\frac{(\sin x + x\cos x)'}{(2xe^{-x^2})'} = \lim_{x\to 0}\frac{2\cos x - x\sin x}{2e^{-x^2} - 4x^2e^{-x^2}} = \frac{2}{2} = 1.$$

The $\infty - \infty$ indeterminate form

If $f(x) \to \infty$ and $g(x) \to \infty$ as $x \to a$, then $\lim_{x\to a}(f(x) - g(x))$ has an $\infty - \infty$ indeterminate form, since $\infty - \infty$ is undefined. In the next example we show that this limit can sometimes be evaluated using L'Hôpital's rule.

Example 4.4.6. Use L'Hôpital's rule to show that

$$\lim_{x\to 0+}\left(\frac{1}{x^2} - \frac{1}{\sin x}\right) = +\infty.$$

Solution. The function $\frac{1}{x^2} - \frac{1}{\sin x}$ is an $\infty - \infty$ indeterminate form as $x \to 0^+$. It can be written as a fraction which is a $\frac{0}{0}$ indeterminate form as $x \to 0^+$ by finding a common denominator as follows:

$$\frac{1}{x^2} - \frac{1}{\sin x} = \frac{\sin x - x^2}{x^2 \sin x}.$$

Applying L'Hôpital's rule, we obtain

$$\lim_{x \to 0^+} \left(\frac{1}{x^2} - \frac{1}{\sin x} \right) = \lim_{x \to 0^+} \frac{\sin x - x^2}{x^2 \sin x} = \lim_{x \to 0^+} \frac{\cos x - 2x}{2x \sin x + x^2 \cos x} = +\infty,$$

because, as $x \to 0^+$, the numerator approaches 1 and the denominator is positive and approaches 0.

The $0 \times \infty$ indeterminate form

If $f(x) \to \infty$ and $g(x) \to 0$ as $x \to a$, then $f(x)g(x) = f(x) \times g(x)$ as $x \to \infty$ is also an indeterminate form, since $0 \times \infty$ is undefined. In the next example, we show that this limit can sometimes be evaluated using L'Hôpital's rule.

Example 4.4.7. Evaluate the limit

$$\lim_{x \to \infty} x \arctan\left(\frac{1}{x} \right).$$

Solution. This function has an $\infty \times 0$ indeterminate form as $x \to \infty$. We can convert it to a form suitable for L'Hôpital's rule by rewriting the limit as

$$\lim_{x \to \infty} \frac{\arctan \frac{1}{x}}{\frac{1}{x}} = \lim_{t \to 0} \frac{\arctan t}{t} = \lim_{t \to 0} \frac{\frac{1}{1+t^2}}{1} = 1.$$

The $\frac{\infty}{\infty}$ indeterminate form

If $f(x) \to \infty$ and $g(x) \to \infty$ as $x \to a$, then $\frac{f(x)}{g(x)}$ is of an $\frac{\infty}{\infty}$ indeterminate form as $x \to a$, since $\frac{\infty}{\infty}$ is also undefined. The next theorem says that this limit can sometimes be evaluated using the same process as L'Hôpital's rule. The proof of the theorem is not given here.

Theorem 4.4.3. *Assume that f and g are differentiable on some interval containing a (except possibly at a), $g'(x) \neq 0$ on that interval, and $\frac{f(x)}{g(x)}$ is an $\frac{\infty}{\infty}$ indeterminate form as $x \to a$. If the limit of the ratio of the derivatives satisfies*

$$\lim_{x \to a} \frac{f'(x)}{g'(x)} = L, \quad \text{where } L \text{ is finite or } \pm\infty,$$

then

$$\lim_{x \to a} \frac{f(x)}{g(x)} = \lim_{x \to a} \frac{f'(x)}{g'(x)} = L.$$

NOTE. This theorem could also be extended to the case where $a = \pm\infty$.

Example 4.4.8. Use L'Hôpital's rule to evaluate the limit

$$\lim_{x \to \infty} \frac{\ln x}{x}.$$

Solution. We have

$$\lim_{x \to \infty} \frac{\ln x}{x} = \lim_{x \to \infty} \frac{\frac{1}{x}}{1} = 0.$$

Example 4.4.9. Evaluate $\lim_{x \to 0^+} x \ln x$.

Solution. The function $x \ln x$ is an $\infty \cdot 0$ indeterminate form as $x \to 0^+$, but if we write it as $x \ln x = \frac{\ln x}{\frac{1}{x}}$, it is now an $\frac{\infty}{\infty}$ indeterminate form as $x \to 0^+$. Hence,

$$\lim_{x \to 0^+} x \ln x = \lim_{x \to 0^+} \frac{\ln x}{\frac{1}{x}} = \lim_{x \to 0^+} \frac{1/x}{-1/x^2} = \lim_{x \to 0^+} (-x) = 0.$$

The 0^0, ∞^0, and 1^∞ indeterminate forms

If $f(x) \to 0$, $g(x) \to 0$, $h(x) \to \infty$, and $k(x) \to 1$ as $x \to a$, then $f(x)^{g(x)}$ is a 0^0 *indeterminate form as* $x \to a$, $h(x)^{g(x)}$ is an ∞^0 *indeterminate form as* $x \to a$, and $k(x)^{h(x)}$ is a 1^∞ *indeterminate form as* $x \to a$, since 0^0, ∞^0, and 1^∞ are all undefined.

The limits of the functions defining these three indeterminate forms may or may not exist as $x \to a$, but can sometimes be evaluated using L'Hôpital's rule. Expressions like $f(x)^{g(x)}$ where $f(x) > 0$ can be written as

$$f(x)^{g(x)} = e^{g(x) \ln f(x)}.$$

If we can show, perhaps with the use of L'Hôpital's rule, that

$$\lim_{x \to a} g(x) \ln f(x) = L, \quad \text{where } L \text{ is finite,}$$

then

$$\lim_{x \to a} f(x)^{g(x)} = \lim_{x \to a} e^{g(x) \ln f(x)} = e^L.$$

The last equality holds because the exponential function is continuous.

Example 4.4.10. Show that $\lim_{x \to 0^+} x^x = 1$.

Solution. The function x^x is a 0^0 indeterminate form as $x \to 0^+$. Using the result of Example 4.4.9, we have

$$\lim_{x \to 0^+} x^x = \lim_{x \to 0^+} e^{x \ln x} = e^0 = 1.$$

Another strategy is to set $y = f(x)^{g(x)}$ and take logarithms of both sides, $\ln y = g(x) \ln f(x)$, as in the next example.

Example 4.4.11. Show that, for any real number c,

$$\lim_{x \to \infty} \left(1 + \frac{c}{x}\right)^x = e^c.$$

Solution. The function $(1 + \frac{c}{x})^x$ is a 1^∞ indeterminate form as $x \to \infty$. Setting $y = (1 + \frac{c}{x})^x$ and taking the natural logarithm of both sides gives

$$\ln y = x \ln\left(1 + \frac{c}{x}\right).$$

As $x \to \infty$, the expression on the right is an $\infty \cdot 0$ indeterminate form, that we convert to a $\frac{0}{0}$ indeterminate form as follows:

$$\lim_{x \to \infty} \ln y = \lim_{x \to \infty} x \ln\left(1 + \frac{c}{x}\right)$$

$$= \lim_{x \to \infty} \frac{\ln(1 + \frac{c}{x})}{\frac{1}{x}}$$

$$= \lim_{x \to \infty} \frac{\frac{1}{(1+\frac{c}{x})}\left(-\frac{c}{x^2}\right)}{-\frac{1}{x^2}} = \lim_{x \to \infty} \frac{\frac{1}{(1+\frac{c}{x})}c}{1} = c.$$

Hence,

$$\lim_{x \to \infty} \left(1 + \frac{c}{x}\right)^x = \lim_{x \to \infty} y = \lim_{x \to \infty} e^{\ln y} = e^c.$$

Example 4.4.12. Show that, for any real number c,

$$\lim_{x \to 0} \left(1 + \frac{c}{x}\right)^x = 1.$$

Solution. The function $(1 + \frac{c}{x})^x$ is an ∞^0 indeterminate form as $x \to 0$. Following the method of the previous example, set $y = (1 + \frac{c}{x})^x$ and take the natural logarithm of both sides to obtain

$$\ln y = x \ln\left(1 + \frac{c}{x}\right).$$

The expression on the right is a $0 \cdot \infty$ indeterminate form as $x \to 0$, so, using the method of Example 4.4.9, we find

$$\lim_{x \to 0} \ln y = \lim_{x \to 0} x \ln\left(1 + \frac{c}{x}\right) = \lim_{x \to 0} \frac{\ln(1 + \frac{c}{x})}{\frac{1}{x}}$$

$$= \lim_{x \to 0} \frac{\frac{1}{(1+\frac{c}{x})}\left(-\frac{c}{x^2}\right)}{-\frac{1}{x^2}}$$

$$= 0.$$

Hence,

$$\lim_{x \to 0} \left(1 + \frac{c}{x}\right)^x = \lim_{x \to 0} y = \lim_{x \to 0} e^{\ln y} = e^0 = 1.$$

4.5 Taylor's theorem

4.5.1 The error analysis for the linear approximation

In the previous chapter, we saw that the tangent line approximation, or the lineariza-tion for a differentiable function $f(x)$ at $x = a$, is

$$f(x) \approx f(a) + f'(a)(x - a).\tag{4.2}$$

If we denote the difference between the function and its linearization at $x = a$ by $R(x)$, we have

$$f(x) = f(a) + f'(a)(x - a) + R(x), \quad \text{or}$$
$$R(x) = f(x) - f(a) - f'(a)(x - a).$$

Because

$$\lim_{x \to a} \frac{R(x)}{x - a} = \lim_{x \to a} \frac{f(x) - f(a) - f'(a)(x - a)}{x - a}$$
$$= \lim_{x \to a} \left[\frac{f(x) - f(a)}{x - a} - f'(a) \right] = 0,$$

$R(x)$ is negligible with respect to $x - a$ when $x \to a$. Using the small o notation, we have $R(x) = o(x - a)$. Therefore,

$$f(x) = f(a) + f'(a)(x - a) + o(x - a).\tag{4.3}$$

This gives some basic information – even though we do not know the exact error, we know that the error would be much smaller than $\Delta x = x - a$. This agrees with our intuition that, as x approaches a, the error in a linear approximation approaches zero. The exact error was given by Example 4.4.2, where we found

$$R(x) = \frac{f''(\xi)}{2}(x - a)^2 \quad \text{for some } \xi \text{ between } a \text{ and } x.$$

This means that, for some ξ between a and x,

$$f(x) = f(a) + f'(a)(x - a) + \frac{f''(\xi)}{2}(x - a)^2.\tag{4.4}$$

Observe that

$$0 < \frac{\xi - a}{x - a} < 1.$$

Let $\theta = \frac{\xi - a}{x - a}$ and $\xi = a + \theta(x - a)$. Then the exact error when we use the tangent line approximation can also be written

$$R(x) = \frac{f''(a + \theta(x - a))}{2}(x - a)^2 \quad \text{for some } \theta \text{ between } 0 \text{ and } 1.$$

Therefore, for some θ between 0 and 1,

$$f(x) = f(a) + f'(a)(x - a) + \frac{f''(a + \theta(x - a))}{2}(x - a)^2. \tag{4.5}$$

Example 4.5.1. Estimate the error if we use the linearization at $x = 0$ to approximate the number $\sin(0.02)$.

Solution. The derivative of $\sin x$ at $x = 0$ is 1 and the tangent line to $\sin x$ at $x = 0$ is $f(x) = x$. The error is given by

$$|\sin x - x| = \left| \frac{-\sin(0 + \theta x)}{2}(x - 0)^2 \right|, \quad 0 < \theta < 1.$$

The tangent line estimate of $\sin 0.02$ is 0.02, so the error bound is

$$|\sin(0.02) - 0.02| = \left| \frac{-\sin(\theta \times 0.02)}{2}(0.02 - 0)^2 \right| \leq \frac{1}{2} \times 0.02^2 = 0.0002.$$

NOTE. Using a calculator, to nine decimal places, the value of $\sin 0.02$ is $0.019\,998\,667$. The error is therefore

$$|0.019\,998\,667 - 0.02| = 0.000\,000\,133\,3 < 0.0002.$$

4.5.2 The quadratic approximation

Now, let us look at Figure 4.5.1, showing a linear and a "curved" approximation to e^x at $x = 0$. Which one is a better approximation? The "curved" approximation is clearly better. But why is it better? And how do we find such a "curved" approximation?

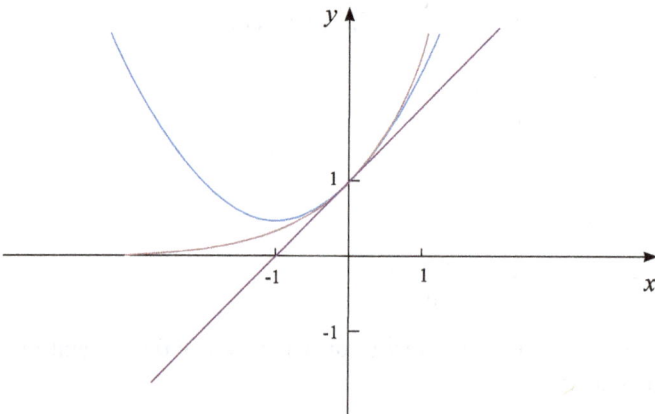

Figure 4.5.1: Linear and quadratic approximation to $y = e^x$ at $x = 0$.

If we think about two runners running along a straight line, they may start from the same position with the same initial velocity, but will they always have the same position? They will initially, but perhaps not for very long because their accelerations may be different! The second derivative makes an important difference in their positions. Now think about a quadratic curve $P(x) = c_0 + c_1(x-a) + c_2(x-a)^2$ which meets a twice differentiable function $f(x)$ at the point $x = a$. Assume that, at this point, the two curves have the same slope (first derivative) and the same second derivative. We summarize this information as follows:

$$f(x) \text{ meets } P(x) \text{ at } x = a, \quad f(a) = P(a),$$
$$\text{they have the same slope at } x = a, \quad f'(a) = P'(a),$$
$$\text{they have the same second derivative at } x = a, \quad f''(a) = P''(a).$$

On the other hand, if $P(x) = c_0 + c_1(x-a) + c_2(x-a)^2$, then

$$P(a) = c_0, \quad P'(a) = c_1, \quad \text{and} \quad P''(a) = 2c_2.$$

Therefore, the coefficients of $P(x)$ are

$$c_0 = f(a), \quad c_1 = f'(a) \quad \text{and} \quad c_2 = \frac{f''(a)}{2}.$$

Thus, the quadratic curve which approximates $f(x)$ must be

$$P(x) = f(a) + f'(a)(x-a) + \frac{f''(a)}{2}(x-a)^2$$

and

$$f(x) \approx f(a) + f'(a)(x-a) + \frac{f''(a)}{2}(x-a)^2. \tag{4.6}$$

Example 4.5.2. Find the quadratic approximation for $f(x) = \ln x$ at $a = 1$.

Solution. First compute the coefficients of the quadratic polynomial as follows:

$$f(1) = \ln 1 = 0, \quad f'(1) = \frac{1}{x}\Big|_{x=1} = 1, \quad f''(1) = -\frac{1}{x^2}\Big|_{x=1} = -1.$$

The quadratic approximation is given by

$$P(x) = 0 + 1(x-1) - \frac{1}{2}(x-1)^2$$

and then

$$f(x) \approx 0 + 1(x-1) - \frac{1}{2}(x-1)^2.$$

Figure 4.5.2 shows the linear and quadratic approximation for $f(x) = \ln x$ at $x = 1$.

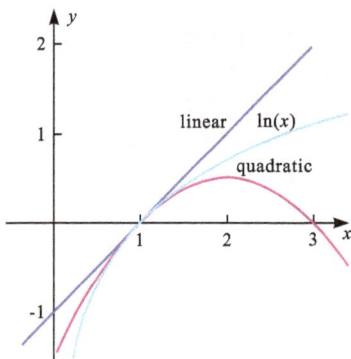

Figure 4.5.2: Linear and quadratic approximation to $y = \ln x$ at x=1.

Error analysis with a quadratic approximation

As for the error analysis with a quadratic approximation, we assume that $f(x)$ is twice differentiable on an interval **I** containing a and that $f''(x)$ is continuous at $x = a$. Again we let $R(x) = f(x) - P(x)$, that is, $f(x) = P(x) + R(x)$. Then

$$f(x) = f(a) + f'(a)(x - a) + \frac{f''(a)}{2}(x - a)^2 + R(x).$$

Now we evaluate the following limit by using L'Hôpital's rule. We have

$$\lim_{x \to a} \frac{R(x)}{(x - a)^2} = \lim_{x \to a} \frac{f(x) - f(a) - f'(a)(x - a) - \frac{f''(a)}{2}(x - a)^2}{(x - a)^2}$$

$$= \lim_{x \to a} \frac{f'(x) - f'(a) - f''(a)(x - a)}{2(x - a)}$$

$$= \lim_{x \to a} \frac{f''(x) - f''(a)}{2} = 0.$$

The last equality holds because $f''(x)$ is continuous at $x = a$. The fact that the limit is 0 means $R(x)$ is negligible with respect to $(x - a)^2$. Thus, $R(x) = o((x - a)^2)$. Therefore,

$$f(x) = f(a) + f'(a)(x - a) + \frac{f''(a)}{2}(x - a)^2 + o((x - a)^2). \qquad (4.7)$$

This agrees with our intuition that, near $x = a$, a quadratic approximation is better than a linear approximation.

To find the exact error, we assume f can be differentiated at least three times on **I**. We use a similar argument as used for linear approximation, but this time we apply the extended value theorem three time with functions $R(x)$ and $g(x) = (x - a)^3$, $R'(x)$

and $g'(x)$, and $R''(x)$ and $g''(x)$, respectively, to obtain

$$\frac{R(x) - R(a)}{g(x) - g(a)} = \frac{R'(\xi_1)}{g'(\xi_1)} \quad \text{for some } \xi_1 \text{ between } a \text{ and } x,$$

$$\frac{R'(\xi_1) - R'(a)}{g'(\xi_1) - g'(a)} = \frac{R''(\xi_2)}{g''(\xi_2)} \quad \text{for some } \xi_2 \text{ between } a \text{ and } \xi_1,$$

$$\frac{R''(\xi_2) - R''(a)}{g''(\xi_2) - g''(a)} = \frac{R'''(\xi)}{g'''(\xi)} \quad \text{for some } \xi \text{ between } a \text{ and } \xi_2.$$

Now evaluating $R(a)$, $g(a)$, $R'(a)$, $g'(a)$, $R''(a)$, and $g''(a)$ (all of them are zeros and $g'''(x) = 3!$), the above equations give

$$R(x) = \frac{f'''(\xi)}{3!}(x - a)^3 \quad \text{for some } \xi \text{ between } a \text{ and } x.$$

Thus, for some ξ between a and x, we have

$$f(x) = f(a) + f'(a)(x - a) + \frac{f''(a)}{2}(x - a)^2 + \frac{f'''(\xi)}{3!}(x - a)^3. \tag{4.8}$$

As shown in the previous section, we have

$$R(x) = \frac{f'''(a + \theta(x - a))}{3!}(x - a)^3 \quad \text{for some } \theta \text{ between } 0 \text{ and } 1.$$

Then we have

$$f(a) + f'(a)(x - a) + \frac{f''(a)}{2}(x - a)^2 + \frac{f'''(a + \theta(x - a))}{3!}(x - a)^3. \tag{4.9}$$

Example 4.5.3. Estimate the error in approximating $e^{0.02}$ when:
1. using a linear approximation at $x = 0$;
2. using a quadratic approximation at $x = 0$.

Solution. Let $f(x) = e^x$. Then $f(0) = 1$, $f'(0) = 1$, and $f''(0) = 1$, so the tangent line to e^x at $x = 0$ is $P_1(x) = 1 + x$. The quadratic approximation is

$$P_2(x) = 1 + x + \frac{1}{2}x^2.$$

The error for the linear approximation is given by

$$\left| e^x - (1 + x) \right| = R(x) = \left| \frac{e^{0 + \theta x}}{2}(x - 0)^2 \right|, \quad 0 < \theta < 1,$$

so

$$\left| e^{0.02} - (1 + 0.02) \right| - R(0.02) = \left| \frac{e^{0 + \theta \times 0.02}}{2}(0.02 - 0)^2 \right|$$

$$\leqslant \frac{e}{2}(0.02)^2 < \frac{2.72}{2} \times (0.02)^2 = 0.000\,544.$$

When using a quadratic approximation, the error is given by

$$\left| e^x - \left(1 + x + \frac{x^2}{2} \right) \right| = R(x) = \left| \frac{e^{0+\theta x}}{3!} (x-0)^3 \right|, \quad 0 < \theta < 1,$$

so

$$\left| e^{0.02} - \left(1 + 0.02 + \frac{0.02^2}{2} \right) \right| = R(0.02) = \left| \frac{e^{\theta \times 0.02}}{3!} (0.02 - 0)^3 \right|$$

$$< \frac{2.72}{3!} \times (0.02)^3 < 0.000\,003\,626\,7.$$

This confirms algebraically that the quadratic approximation is better than the linear one in this situation when $x \approx a$.

4.5.3 Taylor's theorem

The ideas discussed above could easily be extended in order to approximate a function with a polynomial of degree n, for some positive integer n, assuming that the function has an nth derivative. That is, we try an nth degree polynomial approximation to $f(x)$ at $x = a$ of the form

$$P(x) = c_0 + c_1(x-a) + c_2(x-a)^2 + \cdots + c_n(x-a)^n.$$

We attempt to make sure that the approximation to $f(x)$ is accurate by choosing the coefficients c_k so that the derivatives of $f(x)$ and $P(x)$ are the same at $x = a$. That is, we require

$$P(a) = f(a), \quad P'(a) = f'(a), \quad P''(a) = f''(a), \quad \ldots, \quad P^{(n)}(a) = f^{(n)}(a).$$

That means $P^{(k)}(a) = f^{(k)}(a)$ for $k = 0, 1, 2, \ldots, n$. Under these assumptions, by substituting $x = a$ in the approximation and then differentiating $P(x)$ repeatedly and substituting a for x each time, it is easy to show that the coefficients are $c_0 = f(a)$, $c_1 = \frac{f'(a)}{1!}$, $c_2 = \frac{f''(a)}{2!}$, ..., and in general $c_k = \frac{f^{(k)}(a)}{k!}$, for $k = 0, 1, 2, \ldots, n$. Hence, the required approximating polynomial is

$$P_n(x) = f(a) + \frac{f'(a)}{1}(x-a) + \frac{f''(a)}{2!}(x-a)^2 + \cdots + \frac{f^{(n)}(a)}{n!}(x-a)^n$$

and this polynomial is called the *nth degree Taylor polynomial of f centered at a*.

Similar to the linear and quadratic approximation, we have the following useful theorem.

Theorem 4.5.1 (Taylor's theorem). *If $f(x)$ has $(n+1)$th order derivatives in some interval* **I** *containing a, then, for any x in this interval,*

$$f(x) = P_n(x) + R_n(x),$$

Brook Taylor FRS (1685–1731) was an English mathematician who is best known for Taylor's theorem and the Taylor series. The concept of a Taylor series was discovered by the Scottish mathematician James Gregory and formally introduced by the English mathematician Brook Taylor in 1715. http://en.wikipedia.org/wiki/Brook_Taylor

where the Taylor polynomial of degree n, $P_n(x)$, is

$$P_n(x) = f(a) + \frac{f'(a)}{1}(x - a) + \frac{f''(a)}{2!}(x - a)^2 + \cdots + \frac{f^{(n)}(a)}{n!}(x - a)^n \qquad (4.10)$$

and the remainder term $R_n(x)$ is given by

$$R_n(x) = \frac{f^{(n+1)}(c)}{(n + 1)!}(x - a)^{n+1} \quad \text{for some c between a and x,} \qquad (4.11)$$

$$R_n(x) = o((x - a)^n), \qquad (4.12)$$

or

$$R_n(x) = \frac{f^{(n+1)}(a + \theta(x - a))}{(n + 1)!}(x - a)^{n+1} \quad \text{for some } \theta \text{ between 0 and 1} \qquad (4.13)$$

(the value of θ depends on the values of x, a, and n).

Proof. This theorem can easily be proved by replicating the proofs that we have given for $n = 1$ and $n = 2$. □

The remainder $R_n(x)$ of the form $R_n(x) = o((x - a)^n)$ is called the *Peano remainder*, while $R_n(x) = \frac{f^{(n+1)}(c)}{(n+1)!}(x - a)^{n+1}$ is called the *Lagrange remainder*. It gives the difference between the function $f(x)$ and its Taylor polynomial of degree n centered at a. If we get an upper bound M for $f^{(n+1)}(c)$, then an error bound is given by

$$|R_n(x)| = \left| \frac{f^{(n+1)}(c)}{(n + 1)!}(x - a)^{n+1} \right| \leq \frac{M}{(n + 1)!}|x - a|^{n+1}. \qquad (4.14)$$

Example 4.5.4. Find the Taylor polynomial of degree 4 centered at $x = 1$ for $f(x) = \sqrt{x}$ and use it to estimate $\sqrt{1.5}$.

Solution. We first compute the coefficients for this Taylor polynomial. We have

$$f(x) = \sqrt{x}, \quad f'(x) = \frac{1}{2}x^{-\frac{1}{2}}, \quad f''(x) = -\frac{1}{4}x^{-\frac{3}{2}}, \quad f'''(x) = \frac{3}{8}x^{-\frac{5}{2}},$$

and $f^{(4)}(x) = -\frac{15}{16}x^{-\frac{7}{2}}$. Therefore,

$$f(1) = 1, \quad f'(1) = \frac{1}{2}, \quad f''(1) = -\frac{1}{4}, \quad f'''(1) = \frac{3}{8}, \quad \text{and} \quad f^{(4)}(1) = -\frac{15}{16}.$$

The Taylor polynomial of degree 4 centered at $x = 1$ of $f(x) = e^x$ is then given by

$$P_4(x) = 1 + \frac{\frac{1}{2}}{1}(x-1) + \frac{-\frac{1}{4}}{2!}(x-1)^2 + \frac{\frac{3}{8}}{3!}(x-1)^3 + \frac{-\frac{15}{16}}{4!}(x-1)^4$$

$$= 1 + \frac{(x-1)}{2} - \frac{1}{8}(x-1)^2 + \frac{1}{16}(x-1)^3 - \frac{5}{128}(x-1)^4.$$

Hence,

$$f(1.5) \approx P_4(1.5)$$

$$= 1 + \frac{(1.5-1)}{2} - \frac{(1.5-1)^2}{8} + \frac{(1.5-1)^3}{16} - \frac{5(1.5-1)^4}{128}$$

$$= 1.224\,1.$$

To estimate the error, we differentiate $f^{(4)}(x)$ again to obtain

$$f^{(5)}(x) = \frac{105}{32}x^{-\frac{9}{2}}.$$

Then

$$|R_4(x)| = \left| \frac{\frac{105}{32}(1+\theta(x-1))^{-\frac{9}{2}}}{(4+1)!}(x-1)^{4+1} \right|$$

$$= \left| \frac{105}{32 \times 5!}(1+\theta(x-1))^{-\frac{9}{2}}(x-1)^5 \right|, \quad 0 < \theta < 1,$$

$$\leqslant \frac{105}{32 \times 5!}|x-1|^5,$$

so

$$|R_4(1.5)| = \frac{105}{32 \times 5!}|1.5-1|^5 \leqslant 8.545 \times 10^{-4}.$$

Figure 4.5.3 shows the graphs of $\sqrt{x}$ and its Taylor polynomial of degree 4 centered at $x = 1$. The value of $\sqrt{1.5}$ given by a TI-89 graphing calculator is $1.224\,744\,871\,39$ and

$$|1.224\,1 - 1.224\,744\,871\,39| = 6.449 \times 10^{-4} \leqslant 8.545 \times 10^{-4}.$$

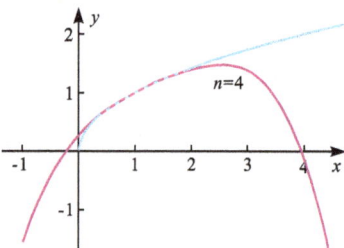

Figure 4.5.3: Graphs of $y = \sqrt{x}$ and its Taylor polynomial of degree 4 at $x = 1$.

To compare with the linear and quadratic approximations, we compute

$$P_1(1.05) = 1 + \frac{(1.5-1)}{2}\bigg|_{x=1.5} = 1.25$$

and

$$P_2(1.05) = 1 + \frac{(1.5-1)}{2} - \frac{1}{8}(1.5-1)^2\bigg|_{x=1.5} = 1.2188.$$

Obviously, the corresponding errors are larger.

NOTES. 1. Taylor's theorem is sometimes called Taylor's formula or Taylor's expansion.

2. It can be shown that the remainder $R_n(x) \to 0$ as $n \to \infty$ for most infinitely differentiable functions. This implies that we can make the approximation as good as we want by using a Taylor polynomial for f of sufficiently large degree.

Maclaurin's formula

When $a = 0$, Taylor's theorem becomes simply

$$f(x) = f(0) + \frac{f'(0)}{1}x + \frac{f''(0)}{2!}x^2 + \cdots + \frac{f^{(n)}(0)}{n!}x^n + R_n(x), \qquad (4.15)$$

$$R_n(x) = \frac{f^{(n+1)}(\theta x)}{(n+1)!}x^{n+1}, \quad \text{where } 0 < \theta < 1, \qquad (4.16)$$

where θ is a number depending on the values of x and n. This is given a special name: the *Maclaurin formula* or *Maclaurin expansion*.

Using the small o notation, we write the remainder of the Maclaurin formula as

$$R_n = \frac{f^{(n+1)}(\theta x)}{(n+1)!}x^{n+1} = o(x^n). \qquad (4.17)$$

Example 4.5.5. Find the Maclaurin formula for the function $f(x) = e^x$ and compute the error in using the 11 terms of the polynomial $(n = 10)$ to approximate e.

Solution. Since $f(x) = f'(x) = \cdots = f^{(n)}(x) = e^x$, we have

$$f(0) = f'(0) = f''(0) = \cdots = f^{(n)}(0) = 1,$$

so the Maclaurin expansion for f is

$$e^x = 1 + x + \frac{x^2}{2!} + \frac{x^3}{3!} + \cdots + \frac{x^n}{n!} + \frac{e^{\theta x}}{(n+1)!}x^{n+1}, \quad \text{where } 0 < \theta < 1.$$

The nth degree polynomial approximation for e^x is

$$e^x \approx 1 + x + \frac{x^2}{2!} + \frac{x^3}{3!} + \cdots + \frac{x^n}{n!}$$

and the error (remainder) for this approximation satisfies

$$|R_n(x)| = \left| \frac{e^{\theta x}}{(n+1)!} x^{n+1} \right| < \frac{e^x}{(n+1)!} |x|^{n+1}.$$

When $x = 1$ and $n = 10$, we obtain the following approximation for the transcendental number e:

$$e \approx 1 + 1 + \frac{1}{2!} + \frac{1}{3!} + \frac{1}{4!} + \frac{1}{5!} + \frac{1}{6!} + \frac{1}{7!} + \frac{1}{8!} + \frac{1}{9!} + \frac{1}{10!}$$
$$\approx 2.718\,281\,8.$$

The error $|R_n|$ satisfies

$$|R_n| < \frac{e}{11!} 1^{n+1} < 10^{-7}.$$

Figure 4.5.4 shows the graphs of $y = e^x$ and some of its Taylor polynomials.

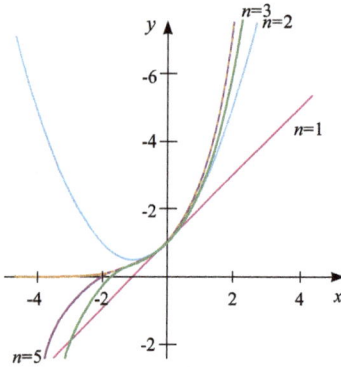

Figure 4.5.4: Graphs of $y = e^x$ and its Taylor polynomial of degree 1, 2, 3 and 5.

In a similar manner, we can find more Maclaurin expansions, such as

$$\sin x = x - \frac{x^3}{3!} + \frac{x^5}{5!} - \frac{x^7}{7!} + \cdots + (-1)^{n-1} \frac{x^{2n-1}}{(2n-1)!} + R_{2n},$$

$$\cos x = 1 - \frac{x^2}{2!} + \frac{x^4}{4!} - \frac{x^6}{6!} + \cdots + (-1)^n \frac{x^{2n}}{(2n)!} + R_{2n+1},$$

$$\ln(1+x) = x - \frac{x^2}{2} + \frac{x^3}{3} - \cdots + (-1)^{n-1} \frac{1}{n} x^n + R_n, \quad \text{and}$$

$$(1+x)^a = 1 + ax + \frac{a(a-1)}{2!} x^2 + \cdots$$
$$+ \frac{a(a-1) \cdots (a-n+1)}{n!} x^n + R_n.$$

This of course leads to some useful approximations, such as

$$\sin x = x - \frac{1}{3!} x^3 + o(x^4) \quad \text{and} \quad \cos x = 1 - \frac{x^2}{2!} + o(x^3).$$

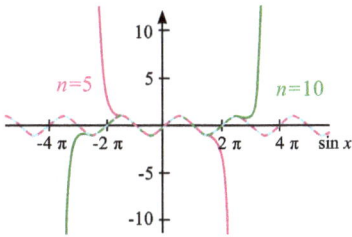

Figure 4.5.5: Graphs of $y = \sin x$ and its Taylor polynomial of degree 5 and 10.

Figure 4.5.5 shows graphs of $y = \sin x$ and its Taylor polynomial of degree 5 and 10. These approximations can also be useful when evaluating limits.

Example 4.5.6. Find $\lim_{x \to 0} \frac{\sin x - x \cos x}{\sin^3 x}$.

Solution. Since

$$\sin x = x - \frac{x^3}{3!} + o(x^4), \quad x \cos x = x - \frac{x^3}{2!} + o(x^4),$$

and $\sin^3 x \sim x^3$ as $x \to 0$, we have

$$\lim_{x \to 0} \frac{\sin x - x \cos x}{\sin^3 x} = \lim_{x \to 0} \frac{x - \frac{x^3}{3!} + o(x^4) - x[1 - \frac{x^2}{2!} + o(x^3)]}{x^3}$$

$$= \lim_{x \to 0} \frac{\frac{1}{3}x^3 + o(x^3)}{x^3} = \frac{1}{3}.$$

4.6 Concave functions and the second derivative test

4.6.1 Concave functions

The sign of the first derivative gives information as regards monotonicity. Figure 4.6.1 (a) and Figure 4.6.1 (b) show the graphs of two increasing functions on (a, b) and both graphs join point A to point B. They look different because the first bends upward on $[a, b]$ and the second bends downward on $[a, b]$. Both functions have posi-

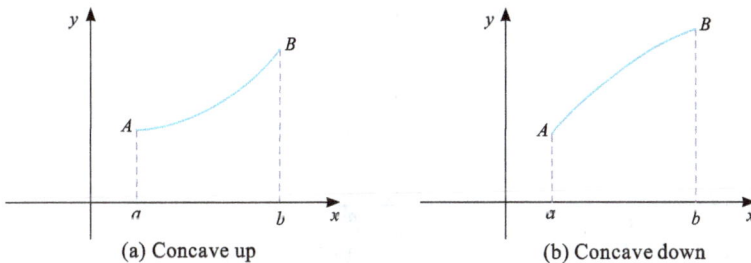

(a) Concave up

(b) Concave down

Figure 4.6.1: Concave up and concave down functions.

tive first derivatives on the interval (a, b). The graph of the function f in Figure 4.6.1 (a) is said to be *concave up* on (a, b) and the graph of the function g in Figure 4.6.1 (b) is said to be *concave down* on (a, b).

As seen in Figure 4.6.2 (a), if the graph is concave up on $[a, b]$, then, for any two points x_1 and x_2 in $[a, b]$, the chord connecting them must lie above the curve. If the graph is concave down, then the chord connecting any two points must lie below the curve. The formal definition of concavity is given below.

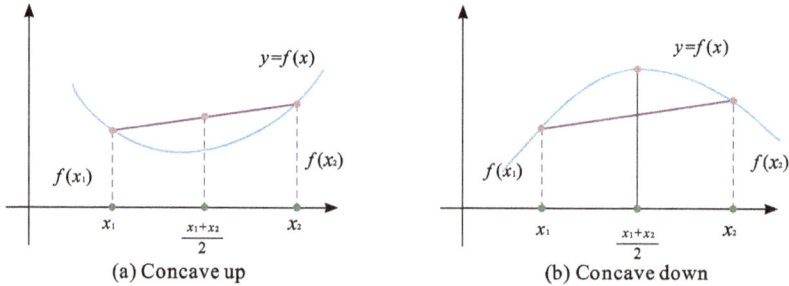

Figure 4.6.2: Concave up/down graphs and their chords.

Definition 4.6.1. Suppose $f(x)$ is continuous on **I** and, for any two values $a \neq b \in I$, we have

$$f\left(\frac{a+b}{2}\right) < \frac{f(a)+f(b)}{2}.$$

Then $f(x)$ is *concave up* on **I**. Similarly, if, for any two values $a \neq b \in I$,

$$f\left(\frac{a+b}{2}\right) > \frac{f(a)+f(b)}{2},$$

then $f(x)$ is *concave down* on **I**.

In Figure 4.6.3 (a), the graph is concave up. Observe that the tangent lines lie below the curve f and their slopes increase as x increases. That means the derivative $f'(x)$ is increasing and therefore the derivative of $f'(x)$ must be positive, i.e., $f''(x) > 0$.

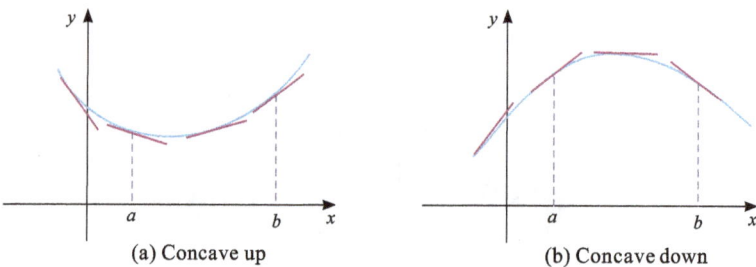

Figure 4.6.3: Concave up/down graphs and their tangents.

In Figure 4.6.3 (b), observe that the tangent lines lie above the curve f and their slopes decrease as x increases. That means the derivative $f'(x)$ is decreasing and therefore the derivative of $f'(x)$ must be negative, i.e., $f''(x) < 0$. The following theorem tells us that, if a function is twice differentiable on an interval, then, indeed, we can use its second derivative to determine its concavity.

Theorem 4.6.1. *Suppose $f(x)$ is a twice differentiable function defined on an interval* I. *Then:*
1. *if $f''(x) > 0$, then $f(x)$ is concave up on* I;
2. *if $f''(x) < 0$, then $f(x)$ is concave down on* I.

Proof. Suppose that a, b are any two points in I and $a < b$. Apply Taylor's theorem to $f(x)$ at the point $x = \frac{a+b}{2}$ to obtain

$$f(x) = f\left(\frac{a+b}{2}\right) + f'\left(\frac{a+b}{2}\right)\left(x - \frac{a+b}{2}\right) + \frac{f''(\xi)}{2!}\left(x - \frac{a+b}{2}\right)^2,$$

where ξ depends on x.

Now evaluating f at $x = a$ and $x = b$, respectively, we have

$$f(a) = f\left(\frac{a+b}{2}\right) + f'\left(\frac{a+b}{2}\right)\left(a - \frac{a+b}{2}\right) + \frac{f''(\xi_1)}{2!}\left(a - \frac{a+b}{2}\right)^2$$

$$= f\left(\frac{a+b}{2}\right) + f'\left(\frac{a+b}{2}\right)\left(\frac{a-b}{2}\right) + \frac{f''(\xi_1)}{2!}\left(\frac{a-b}{2}\right)^2,$$

$$f(b) = f\left(\frac{a+b}{2}\right) + f'\left(\frac{a+b}{2}\right)\left(b - \frac{a+b}{2}\right) + \frac{f''(\xi_2)}{2!}\left(b - \frac{a+b}{2}\right)^2$$

$$= f\left(\frac{a+b}{2}\right) + f'\left(\frac{a+b}{2}\right)\left(\frac{b-a}{2}\right) + \frac{f''(\xi_2)}{2!}\left(\frac{b-a}{2}\right)^2.$$

Notice that $(\frac{a-b}{2}) + (\frac{b-a}{2}) = 0$, so if we add $f(a)$ and $f(b)$ we obtain

$$f(a) + f(b) = 2f\left(\frac{a+b}{2}\right) + \frac{f''(\xi_1)}{2!}\left(\frac{a-b}{2}\right)^2 + \frac{f''(\xi_2)}{2!}\left(\frac{b-a}{2}\right)^2.$$

If $f''(x) > 0$ for every value of x in I, then $f''(\xi_1)$ and $f''(\xi_2)$ are both greater than 0, so

$$f(a) + f(b) > 2f\left(\frac{a+b}{2}\right).$$

This means

$$f\left(\frac{a+b}{2}\right) < \frac{f(a) + f(b)}{2}.$$

Then, by the definition of concave up, we conclude that $f(x)$ is concave up on I.

If $f''(x) < 0$, similar arguments lead to

$$f\left(\frac{a+b}{2}\right) > \frac{f(a)+f(b)}{2}$$

for every point a and b in **I**, so $f(x)$ is concave down on **I**. This completes the proof. □

NOTE. If the function $f(x)$ is twice differentiable on **I** and concave up, then it can be shown that the graph of the function lies above the tangent line at each point $x = a$ in **I**, as shown in Figure 4.6.3 (a). In fact, by Taylor's theorem, we have

$$f(x) = f(a) + f'(a)(x-a) + \frac{f''(\xi)}{2!}(x-a)^2.$$

If $f''(x) > 0$ for every value of x in **I**, then $f''(\xi) > 0$, so $f(x) > f(a) + f'(a)(x-a)$, which means the graph of $f(x)$ is above the line $y = f(a) + f'(a)(x-a)$, which is exactly the line tangent to the graph of $f(x)$ at $x = a$. Similarly, if $f''(x) < 0$ for all x in **I**, then $f(x) < f'(a) + f'(a)(x-a)$. Therefore, the tangent line at $x = a$ is below the graph of $f(x)$.

NOTE. A more general definition of concavity is as follows. Given any points $x_1 < x_2 < \cdots < x_n$ in **I** and any positive numbers $0 < \lambda_1 < \lambda_2 < \cdots < \lambda_n < 1$ with $\sum_{i=1}^{n} \lambda_i = 1$, if $f(\sum_{i=1}^{n} \lambda_i x_i) < \sum_{i=1}^{n} \lambda_i f(x_i)$, then $f(x)$ is said to be concave up on **I**. If $f(\sum_{i=1}^{n} \lambda_i x_i) > \sum_{i=1}^{n} \lambda_i f(x_i)$, then $f(x)$ is said to be concave down on **I**.

Now, what happens if $f''(x)$ is 0 or $f''(x)$ does not exist at a point? The answer is that the function may change its concavity there. We give the following definition.

Definition 4.6.2. A point $(c, f(c))$ on the graph of f is an *inflection point* (or *point of inflection*) of f if f changes its concavity there.

If $(c, f(c))$ is an inflection point, then there is an interval $a < c < b$ such that either $f(x)$ is concave up on (a, c) and concave down on (c, b), or $f(x)$ is concave down on (a, c) and concave up on (c, b). If the function is twice differentiable, then there is also an interval $a < c < b$ such that $f''(x)$ exists and has a constant sign on (a, c) (either positive or negative) and $f''(x)$ exists and has the opposite constant sign on the interval (c, b). This implies that either $f''(c) = 0$ and that f is concave up on one of the intervals, (a, c) or (c, b), and concave down on the other interval. The second derivative does not have to exist at c in order for it to be an inflection point.

Example 4.6.1. Determine the intervals on which $f(x) = x^3$ is concave up or concave down and locate any inflection points.

Solution. In order to use the concavity test, we calculate

$$f'(x) = 3x^2 \quad \text{and} \quad f''(x) = 6x.$$

We see that, on the interval $(0, +\infty)$, $f''(x) > 0$, so f is concave up. On the interval $(-\infty, 0)$, $f''(x) < 0$, so f is concave down. It follows that, at the junction of these two intervals, $x = 0$, the point $(0, f(0)) = (0, 0)$ on the graph is an inflection point (note also that $f''(0) = 0$ at this point). A sign and behavior diagram is shown in Figure 4.6.4.

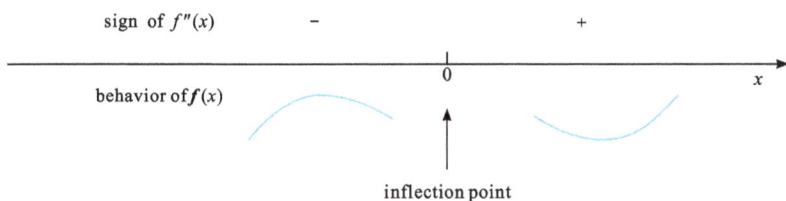

Figure 4.6.4: Sign of $f''(x)$ and behavior of $f(x)$.

Example 4.6.2. Determine the intervals on which

$$f(x) = 3x^4 - 4x^3 + 1$$

is concave up and concave down and locate all of its inflection points.

Solution. We calculate $f'(x) = 12x^3 - 12x^2$ and $f''(x) = 36x^2 - 24x = 36x(x - \frac{2}{3})$. Hence, $f''(x) > 0$ when $x < 0$ and when $x > \frac{2}{3}$. On these two intervals, $f(x)$ is concave up. Similarly, $f''(x) < 0$ when $0 < x < \frac{2}{3}$ and on this interval $f(x)$ is concave down. Since $f(x)$ changes its concavity at the point $x = 0$ (where $y = 1$) and at $x = \frac{2}{3}$ (where $y = \frac{11}{27}$), the inflection points are $(0, 1)$ and $(\frac{2}{3}, \frac{11}{27})$. For reference, we show the graph of $y = 3x^4 - 4x^3 + 1$ in Figure 4.6.5.

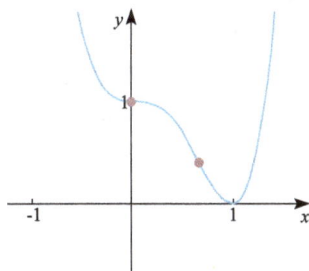

Figure 4.6.5: Graph of $f(x) = 3x^4 - 4x^3 + 1$.

NOTE. If $f(x)$ is twice differentiable and $(c, f(c))$ is an inflection point, then $f''(c) = 0$. However, the converse may not be true, as the next example shows.

Example 4.6.3. Find all inflection points on the graph of $y = x^4$.

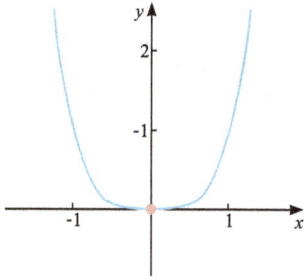

Figure 4.6.6: Graph of $y = x^4$.

Solution. $y' = 4x^3$ and $y'' = 12x^2$, so $y'' = 0$ when $x = 0$. However, we see that y'' is defined for all values of x and it does not change sign at $x = 0$ or any other value of x. The function $f(x) = x^4$ does not have any points of inflection, even though $f''(0) = 0$. In fact, the curve is concave up for all values of x, as shown in Figure 4.6.6.

NOTE. For some functions $f(x)$, an inflection point $(c, f(c))$ occurs at $x = c$ where $f''(c)$ does not exist, as in the next example.

Example 4.6.4. Locate any inflection points on the graph of the function

$$y = \sqrt[3]{x}.$$

Solution. The function is continuous on $(-\infty, +\infty)$. When $x \neq 0$, we have

$$y' = \frac{1}{3x^{\frac{2}{3}}} \quad \text{and} \quad y'' = -\frac{2}{9x^{\frac{5}{3}}}.$$

Observe that $y''(0)$ does not exist, but y'' changes from being positive when $x < 0$ to being negative when $x > 0$. Hence the graph changes from concave up to concave down at $x = 0$, so $(0, 0)$ is a point of inflection. The graph of $y = \sqrt[3]{x}$ is shown in Figure 4.6.7.

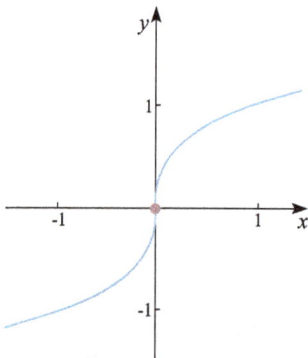

Figure 4.6.7: Graph of $y = \sqrt[3]{x}$.

4.6.2 The second derivative test

Now we show another way to determine the nature of a critical point. For a twice differentiable function $f(x)$, if $f'(a) = 0$, we want to determine the nature of the point $(a, f(a))$. If, in addition, we have $f''(a) > 0$, then we can imagine that there is a neighborhood of $x = a$ such that, on this interval, $f(x)$ is concave up, but $x = a$ is the turning point on that interval, so it would be a local minimum. If, on the other hand, $f''(a) < 0$, then $f(x)$ is concave down on a small interval containing a and we would believe that $f(x)$ reaches a local maximum at $x = a$.

In fact, we have the following theorem.

Theorem 4.6.2 (Second derivative test). *If $f(x)$ has a continuous second derivative and $f'(a) = 0$, then:*
1. *if $f''(a) > 0$, then $f(x)$ has a local minimum at $x = a$;*
2. *if $f''(a) < 0$, then $f(x)$ has a local maximum at $x = a$.*

Proof. First, we notice that, if $f''(x)$ is continuous and $f''(a) > 0$, there is a small interval I containing a, where $f''(x) > 0$ for each $x \in I$. On the other hand, the Taylor expansion at $x = a$ for $f(x)$ is for some ξ depending on x and a. We have

$$f(x) = f(a) + f'(a)(x - a) + \frac{f''(\xi)}{2!}(x - a)^2 \quad \text{for } x \in I.$$

Since $f'(a) = 0$, this becomes

$$f(x) = f(a) + \frac{f''(\xi)}{2!}(x - a)^2.$$

By assumption, $f''(x) > 0$ for each $x \in I$, so $f''(\xi) > 0$. Therefore,

$$f(x) > f(a) \quad \text{for each } x \in R,$$

so $f(a)$ is a local minimum.

A similar argument leads to the conclusion that, if $f''(a) < 0$, $f(x)$ has a local maximum at $x = a$. □

NOTE. In fact, $f''(x)$ does not need to be continuous. The second derivative test holds as long as $f''(a)$ exists. One can deduce this by investigating the change of sign of $f'(x)$ in the limit $\lim_{x \to a} \frac{f'(x) - f'(a)}{x - a} = f''(a)$.

Example 4.6.5. Locate and classify all local extrema of the function

$$f(x) = \frac{1}{3}x^3 - x^2 + \frac{1}{3}.$$

Solution. The function is continuous and differentiable on $(-\infty, \infty)$. Hence, we identify the local extrema using the first or second derivative tests as follows:

$$f'(x) = x^2 - 2x = x(x - 2).$$

The critical points are $x = 0$ and $x = 2$. Then

$$f''(x) = 2x - 2,$$

so

$$f''(0) = -2 < 0, \quad f''(2) = 2 > 0.$$

Therefore, it follows from the second derivative test that f has a local maximum at $x = 0$ and a local minimum at $x = 2$, as seen in Figure 4.6.8.

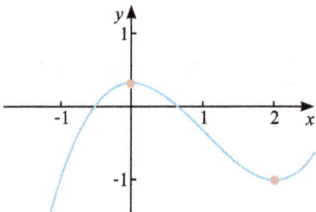

Figure 4.6.8: Graph of $y = \frac{1}{3}x^3 - x^2 + \frac{1}{3}$.

Example 4.6.6. Find all local extreme values of the function

$$f(x) = x - \sqrt{x}, \quad \text{for } x > 0.$$

Solution. The function is continuous and differentiable for $x > 0$, so we have

$$f'(x) = 1 - \frac{1}{2x^{\frac{1}{2}}}.$$

Solving $f'(x) = 0$ gives $x = \frac{1}{4}$. This is the only candidate to be a local extremum of f. We now find the second derivative of f, which is

$$f''(x) = \left(1 - \frac{1}{2x^{\frac{1}{2}}}\right)' = 0 - \frac{1}{2}\left(-\frac{1}{2}\right)x^{-\frac{1}{2}-1} = \frac{1}{4}x^{-\frac{3}{2}},$$

$$f''\left(\frac{1}{4}\right) = \frac{1}{4}\left(\frac{1}{4}\right)^{-\frac{3}{2}} > 0,$$

so we conclude that f has only one local minimum, which occurs at $x = \frac{1}{4}$. This local minimum value is $f(\frac{1}{4}) = \frac{1}{4} - \frac{1}{2} = -\frac{1}{4}$. The graph of f is shown in Figure 4.6.9. Observe that f has no local maximum.

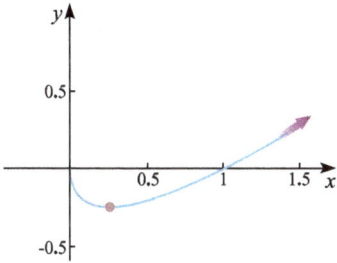

Figure 4.6.9: Graph of $f(x) = x - \sqrt{x}$.

Unfortunately, the second derivative test does not tells us whether $f(a)$ is a local extremum or not when $f''(a) = 0$. Actually, if $f''(a) = 0$, $f(a)$ could be a local maximum, a local minimum or neither. We take, for example, $f(x) = x^3$, $f(x) = x^4$, and $f(x) = -x^4$. For all three of these, the second derivative at $x = 0$ is 0, but $f(x) = x^3$ has no extremum at $x = 0$, $f(x) = x^4$ has a local minimum at $x = 0$, and $f(x) = -x^4$ has a local maximum at $x = 0$. A further result for determining the nature of a stationary point using higher order derivatives is given by the following theorem.

Theorem 4.6.3. *If $f'(a) = 0$, $f''(a) = 0, \ldots, f^{(n-1)}(a) = 0$, and $f^{(n)}(a) \neq 0$, then:*
1. *if n is odd, $f(a)$ is not a local extreme value;*
2. *if n is even and $f^{(n)}(a) > 0$, then $f(x)$ has a local minimum at $x = a$; if $f^{(n)}(a) < 0$, then $f(x)$ has a local maximum at $x = a$.*

Example 4.6.7. Determine the nature of the stationary point of $f(x) = x^3$ and $g(x) = x^4$.

Solution. Since

$$f'(x) = 3x^2 \quad \text{and} \quad g'(x) = 4x^3,$$

both f and g have only one stationary point, $x = 0$. Because

$$f''(x) = 6x, \quad f''(0) = 0 \quad \text{(inconclusive by the second derivative theorem)}.$$

We differentiate f'' again to get the third derivative, $f'''(x) = 6 > 0$. Since 3 is odd, we conclude by Theorem 4.6.3 that $f(x)$ has no local extreme value at $x = 0$. Similarly, since $g''(0) = g'''(0) = 0$ and $g^{(4)}(0) = 24 > 0$, we conclude that $g(x)$ has a local minimum at $x = 0$.

4.7 Extreme values of functions revisited

From Section 4.1 of this chapter, we know how to find the candidates for extreme values of a function and we also know how to find global extreme values for a continuous function defined on a closed interval. From Section 4.3, we know how to investigate whether or not a candidate is a local extreme value. If a function is not continuous or not defined on a closed interval, then it may or may not have local or global extreme values. This will depend on individual cases.

Example 4.7.1. If $f(x) = x^2$, then $f(x) \geqslant f(0)$ because $x^2 \geqslant 0$ for all x. Therefore, $f(0) = 0$ is the absolute (and local) minimum value of f. This corresponds to the fact that the origin is the lowest point on the parabola $y = x^2$. However, there is no highest point on the parabola, so this function has no maximum value.

Example 4.7.2. For the following equation, find all candidates for extrema of f and determine the global extrema:

$$f(x) = 3x^4 - 16x^3 + 18x^2, \quad -1 \leqslant x < 2.$$

Solution. Differentiating f, we obtain

$$f'(x) = 12x^3 - 48x^2 + 36x = 12x(x-1)(x-3).$$

Hence, $f'(x) = 0$ when $x = 0$, $x = 1$, and $x = 3$. The candidates for extrema in this interval $[-1, 2)$ are therefore $x = -1$, $x = 0$, and $x = 1$. On $[-1, 0) \cup (1, 2)$, $f'(x) < 0$ and $f'(x) > 0$ on $(0, 1)$, so we construct a rough diagram in Figure 4.7.1.

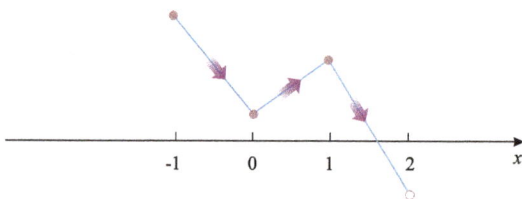

Figure 4.7.1: Rough graph for $f(x) = 3x^4 - 16x^3 + 18x^2$ for $-1 \leqslant x < 2$.

The function f starts at the point $(-1, f(-1))$ and then decreases until $x = 0$. Then it starts to rise until $x = 1$ and it decreases again, so f has a local minimum at $x = 0$ and a local maximum at $x = 1$. Observe that

$$f(-1) = 37, \quad f(0) = 0, \quad \text{and} \quad f(1) = 5.$$

On the interval $[1, 2)$, the value of $f(2)$ is not defined, but we find

$$\lim_{x \to 2} f(x) = \lim_{x \to 2} (3x^4 - 16x^3 + 18x^2)$$
$$= 3(2^4) - 16(2^3) + 18(2^2)$$
$$= -8.$$

This means that, after passing through the point $(1,5)$, the graph of $f(x)$ falls and gets closer and closer to -8 as $x \to 2$, but it never reaches -8 for x in $[1,2)$. Therefore, $f(x)$ has no global minimum, but it has a global maximum at $x = -1$, a local minimum at $x = 0$, and a local maximum at $x = 1$. The graph of $f(x)$ is shown in Figure 4.7.2.

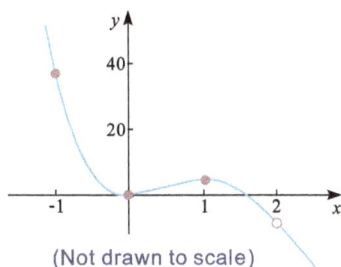

(Not drawn to scale)

Figure 4.7.2: Graph of $f(x) = 3x^4 - 16x^3 + 18x^2$ for $-1 \leqslant x < 2$.

Example 4.7.3. Find the extreme values of

$$f(x) = \begin{cases} 3 - 2x^2, & x \leqslant 1 \\ x + 1, & x > 1. \end{cases}$$

Solution. The function is not continuous at $x = 1$, since

$$\lim_{x \to 1^-} f(x) = \lim_{x \to 1^-} (3 - 2x^2) = 1 \quad \text{but}$$
$$\lim_{x \to 1^+} f(x) = \lim_{x \to 1^-} (x + 1) = 2.$$

However, it is continuous and differentiable anywhere else. The derivative of f is given by

$$f'(x) = \begin{cases} -4x, & x < 1 \\ 1, & x > 1. \end{cases}$$

We find that $f'(x) = 0$ when $x = 0$ and $f'(1)$ does not exist, so the two critical numbers are $x = 0$ and $x = 1$. The domain is unbounded above and unbounded below, so we have to investigate the behavior of f when $x \to \pm\infty$. We have

$$\lim_{x \to \infty} f(x) = \lim_{x \to \infty} (x + 1) = \infty,$$
$$\lim_{x \to -\infty} f(x) = \lim_{x \to -\infty} (3 - 2x^2) = -\infty.$$

Therefore, $f(x)$ is unbounded above and unbounded below, so it has neither a global maximum nor a global minimum. However, at $x = 0$, $f''(0) = -4 < 0$, so $f(x)$ has a local maximum at $x = 0$.

We see that $f'(x)$ changes from negative to positive at $x = 1$, but $f(x) \to 1$ as $x \to 1^-$ and $f(x) \to 2$ as $x \to 1^+$, so f has a jump discontinuity at $x = 1$ (it jumps from $f(1) = 1$ to 2 at $x = 1$), so $f(x)$ has a local minimum value at $x = 1$. The local maximum value is $f(0) = 3 - 2(0^2) = 3$. The local minimum value is $f(1) = 1$. The graph of $f(x)$ is shown in Figure 4.7.3.

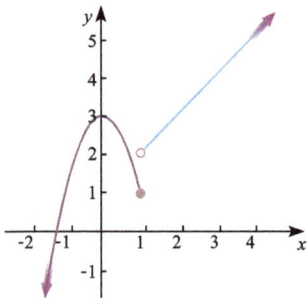

Figure 4.7.3: Graph of $f(x)$ in Example 4.7.3.

Example 4.7.4. The graph of the derivative of a function f defined on a closed interval $[-2, 5]$ is shown in Figure 4.7.4. It consists of a semi-circle and two line segments. Identify any extreme values of f.

Solution. The derivative $f'(x)$ is positive for $-2 < x < 2$ and $2 < x < 4$ and is negative for $4 < x < 5$. There are two stationary points $x = 2$ and $x = 4$, since $f'(2) = f'(4) = 0$, so $f(x)$ increases on the interval $(-2, 2)$, then arrives at a stationary point $(2, f(2))$, and then increases again on the interval $(2, 3)$. Therefore, f does not have a local extreme value

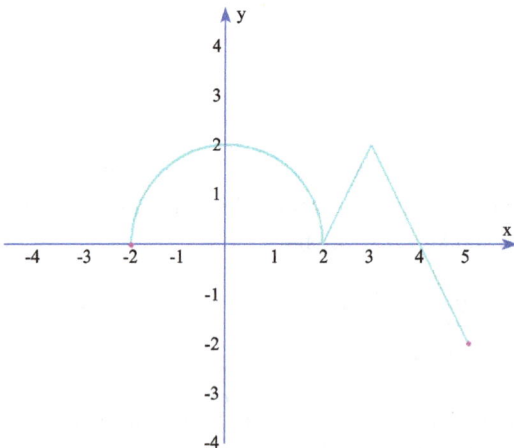

Figure 4.7.4: Graph of $f'(x)$ in Example 4.7.4.

at $x = 2$. However, f increases on the interval $(2, 4)$ and decreases on $(4, 5)$. Therefore, f does have a local extreme value at $x = 4$. Furthermore, this local maximum is also the global maximum of f, since f rises to $(4, f(4))$, then falls, and does not rise again. We cannot make any conclusion at this moment about the global minimum of f. In Chapter 5, we shall be able to determine that f takes its global minimum at one of the endpoints $x = -2$ and $x = 5$.

4.8 Curve sketching

The graph of a function $y = f(x)$ can usually be sketched showing all of its most important features by using the following algorithm.

1. *Domain* It is often useful to start by determining the domain D of f (the set of values of x for which $f(x)$ is defined). Note that, at single points where f is undefined, concavity may change and the function may change from increasing to decreasing or vice versa.

2. *Intercepts* The y-intercept is the point $(0, f(0))$ and this tells us where the curve intersects the y-axis. The x-intercepts are the points, $(c, 0)$, where the graph of f intersects the x-axis. A function may not have any x-intercepts, or it may be impossible to find the x-intercepts if the equation is difficult to solve.

3. *Symmetry* Determine whether the function is even $(f(-x) = f(x)$ for all $x)$, odd $(f(-x) = -f(x)$ for all $x)$, periodic $(f(x + p) = f(x)$ for all x and a fixed $p)$, or none of these. If it is an even function, then we need only to determine the graph for $x \geqslant 0$. We can create the graph for $x < 0$ by reflecting this across the y-axis. If it is an odd function, then we need only determine the graph for $x \geqslant 0$ and then reflect through the point $(0, 0)$, the origin. Finding the reflection of a graph in the origin is equivalent to finding its reflection in the y-axis and then finding the reflection of this in the x-axis. If the function is periodic with period p, then we need only to find the graph for $0 \leqslant x \leqslant p$ or some other interval of length p and we can create the rest of the graph of f by copying this part for other x-values to the left and right.

4. *Asymptotes* Find all horizontal, vertical, and slant asymptotes in order to determine the behavior of the graph for large x- and y-values (see Section 2.2.5).

5. *Intervals of increase or decrease* Use the increasing/decreasing test. That is, compute $f'(x)$ and find the intervals on which $f'(x)$ is positive (f is increasing) and the intervals on which $f'(x)$ is negative (f is decreasing).

6. *Local maximum and minimum values* Find the critical points of f ($f'(x) = 0$ or is undefined) and then use the first derivative test. That is, if $f'(x)$ changes from positive to negative at a critical number c, then $f(c)$ is a local maximum and if it changes from negative to positive, then $f(c)$ is a local minimum. We can also use the second derivative test to identify the local maxima and minima.

7. *Concavity and points of inflection* Compute f'' and use the concavity test. That is, the curve is concave up where $f''(x) > 0$ and concave down where $f''(x) < 0$. Inflection points occur where the concavity changes from up to down or from down to up.

8. *Sketch the curve* Use the information obtained to draw the graph. Sketch the asymptotes as dashed lines. Plot the intercepts, maximum and minimum points, and inflection points. Then make the curve pass through these points, rising and falling on the intervals found in Step 5, with concavity found in Step 7, and approaching the asymptotes for large y- and x-values. In order to position the graph reasonably and accurately you may need to compute some points on the graph for a few additional well-chosen values of x.

The same methods can, to some extent, be used to sketch the graph of an implicitly defined function $F(x, y) = 0$. However, except for symmetry, it is generally much harder to determine the same information, such as domain, intercepts, asymptotes, intervals of increase/decrease, maxima and minima, and concavity.

Example 4.8.1. Sketch the graph of the function $f(x) = x^4 - 2x^3$.

Solution. The domain of the function is $(-\infty, +\infty)$. The y-intercept is $(0, 0)$. This is also an x-intercept; the other x-intercept is $(2, 0)$. There is no symmetry and there are no asymptotes.

The derivative is $f'(x) = 4x^3 - 6x^2 = 2x^2(2x - 3)$. Solving $f'(x) = 0$, we obtain the critical points $x = 0$ and $x = \frac{3}{2}$.

The second derivative is $f''(x) = 12x^2 - 12x = 12x(x - 1)$. Solving $f''(x) = 0$ for x, we obtain $x = 0$ and $x = 1$. These points divide the domain into four intervals. We compute the sign table below.

x	$(-\infty, 0)$	0	$(0, 1)$	1	$(1, \frac{3}{2})$	$\frac{3}{2}$	$(\frac{3}{2}, +\infty)$
$f'(x)$	$-$	0	$-$	-2	$-$	0	$+$
$f''(x)$	$+$	0	$-$	0	$+$	$+$	$+$
$f(x)$	↘ decreasing	0	↘ decreasing	-1	↘ decreasing	$-\frac{27}{16}$	↗ increasing
Concave	Up		Down		Up		Up

This shows a local (and absolute) minimum at $x = \frac{3}{2}$, $y = -\frac{27}{16}$ and the points of inflection $(1, -1)$ and $(0, 0)$. The graph is shown in Figure 4.8.1. Observe that we have calculated the point $(-1, 3)$ for additional accuracy.

Example 4.8.2. Find the asymptotes of the graph of the function

$$f(x) = \frac{(x - 3)^2}{4(x - 1)}.$$

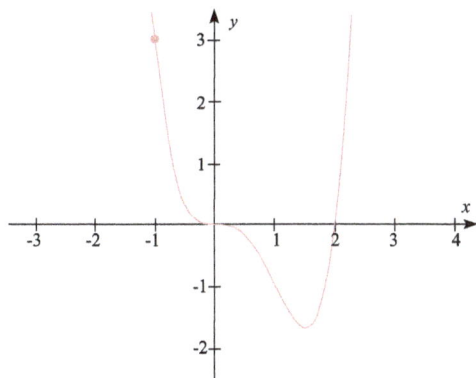

Figure 4.8.1: Graph of $f(x) = x^4 - 2x^3$.

Solution. Since $\lim_{x \to 1^-} f(x) = -\infty$ and $\lim_{x \to 1^+} f(x) = \infty$, we know that $x = 1$ is a vertical asymptote (approached on the left through negative y-values and on the right through positive y-values). We have

$$\lim_{x \to \pm\infty} \frac{f(x)}{x} = \lim_{x \to \pm\infty} \frac{(x-3)^2}{4(x-1)x} = \frac{1}{4},$$

$$\lim_{x \to \pm\infty} \left[f(x) - \frac{x}{4} \right] = \lim_{x \to \pm\infty} \left[\frac{(x-3)^2}{4(x-1)} - \frac{x}{4} \right] = -\frac{5}{4}.$$

This gives the slant asymptote $y = \frac{1}{4}x - \frac{5}{4}$ as $x \to \infty$ and as $x \to -\infty$. There are no horizontal asymptotes. The graph is shown later as part of Example 4.8.3 in Figure 4.8.2.

Example 4.8.3. Sketch the graph of $f(x) = \frac{(x-3)^2}{4(x-1)}$.

Solution. The domain is $(-\infty, 1) \cup (1, \infty)$ and the intercepts are the points $(3, 0)$ and $(0, -\frac{9}{4})$. There is no symmetry and, as previously discussed in Example 4.8.2, the vertical asymptote is $x = 1$, the slant asymptote is $y = \frac{1}{4}x - \frac{5}{4}$, and there is no horizontal asymptote. We compute the derivatives

$$f'(x) = \frac{(x-3)(x+1)}{4(x-1)^2} \quad \text{and} \quad f''(x) = \frac{2}{(x-1)^3}.$$

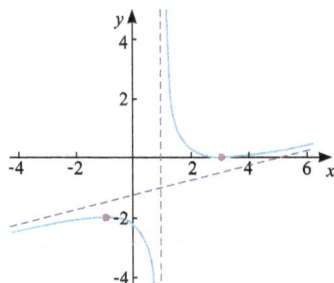

Figure 4.8.2: Graph of $f(x) = \frac{(x-3)^2}{4(x-1)}$ and its asymptotes.

Then $f'(x) = 0$ when $x = -1$ and when $x = 3$, but $f'(x)$ and $f''(x)$ are undefined at $x = 1$ (where there is a change of concavity). We set up the sign table.

x	$(-\infty, -1)$	-1	$(-1, 1)$	1	$(1, 3)$	3	$(3, +\infty)$
$f'(x)$	+	0	–	dne	–	0	+
$f''(x)$	–	–	–	dne	+	+	+
$f(x)$	↗ increasing	-2	↘ decreasing	dne	↘ decreasing	0	↗ increasing
Concave	Down		Down		Up		Up

This shows a local minimum at $(3, 0)$, a local maximum at $(-1, -2)$, and a change of concavity at $x = 1$ (however, it is not a point of inflection because $f(1)$ is not defined). The graph is shown in Figure 4.8.2.

4.9 Solving equations numerically

Any equation of the single variable x can be rearranged so that it takes the form $f(x) = 0$. It is often important to find specific values of x such that $f(x) = 0$. We give a specific name to these values.

Definition 4.9.1. A value of x for which $f(x)$ evaluates to 0 is called a *root* of the equation $f(x) = 0$ (or a *zero* of the function f). The *solution* of the equation $f(x) = 0$ is the set of all of its roots.

For any particular function f, a root $x = c$ satisfying $f(c) = 0$ may exist, but we may be unable to find its exact value. In this case we often try to find another specific x-value, $x = d$, that approximates the root: $f(d) \approx 0$, while $d \approx c$. In this section we give a brief introduction to methods for finding such approximate roots.

The *sign-change rule*: recall the intermediate value theorem from Section 2.6.3 in Chapter 2: if a function f is continuous on an interval $a \leqslant x \leqslant b$ of its domain and $f(a) \times f(b) < 0$, then $f(x) = 0$ has at least one root between a and b.

4.9.1 Decimal search

By this method, we find an interval containing one root of $f(x) = 0$, perhaps using graphical means, and then subdivide the interval into two parts, locating the root in one of these subintervals using the sign change rule described above. This method is instructive but very inefficient, so we only show it in the form of an example.

Example 4.9.1. Show that the equation $xe^x = 1$ has one root and find an approximation to this root correct to two decimal places (that is, if we round the approximate root to two decimal places, then it will be the same as the actual root rounded to two decimal places).

Solution. We first investigate the equation graphically, rewriting the equation as $e^x = \frac{1}{x}$ and plotting on the same graph both $y = e^x$ and $y = \frac{1}{x}$, as shown in Figure 4.9.1.

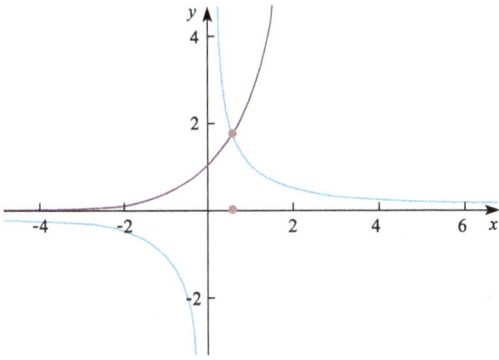

Figure 4.9.1: Graphs of $y = e^x$ and $y = \frac{1}{x}$.

The roots of $xe^x = 1$ are exactly the x-values of the points of intersection of the two curves $y = \frac{1}{x}$ and $y = e^x$. The graph in Figure 4.9.1 indicates that there is only one root of the equation $e^x = \frac{1}{x}$, which lies between 0 and 1. Note that there cannot be another root to the left of the part of the graph shown, because, as $x \to -\infty$, $\frac{1}{x}$ is negative and e^x is positive ($y = 0$ is a horizontal asymptote for both). Similarly, there is no root to the right of the part shown; as $x \to \infty$, $\frac{1}{x}$ is decreasing and e^x is increasing. Let $f(x) = e^x - \frac{1}{x}$, so that we need to find an approximation to the root of $f(x) = 0$ within two decimal places of the actual root.

1. We arbitrarily choose $x = 0.3$ inside the original interval $(0,1)$. Since $f(0.3) \approx -1.98335$ and $f(1) \approx 1.7183 > 0$, by the sign-change rule, the root lies in the interval $(0.3,1)$. Notice that there is no sign change for the other subinterval $(0,0.3)$ since $f(0) = -1$ and $f(0.3) \approx -1.983$. This will always be the case – there can only be a sign change for one of the two subintervals.

2. We arbitrarily choose $x = 0.6$ inside the previous interval $(0.3,1)$. Since $f(0.3) \approx -1.9835 < 0$ and $f(0.6) \approx 0.15545 > 0$, by the sign-change rule, the root lies in the interval $(0.3,0.6)$.

3. We arbitrarily choose $x = 0.5$ inside the previous interval $(0.3,0.6)$. Since $f(0.5) \approx -0.35313 < 0$ and $f(0.6) \approx 0.15545 > 0$, by the sign-change rule, the root lies in the interval $(0.5,0.6)$.

4. We arbitrarily choose $x = 0.55$ inside the previous interval $(0.5,0.6)$. Since $f(0.55) \approx -0.08449 < 0$ and $f(0.6) \approx 0.15545 > 0$, by the sign-change rule, the root lies in the interval $(0.55,0.6)$.

5. We arbitrarily choose $x = 0.57$ inside the previous interval $(0.55, 0.6)$. Since $f(0.55) \approx -0.0849 < 0$ and $f(0.57) \approx 0.014 > 0$, by the sign-change rule, the root lies in the interval $(0.55, 0.57)$.
6. We arbitrarily choose $x = 0.56$ inside the previous interval $(0.55, 0.57)$. Since $f(0.56) \approx -0.035 < 0$ and $f(0.57) \approx 0.014 > 0$, by the sign-change rule, the root lies in the interval $(0.56, 0.57)$.
7. We arbitrarily choose $x = 0.565$ inside the previous interval $(0.56, 0.57)$. Since $f(0.565) \approx -0.01 < 0$ and $f(0.57) \approx 0.014 > 0$, by the sign-change rule, the root lies in the interval $(0.565, 0.57)$.
8. Both endpoints of the previous interval, rounded to two decimal places, have the same value, 0.57. Hence, the actual root is $x = 0.57$, rounded to two decimal places.

4.9.2 Newton's method

Newton's method is a procedure or algorithm, superior to the decimal search method, for approximating the roots of an equation $f(x) = 0$ (or equivalently, the zeros of the function f). As with the decimal search method, we start with an approximate value $x = a$ for the root, perhaps found by graphical methods. Then we find a better approximation by drawing the tangent to $y = f(x)$ at $x = a$ and taking the next approximation as the point where this tangent crosses the x-axis. The equation of this tangent line is

$$y - f(a) = f'(a)(x - a)$$

and this line meets the x-axis at point $x = a^*$ satisfying

$$0 - f(a) = f'(a)(a^* - a),$$
$$a^* = a - \frac{f(a)}{f'(a)}.$$

If the first approximation $x = a$ is reasonably good, then the value of a^* will be much closer to the root than a. We keep applying this formula, each time replacing a by a^*, until an a^* is found that is sufficiently close to the required zero of f. Figure 4.9.2 illustrates this idea. We write this algorithm more precisely in the following way.

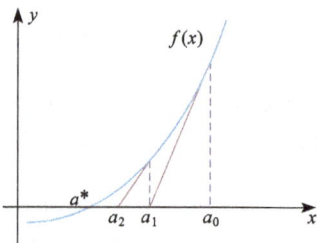

Figure 4.9.2: Illustration for Newton's method.

Newton's algorithm for finding approximate roots

Step 1. Determine an approximate value a_0 of a zero of f. Let $n = 0$.

   Step 2. Calculate

$$a_{n+1} = a_n - \frac{f(a_n)}{f'(a_n)}. \tag{4.18}$$

   Step 3. If $f(a_n)$ is sufficiently close to 0 (use the sign-change rule to check the accuracy), then stop – the required approximation is a_n; otherwise, increase n by 1 and repeat Step 2.

Example 4.9.2. Determine an approximate value for $\sqrt{2}$, accurate to 10^{-10}, by applying Newton's Method.

Solution. The required value $\sqrt{2}$ is the positive zero of the function $f(x) = x^2 - 2$. Take $a_0 = 1.5$ as the first guess, basing this on the numerical evidence that $a = 1$ is too small ($1^2 < 2$) and $a = 2$ is too large ($2^2 > 2$). Since $f'(x) = 2x$, the iterative formula is

$$a_{n+1} = a_n - \frac{f(a_n)}{f'(a_n)} = a_n - \frac{a_n^2 - 2}{2a_n},$$

$$a_{n+1} = \frac{a_n^2 + 2}{2a_n}.$$

Hence,

$$a_1 = \frac{a_0^2 + 2}{2a_0} \approx 1.416\,666\,666\,67.$$

We test whether or not a_1 is within 10^{-10} of the actual root, using the sign-change rule.

$$f(1.416\,666\,666\,67 + 10^{-10}) \approx 6.944\,4 \times 10^{-3} > 0 \quad \text{and}$$
$$f(1.416\,666\,666\,67 - 10^{-10}) \approx 6.944\,4 \times 10^{-3} > 0.$$

There is no sign change so the root is not within 10^{-10} of a_1. We repeat the process and obtain

$$a_2 = \frac{a_1^2 + 2}{2a_1} \approx 1.414\,215\,686\,28.$$

Now the test gives

$$f(1.414\,215\,686\,28 + 10^{-10}) \approx 6.007\,6 \times 10^{-6} > 0 \quad \text{and}$$
$$f(1.414\,215\,686\,28 - 10^{-10}) \approx 6.007 \times 10^{-6} > 0.$$

There is still no sign change so the root is not within 10^{-10} of a_2. We repeat the process and obtain

$$a_3 = \frac{a_2^2 + 2}{2a_2} \approx 1.414\,213\,562\,37.$$

Now the test gives

$$f(1.414\,213\,562\,37 + 10^{-10}) \approx 2.740\,9 \times 10^{-10} > 0 \quad \text{and}$$
$$f(1.414\,213\,562\,37 - 10^{-10}) \approx -2.916\,0 \times 10^{-10} < 0.$$

There is now a sign change so the actual root is between $1.414\,213\,562\,37 - 10^{-10}$ and $1.414\,213\,562\,37 + 10^{-10}$, so the result is within 10^{-10} of the last approximation $a_3 = 1.414\,213\,562\,37$, as required.

NOTES. 1.   The number of correct digits in this calculation approximately doubles with each iteration. This is typical of Newton's method. An accurate approximation of $\sqrt{2}$ to 15 places $\sqrt{2} = 1.414\,213\,562\,373\,10$ and this coincides exactly with our approximation in the first 11 decimal places. Hence, our approximation is actually more accurate than 10^{-10} and the process only required three iterations!

2.   There are some cases where Newton's method fails, because either the iterative sequence does not converge or it converges to another root which we do not want. Some examples are shown in Figure 4.9.3 and Figure 4.9.4.

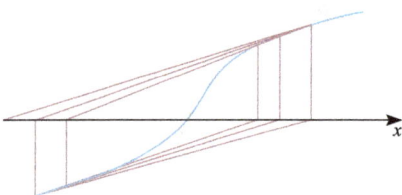

Figure 4.9.3: Newton's method fails: parallel tangents.

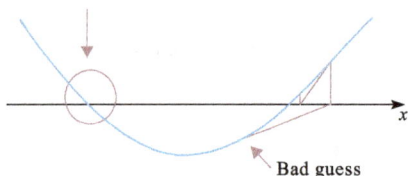

Figure 4.9.4: Newton's method fails: find a wrong root.

4.10 Curvatures and the differential of the arc length

Intuitively, *curvature* is the amount by which a geometric object deviates from being flat, or straight in the case of a line. It is the measure of how sharply the curve bends. For example, a small circle bends more sharply than a bigger circle. All the points on the same circle should have the same curvature. The curvature at every point on a line is zero. We denote the curvature by the letter K. Before we can define curvature, we must introduce the concept of arc length.

Definition 4.10.1. The arc length function $s = s(x)$ associated with the graph of a continuous function $f(x)$ on $[a, b]$ is defined as an increasing, nonnegative function that is the length of the segment from initial point $(a, f(a))$ to the terminal point $(x, f(x))$ for any x in $[a, b]$. The length $s(a)$ is defined to be zero.

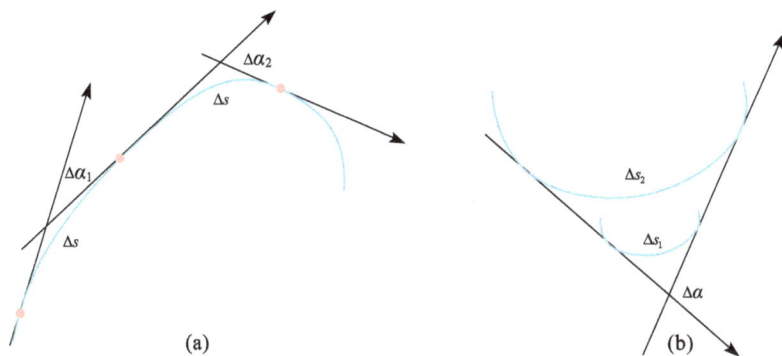

Figure 4.10.1: Curvature and arc length.

As seen in Figure 4.10.1 (a) and Figure 4.10.1 (b), for the same change in arc length Δs, the more sharply the curve bends, the bigger the change in the angle α, which is the angle between the tangent and the positive x-axis measured in counterclockwise direction. Or, if Δs is constant, the more sharply the curve bends, the larger the change in the angle $\Delta \alpha$. On the other hand, if $\Delta \alpha$ is fixed, the smaller Δs is, the more sharply the curve bends, so, as the definition of the average rate of change, the "average curvature" over an interval could be defined as $\frac{\Delta \alpha}{\Delta s}$. The curvature at a specific point can be defined as

$$\lim_{\Delta s \to 0} \left| \frac{\Delta \alpha}{\Delta s} \right|,$$

provided that the limit exists. *The curvature K at a point is the absolute value of the derivative of α with respect to s, which is $\frac{d\alpha}{ds}$.*

To find $\frac{d\alpha}{ds}$, we consider two points, P and Q, on a curve C. Let P have coordinates (x, y) and let Q be another point nearby with coordinates $(x + \Delta x, y + \Delta y)$. Let s be the arc length function for this curve, so that Δs is the length of the arc PQ. Because Δs is very small, it is reasonable to approximate the arc by the hypotenuse of the right angle PQM, as shown in Figure 4.10.2 (it is indeed true that $\lim_{\Delta x \to 0} \frac{\Delta s}{|PQ|} = 1$). Hence, using the Pythagorean theorem, we have

$$(\Delta s)^2 \approx |PQ|^2 = (\Delta x)^2 + (\Delta y)^2.$$

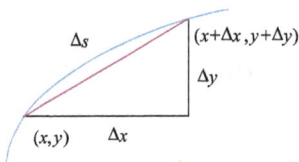

Figure 4.10.2: Arc length element.

Dividing each term by $(\Delta x)^2$, we obtain

$$\frac{(\Delta s)^2}{(\Delta x)^2} \approx 1 + \left(\frac{\Delta y}{\Delta x}\right)^2.$$

When we take the limit as $\Delta x \to 0$, we have

$$\left(\frac{ds}{dx}\right)^2 = 1 + \left(\frac{dy}{dx}\right)^2.$$

Therefore,

$$\frac{ds}{dx} = \sqrt{1 + \left(\frac{dy}{dx}\right)^2} \tag{4.19}$$

or

$$ds = \sqrt{1 + y'^2}\, dx. \tag{4.20}$$

On the other hand, $\frac{dy}{dx} = \tan \alpha$. Using implicit differentiation, we have $\sec^2 \alpha \frac{d\alpha}{dx} = y''$, so $\frac{d\alpha}{dx} = y'' \cos^2 \alpha$. Notice that $\tan \alpha = y'$, so

$$\cos \alpha = \pm \frac{1}{\sqrt{1 + (\frac{dy}{dx})^2}}.$$

Therefore,

$$\frac{d\alpha}{ds} = \frac{\frac{d\alpha}{dx}}{\frac{ds}{dx}} = \frac{y'' \cos^2 \alpha}{\sqrt{1 + y'^2}} = \frac{y''}{\sqrt{1 + y'^2}}\left(\frac{\pm 1}{\sqrt{1 + y'^2}}\right)^2 = \frac{y''}{(1 + y'^2)^{\frac{3}{2}}}.$$

Hence, the curvature K is

$$K = \left|\frac{d\alpha}{ds}\right| = \frac{|y''|}{(1 + y'^2)^{\frac{3}{2}}}. \tag{4.21}$$

We now check whether or not this definition of curvature agrees with our initial thoughts of curvature in a few examples.

Example 4.10.1. Find the curvature for each point on the straight line $y = mx + b$, where m, b are constants.

Solution. We find the first and second derivatives first:

$$y' = m \quad \text{and} \quad y'' = 0.$$

Since $y'' = 0$ at each point on the line, by the above definition, we conclude that the curvature K at each point (x, y) on the line $y = mx + b$ is 0.

If the curve is given by parametric equations

$$\begin{cases} x = \phi(t) \\ y = \psi(t), \end{cases}$$

then

$$\frac{dy}{dx} = \frac{dy/dt}{dx/dt} = \frac{\psi'(t)}{\phi'(t)} \quad \text{and} \quad \frac{d^2y}{dx^2} = \frac{\psi''(t)\phi'(t) - \psi'(t)\phi''(t)}{[\phi'(t)]^3},$$

so

$$K = \left| \frac{y''}{(1 + y'^2)^{3/2}} \right| = \frac{\left| \frac{\psi''(t)\phi'(t) - \psi'(t)\phi''(t)}{[\phi'(t)]^3} \right|}{(1 + (\frac{\psi'(t)}{\phi'(t)})^2)^{\frac{3}{2}}} = \frac{\left| \frac{\psi''(t)\phi'(t) - \psi'(t)\phi''(t)}{[\phi'(t)]^3} \right|}{(\frac{[\phi'(t)]^2 + [\psi'(t)]^2}{(\phi'(t))^2})^{\frac{3}{2}}}$$

$$= \left| \frac{\psi''(t)\phi'(t) - \psi'(t)\phi''(t)}{([\phi'(t)]^2 + [\psi'(t)]^2)^{3/2}} \right|, \tag{4.22}$$

or simply

$$K = \left| \frac{y''x' - y'x''}{((x')^2 + (y')^2)^{3/2}} \right|, \quad \text{where the prime is with respect to the parameter } t.$$

Example 4.10.2. Find the curvature for each point on the circle $x^2 + y^2 = R^2$, where R is constant.

Solution. The parametric equations for this circle are

$$\begin{cases} x = R\cos t \\ y = R\sin t, \end{cases}$$

so

$$x' = -R\sin t, \quad x'' = -R\cos t, \quad y' = R\cos t, \quad \text{and} \quad y'' = -R\sin t.$$

The curvature K at each point is given by

$$K = \left| \frac{y'' x' - y' x''}{((x')^2 + (y')^2)^{3/2}} \right| = \frac{|(-R\sin t)(-R\sin t) - (R\cos t)(-R\cos t)|}{((-R\sin t)^2 + (R\cos t)^2)^{3/2}}$$

$$= \frac{R^2 \sin^2 t + R^2 \cos^2 t}{(R^2 \sin^2 t + R^2 \cos^2 t)^{3/2}}$$

$$= \frac{R^2}{(R^2)^{3/2}} = \frac{1}{R}.$$

Again, this agrees with our observation that the smaller the radius R of a circle, the more sharply the curve bends, so the larger the curvature. In addition, it is clear that every point on a circle has the same curvature.

Osculating circle and radius of curvature

Among all circles that are tangent to a given curve at a specific point, the *osculating circle* is the tangent circle that most closely fits the curve, as shown in Figure 4.10.3. This means the osculating circle and the curve meet at the point, have the same first and second derivatives at the point, and, of course, have the same curvature there. The center and radius of the osculating circle at a given point are called the *center of curvature* and the *radius of curvature* of the curve at that point. The coordinates of the center (ξ, η) are given by

$$\begin{cases} \xi = x - \frac{y'(1+y'^2)}{y''} \\ \eta = y + \frac{1+y'^2}{y''} \end{cases}$$

and the radius of the curvature is given by $R = \frac{1}{K}$.

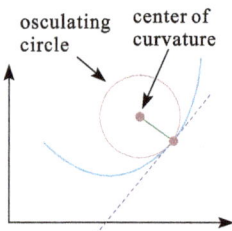

osculating circle   center of curvature

Figure 4.10.3: Osculating circles.

Example 4.10.3. Find the curvature, center of curvature, and radius of curvature at the point $(-1, -2)$ on the parabola $y = x^2 + 2x - 1$.

Solution. We find the first and second derivatives for the function first:

$$y' = 2x + 2 \quad \text{and} \quad y'' = 2,$$

so $y'(-1) = 0$ and $y''(-1) = 2$. Therefore, the curvature of the curve at $(-1-2)$ is

$$K = \frac{2}{(1+0^2)^{3/2}} = 2$$

and the radius of curvature R at this point is $R = \frac{1}{K} = \frac{1}{2}$.
The coordinates of the center (ξ, η) are given by

$$\xi = x - \frac{y'(1+y'^2)}{y''} = -1 - 0 \times \frac{1+0^2}{2} = -1$$

and

$$\eta = y + \frac{1+y'^2}{y''} = -2 + \frac{1+0^2}{2} = -\frac{3}{2}.$$

Thus, the center of curvature is $(-1, -\frac{3}{2})$. Figure 4.10.4 shows the graph.

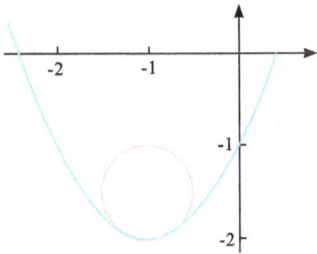

Figure 4.10.4: Graph of $y = x^2 + 2x - 1$ and its osculating circle at $x = -1$.

4.11 Exercises

1. Find the critical points for the following functions and find the absolute (global) extreme values on the given intervals:
   (a) $f(x) = 3x^4 - 8x^3 - 18x^2 + 1, [-2, 5]$;   (b) $y = 5x^{3/5} - x^{5/3}, [-4.2]$;
   (c) $y = \begin{cases} x^2, & -1 \leqslant x < 0 \\ 2-x^2, & 0 \leqslant x \leqslant 1; \end{cases}$ (d) $s(t) = |t^3 - 9t|, [-4, 4]$;
   (e) $r = \theta^2 \sqrt{3 - \theta}, [-1, 3]$.
2. (**Optimization**) A rectangle is inscribed between the parabolas $y = 4x^2$ and $y = 30 - x^2$ (two of its four vertices are on one parabola, the other two vertices are on the other parabola). What is the maximum area of such a rectangle?
3. (**Optimization**) A rectangle is inscribed under the arch of the curve $y = 4\cos\frac{x}{2}$ from $x = -\pi$ to $x = \pi$. Find the length and width of the rectangle with the largest area.
4. Show that the equation $x^5 + x - 1 = 0$ has exactly one root.

5. If $a_0 + \frac{a_1}{2} + \cdots + \frac{a_n}{n+1} = 0$, show that $f(x) = a_0 + a_1 x + \cdots + a_n x^n$ has at least one root in $(0, 1)$.

6. If $f(x)$ is continuous on $[0, a]$ and differentiable on $(0, a)$, while $f(a) = 0$, then show that there is a number $\xi \in (0, a)$ such that $f(\xi) + \xi f'(\xi) = 0$.

7. If $f(x)$ is continuous on $[0, 1]$ and differentiable on $(0, 1)$, while $f(0) = f(1) = 0$ and $f(\frac{1}{2}) = 1$, show that:
   (a) there is a number $c \in (\frac{1}{2}, 1)$ such that $f(c) = c$;
   (b) for any real number λ, there is a number $d \in (0, c)$ such that $f'(d) - \lambda(f(d) - d) = 1$.

8. Use the mean value theorem to prove the following inequalities:
   (a) $\frac{b-a}{b} < \ln \frac{b}{a} < \frac{b-a}{a}$ $(b > a > 0)$;
   (b) $ny^{n-1}(x - y) < x^n - y^n < nx^{n-1}(x - y)$ $(n > 1$ and $x > y > 0)$.

9. Prove the following identities:
   (a) $3 \arccos x - \arccos(3x - 4x^3) = \pi$, $x \in [-\frac{1}{2}, \frac{1}{2}]$;
   (b) $2 \arctan x + \arcsin \frac{2x}{1+x^2} = \pi$, for $x \in [1, \infty)$;
   (c) $\arctan \frac{1+x}{1-x} = -\frac{3\pi}{4} + \arctan x$, for $x > 1$.

10. Graph the function $f(x) = \sin x \sin(x + 2) - \sin^2(x + 1)$. What do you find? Explain why.

11. Find all the antiderivatives for each of the following functions:
   (a) $y = x^3$;   (b) $y = \sqrt{x}$;   (c) $s = \frac{1}{\sqrt{1-t^2}}$;
   (d) $\rho = \sin 2t$;   (e) $k = e^{-3\theta}$;   (f) $g = \sec y \tan y$;
   (g) $E(m) = mc^2$;   (h) $KE(v) = \frac{5}{2}v^2$; (i) $V = 2 \cdot 3^x$;
   (j) $y = \sin x - 3 \cos x + \cot 2x \csc 2x + \frac{1}{1+x^2} + 2$.

12. Assume f is an odd function and is differentiable everywhere. Show that, for any number $b > 0$, there is a number $\xi \in (-b, b)$ such that $f'(\xi) = \frac{f(b)}{b}$.

13. Suppose $g(x)$ is twice differentiable everywhere and $g(0) = 0, g(1) = 1$, and $g(2) = 2$. Show that there is a number $c \in (0, 2)$ such that $g''(c) = 0$.

14. (**Darboux's theorem: intermediate value theorem for derivatives**) If a and b are any two points in an interval $\mathbf{I}$ on which a function $f(x)$ is differentiable, then $f'(x)$ takes on every value between $f'(a)$ and $f'(b)$.
   Use this theorem or any other method to show that, if a function $f(x)$ is continuous on $[0, 3]$ and differentiable on $(0, 3)$, while $f(0) + f(1) + f(2) = 3$ and $f(3) = 1$, there is a number $\eta \in (0, 3)$ such that $f'(\eta) = 0$. *Group activity: prove Darboux's theorem.

15. Find critical points for each of the following functions, determine the intervals of increase and of decrease for each function, and find any local extreme values:
   (a) $y = 2x^3 - 6x^2 - 18x + 7$;   (b) $y = \frac{x}{x^2+4}$;
   (c) $y = \frac{\ln x}{x}$;   (d) $s = \cos^4 t + \sin^4 t$, $0 < t < \frac{\pi}{2}$.

16. Find the x-values where a continuous function $f(x)$ takes on local extreme values if:

(a) $f'(x) = x^2(x-1)(x+2)^3$; (b) $f'(x) = x(x+1)^3(x+2)^2$;

(c) $f'(x) = \frac{(x-1)(x+1)^2}{x}$.

17. The graph of the derivative of a function f is shown below. Find the x-value(s) where f takes a local maximum value and, also, find the x-value(s) where f takes a local minimum value.

Graph of $f'(x)$

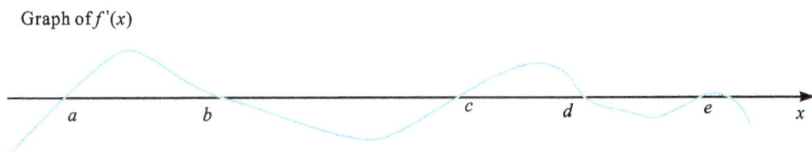

Question 17

18. Prove the following inequalities:
    (a) $\frac{2x}{\pi} < \sin x < x$ for $x \in (0, \frac{\pi}{2})$;
    (b) $\tan x > x + \frac{1}{3}x^3$ for $x \in (0, \frac{\pi}{2})$;
    (c) $1 + x \ln(x + \sqrt{1+x^2}) > \sqrt{1+x^2}$ $(x > 0)$;
    (d) $\sin x > x - \frac{x^3}{3}$ for $x > 0$;
    (e) $(\frac{\sin x}{x})^3 \geqslant \cos x$ for $0 < |x| < \frac{\pi}{2}$;
    (f) $1 - (1 - \frac{x}{n})^n e^x \leqslant \frac{x^2}{n}$ for $x \leqslant n$, where n is a positive integer.

19. Let $b > a > 0$. Assume the function $f(x)$ is continuous on $[a, b]$ and differentiable on (a, b). Prove that there exists a number $\xi \in (a, b)$ such that $2\xi(f(b) - f(a)) = (b^2 - a^2)f'(\xi)$.

20. If $b > a > 0$ and the function $f(x)$ is continuous on $[a, b]$ and differentiable on (a, b), then show that there is a number $c \in (a, b)$ such that $f(b) - f(a) = cf'(c) \ln \frac{b}{a}$.

21. Let $f(x)$ be continuous on $[a, b]$ and differentiable on (a, b) $(ab > 0)$. Show that there is a number η such that

$$\frac{1}{b-a} \begin{vmatrix} a & b \\ f(a) & f(b) \end{vmatrix} = \eta f'(\eta) - f(\eta).$$

NOTE: The 2×2 determinant $\begin{vmatrix} x & y \\ z & w \end{vmatrix} = xw - yz$.

22. Use L'Hôpital's rule to find each of the following limits:
    (a) $\lim_{x \to 1} \frac{x^7-1}{x^3-1}$;
    (b) $\lim_{t \to 0} \frac{e^{3t}-1}{2t}$;
    (c) $\lim_{x \to 0} \frac{x+\tan x}{\sin x}$;
    (d) $\lim_{x \to -2} \frac{x+2}{x^2+3x+2}$;
    (e) $\lim_{x \to 0} \frac{\sin x - \tan x}{x^3}$;
    (f) $\lim_{\theta \to 0} \frac{5^\theta - 3^\theta}{\theta}$;
    (g) $\lim_{x \to 0} \frac{(1+x)^{\frac{1}{x}}-e}{x}$;
    (h) $\lim_{x \to 0} \frac{1-\cos x^2}{x^3 \sin x}$;
    (i) $\lim_{x \to 0}(\csc x - \cot x)$;
    (j) $\lim_{x \to \infty}(xe^{1/x} - x)$;
    (k) $\lim_{x \to 1}(\frac{1}{\ln x} - \frac{1}{x-1})$;
    (l) $\lim_{y \to 0}(\frac{1}{y} - \frac{1}{e^y-1})$;
    (m) $\lim_{x \to 0^+} \sin x \ln x$;
    (n) $\lim_{x \to \pi}(\pi - x)\tan \frac{x}{2}$;
    (o) $\lim_{x \to -\infty} x^2 e^x$;
    (p) $\lim_{x \to 0^+} \sqrt{x} \ln x$;
    (q) $\lim_{x \to \infty} \frac{2x^2+7}{3x^3-2x^5}$;
    (r) $\lim_{x \to \infty} \frac{x^n}{e^{ax}}$, $a > 0$;
    (s) $\lim_{t \to 0^+} t^{1/t}$;
    (t) $\lim_{x \to \infty}(1 + \frac{3}{x} + \frac{5}{x^2})^x$;
    (u) $\lim_{x \to 0^+}(\tan 2x)^x$;
    (v) $\lim_{x \to 0}(\cos 3x)^{5/x}$;
    (w) $\lim_{x \to \infty}(\frac{x}{1+x})^x$;
    (x) $\lim_{x \to \infty}(e^x + x)^{1/x}$.

23. Find $\lim_{x\to\infty} \frac{x+\sin x}{x+\cos x}$. Does L'Hôpital's rule apply to this limit? Explain.

24. For any value of $x > 0$, by the mean value theorem, there is a number $\xi \in (0, x)$ (depending on x) such that $e^x - e^0 = e^\xi (x - 0)$. Let $\theta = \frac{\xi}{x}$. Then θ is a function of x and $0 < \theta < 1$. Find $\lim_{x\to 0} \theta$.

25. Find the Taylor polynomial of degree 3 and the corresponding Lagrange remainder for each of the following functions at the given point:
    (a) $f(x) = \sin x$, $a = \frac{\pi}{2}$; (b) $f(x) = xe^x$, $a = 0$; (c) $f(x) = \cos x$, $x = \pi$.

26. Use the Taylor polynomial of degree 2 for the function $\sqrt{x}$ at $x = 4$ to estimate the number $\sqrt{4.3}$. Also find an error bound for this approximation using the Lagrange remainder.

27. Use suitable Taylor/MacLaurin expansions to evaluate each of the following limits:
    (a) $\lim_{x\to 0} \frac{\cos x - e^{-\frac{x^2}{2}}}{x^2(x+\ln(1-x))}$; (b) $\lim_{x\to+\infty}(\sqrt[3]{x^3 + 3x^2} - \sqrt[4]{x^4 - 2x^3})$.

28. (*Group activity) If $f(x)$ is twice differentiable on $[a, b]$ and $f'(a) = f'(b) = 0$, then:
    (a) write out the Taylor expansion for $f(x)$ at $x = a$;
    (b) write out the Taylor expansion for $f(x)$ at $x = b$;
    (c) by considering the value of f at $\frac{a+b}{2}$ or by any other method, show that there is a number $c \in (a, b)$ such that

$$|f''(c)| \geq \frac{4}{(b-a)^2}|f(b) - f(a)|.$$

29. If $f(x)$ is twice differentiable at $x = 0$ and $\lim_{x\to 0}(\frac{\sin 2x}{x^3} + \frac{f(x)}{x^2}) = 0$, then, by considering the MacLaurin expansion for $\sin x$ at 0:
    (a) show that $xf(x) = -2x + \frac{(2x)^3}{3!} + o(x^3)$;
    (b) find $\lim_{x\to 0} \frac{f(x)+2}{x^2}$.

30. (*Group activity) Assume that the function $f(x)$ has a continuous second derivative for all $x \in \mathbf{R}$. If, for any values of h and x, there is a number θ between 0 and 1 such that $f(x + h) = f(x) + hf'(x + \theta h)$, while $f''(x) \neq 0$, prove that $\lim_{h\to 0} \theta = \frac{1}{2}$.

31. Determine the interval(s) where each of the following functions is concave up and concave down and find any inflection points:
    (a) $f(x) = 4x^3 + 21x^2 + 36x$; (b) $s(t) = t^{1/3}(t - 4)$;
    (c) $r(x) = \frac{x}{x^2+1}$; (d) $k(\theta) = e^{\arctan \theta}$.

32. Use the second derivative test to determine the nature of the stationary points for each of the following functions:
    (a) $f(x) = 4x^3 + 21x^2 + 36x - 20$; (b) $r'(t) = (t - 1)(t + 2)^3$.

33. The graph of the derivative of a function f is shown below. Find the value(s) of x at which f has an inflection point.

Graph of $f'(x)$

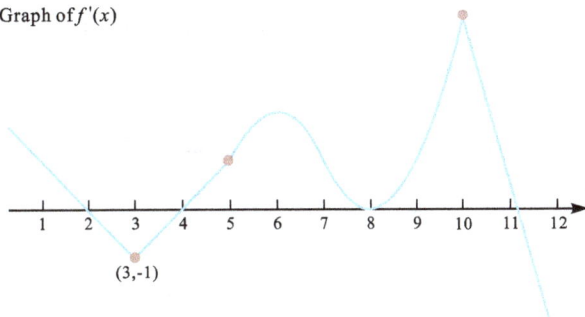

$(3,-1)$

Question 33

34. By considering the concavity of a suitable function on a closed interval, prove each of the following inequalities:

   (a) $\frac{x^n+y^n}{2} > (\frac{x+y}{2})^n$, $x > 0$, $y > 0$, $x \neq y$ and $n > 1$;

   (b) $\frac{n}{\frac{1}{x_1}+\frac{1}{x_2}+\cdots+\frac{1}{x_n}} \leqslant \sqrt[n]{x_1 x_2 \cdots x_n} \leqslant \frac{x_1+x_2+\cdots+x_n}{n}$, where all $x_i > 0$.

35. (**Airplane landing path**) Suppose an airplane is flying at altitude H meters when it begins its descent to an airport runway. The horizontal ground distance between the runway and the airplane is h meters. Assume the landing path of the airplane is a cubic function, $y = ax^3 + bx^2 + cx + d$. By setting up a coordinate system and considering the values of $\frac{dy}{dx}$ at the point where the airplane starts to descend and the point where it lands on the runway, find (a) the constants a, b, c, and d and (b) any inflection point.

36. Find all local extreme values for each of the following functions and discuss the global extreme values:

   (a) $f(x) = xe^{-x}$; (b) $r(t) = \frac{\ln t}{t}$;

   (c) $y = \arctan x - \frac{1}{2}\ln(x^2 + 1)$; (d) $s(\theta) = 4\theta - \tan\theta$, $0 \leqslant \theta < \pi$;

   (e) $y = \sqrt{1 - \cos x}$; (f) $f(x) = \frac{x}{1+x^2}$.

37. (An example of the **Banach contraction mapping principle**) Assume $f'(x) = \frac{x}{1+x^2}$. Then:

   (a) show that, for any numbers a and b, $2|f(a) - f(b)| \leqslant |a - b|$;

   (b) prove that the iterative sequence $x_{n+1} = f(x_n)$ converges to a fixed point x^* of f, no matter what the initial value x_0 is.

38. Use calculus to prove that the minimum value of the function $f(x) = x + \frac{1}{x}$, $x \geqslant 1$, is 2 and then show that

$$\frac{(a^2 + 1)(b^2 + 1)(c^2 + 1)(d^2 + 1)}{abcd} \geqslant 16$$

for any positive integers a, b, c, and d.

39. Find the highest and lowest points on the curve defined by $x^2 + xy + y^2 = 12$.

40. Find the largest term in the sequence $\{\sqrt[n]{n}\}|_{n=1}^{\infty}$.

41. Assume $f(x)$ is twice differentiable on $[0, \infty)$ and $f(0) = \lim_{x \to \infty} f(x) = 0$. If, for all $x > 0$, $f''(x) + e^{f'(x)} = e^{f(x)}$, show that $f(x) \equiv 0$.

42. Sketch a possible graph of a function with the properties given in the table.

x	$(-\infty, 1)$	1		$(1, 3)$	3	$(3, \infty)$
$f(x)$		$\lim_{x \to 1^-} f(x) = \infty$ and $\lim_{x \to 1^+} f(x) = -\infty$			1	
$f'(x)$	+			+	0	+
$f''(x)$	+			$-$		+

43. Sketch a possible graph of a function with the following properties:
    (a) $f(0) = 0$, $f'(0) = f'(2) = f'(4) = 0$;
    (b) $f'(x) > 0$ for $x < 0$ and $2 < x < 4$, $f'(x) < 0$ for $0 < x < 2$ and $x > 4$;
    (c) $f''(x) > 0$ for $1 < x < 3$ and $f''(x) < 0$ for $x < 1$ and $x > 3$.

44. Sketch the graph of each of the following functions:
    (a) $y = e^{-(x-1)^2}$;   (b) $s(t) = \frac{1}{5}(t^4 - 6t^2 + 8t + 7)$;
    (c) $r(x) = \frac{x}{x^2+1}$;   (d) $k(\theta) = \frac{\theta^2+1}{\theta}$.

45. (**Optimization**) Assume the revenue function $R(x) = 6x$ and the cost function $C(x) = x^3 - 6x^2 + 15$. Determine the value x that maximizes the profit.

46. Show that the equation $x^3 - 3x^2 + 5x - 1 = 0$ has a unique solution in the interval $(0, 1)$. Use the decimal search method to find this root correct to two decimal places.

47. Use Newton's method to estimate the real root of the equation $x^3 + 3x + 1 = 0$. Round your answer to three decimal places.

48. Find the curvature of the curve $y = \ln \sec x$ at each point x in its domain. Then find the radius of curvature and an equation of the osculating circle at the point $(0, 0)$.

49. (**Witch of Agnesi**) Find an equation for the osculating circle of the curve $y = \frac{a^3}{x^2+a^2}$, $a > 0$, at the point $(0, a)$.

50. Find the curvature of the cycloid $\begin{cases} x = a(t - \sin t) \\ y = a(1 - \cos t), \end{cases}$ where $a > 0$, at the point $t = \pi$.

5 The definite integral

In this chapter, you will learn about:
- *definite integrals;*
- *the fundamental theorem of calculus;*
- *numerical integration.*

Integral calculus has a longer history than differential calculus, which allowed us to find properties of a function at a particular point or instant in time, such as the slope of the tangent, the concavity, or the velocity of a moving object (derivative of its position or distance) and its acceleration (derivative of its velocity). Quite often we want to reverse this process. For example, it is relatively easy to measure the acceleration of a moving object as a function of time, or we may know the acceleration (such as the acceleration of gravity at the surface of the earth), but it is often difficult to directly measure the velocity or position of that object. Consequently, if we know the acceleration as a function of time, then we need a process of reversing differentiation to find its velocity and position as a function of time. Integral calculus provides the tools for this. Another example of this is a biologist who knows the rate at which an insect population is increasing, who might wish to know what size the population will be at some future time. In each case, the problem is to find, from a given function f, another function F such that $F' = f$. For obvious reasons, F is called an *antiderivative* of f.

Integral calculus allows one to go further than just reversing differentiation in that it can be used to find overall properties of functions, such as the area bounded by a curve, the volume bounded by a surface, and the length of a curve. It provides a basis for solving problems in many nonmathematical areas, such as computing the mass, moment of inertia, or center of gravity of an object even if it has a variable density. It is used in the computation of the orbits and paths of celestial bodies and space ships. An engineer who measures the variable rate at which water is flowing in a river may use integral calculus to compute the total amount of water that has flowed during a certain time period. The applications of integral calculus are endless, and only a few are discussed in this chapter.

5.1 Definite integrals and properties

5.1.1 Introduction

We often would like to know the answer to problems that concern a whole function, problems that concern the total change over time resulting from some constantly changing process, or many other similar problems. We start this section with two specific problems of this type: computing the area bounded by the graph of a func-

https://doi.org/10.1515/9783110527780-005

tion $y = f(x)$ and computing the distance traveled over a period of time by a moving object that has a variable velocity. The two problems are quite different but we will see that both can be solved using the same process, called *integration*. This process will use the idea of a *definite integral of a function*. The definition of a definite integral does not involve calculus, but we will show, in the fundamental theorem of calculus (FTC), that the definite integral of a function can be computed using antiderivatives as discussed in the previous chapter. We start with the distance and area problems that we introduced in Chapter 1.

The distance problem

We model the movement of a particle over time t, moving along the x-axis, with variable velocity $v(t) \geqslant 0$. We want to know the distance traveled by the particle during time interval $[a, b]$, that is, from $t = a$ to $t = b$. We create a partition P of $[a, b]$ into n subintervals as follows:

$$P : a = t_0 < t_1 < t_2 < \cdots < t_{n-1} < t_n = b, \quad \text{with}$$

$$\Delta t_i = t_i - t_{i-1} \quad \text{and} \quad t_i^* \in [t_{i-1}, t_i] \quad \text{for } i = 1, 2, \ldots, n.$$

If the ith subinterval, $[t_{i-1}, t_i]$, is small, then the speed of the object will not change very much during this time subinterval, so it is approximately $v(t_i^*)$ throughout the subinterval (note that t_i^* is an arbitrary time in the given subinterval). Hence, the change in position from $t = t_{i-1}$ to $t = t_i$ is approximately $v(t_i^*)(t_i - t_{i-1})$ (velocity × time), as seen in Figure 5.1.1. Therefore, the total change in a position from $t = a$ to $t = b$ is approximately

$$\text{distance traveled from } t = a \text{ to } t = b$$

$$\approx \sum_{i=1}^{n} v(t_i^*)(t_i - t_{i-1}) = \sum_{i=1}^{n} v(t_i^*)\Delta t_i.$$

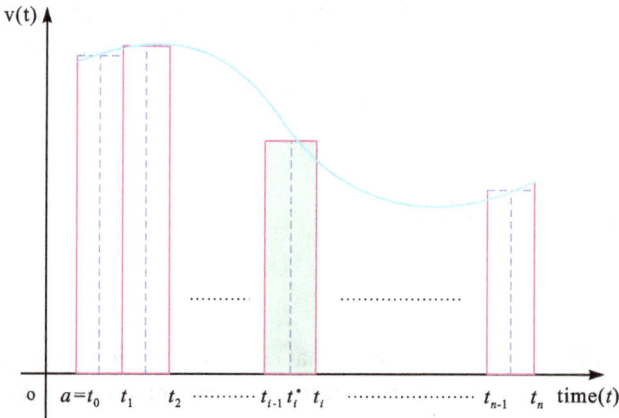

Figure 5.1.1: Changes in position are approximated by small rectangles.

Now we let $n \to \infty$ and $\|P\| = \max_i |\Delta t_i| \to 0$. If the limit

$$\lim_{\|P\| \to 0} \sum_{i=1}^{n} v(t_i^*) \Delta t_i$$

exists for any possible partition of $[a, b]$ and any sample point t_i^* in each subinterval $[t_{i-1}, t_i]$, then it is reasonable to define this limit to be the total distance traveled by the particle during the time interval $[a, b]$. Of course, if this limit exists, it must be unique.

The area problem

We now try to find the area of the region under the graph of a nonnegative function $f(x)$ defined on $[a, b]$, above the x-axis and between the vertical lines $x = a$ and $x = b$. Let

$$P : a = x_0 < x_1 < \cdots < x_n = b$$

be a partition of $[a, b]$ into n subintervals and for $i = 1, 2, \ldots, n$ and denote the width of the subinterval $[x_{i-1}, x_i]$ by $\Delta x_i = x_i - x_{i-1}$. Let $x_i^* \in [x_{i-1}, x_i]$ be any point in the subinterval, as shown in Figure 5.1.2. A product like

$$f(x_i^*)(x_i - x_{i-1}) = f(x_i^*) \Delta x_i$$

is the area of a rectangle with base $[x_{i-1}, x_i]$ (the ith subinterval) of width Δx_i and height $f(x_i^*)$, which is the height of the curve $y = f(x)$ at $x = x_i^*$, as shown in Figure 5.1.2. Intuitively, the following sum of the areas of all such rectangles formed by the partition is the following approximation to the total area A underneath this curve:

$$A \approx \sum_{i=1}^{n} f(x_i^*) \Delta x_i.$$

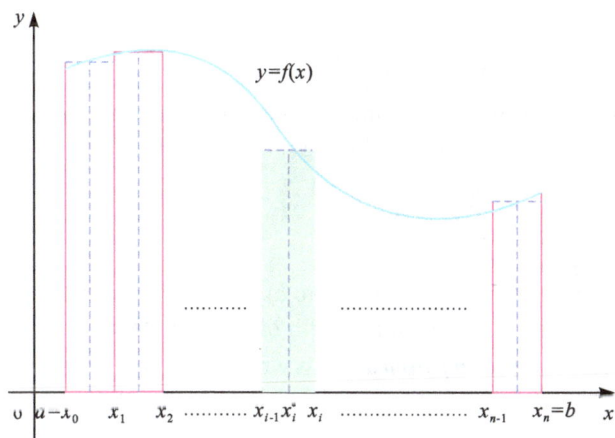

Figure 5.1.2: Area of rectangles approximate the area below the curve.

It is reasonable to suppose that the approximation to the area A will become better and better as the number of rectangles increases and the widths of the rectangles decrease ($\|P\| = \max_i |\Delta x_i| \to 0$), because the narrower the rectangle, the more closely the rectangle approximates the region under the curve above that subinterval. Consequently, we define the area A under $y = f(x)$ between $x = a$ and $x = b$ and above the x-axis as the limit

$$A = \lim_{\|P\| \to 0} \sum_{i=1}^{n} f(x_i^*) \Delta x_i,$$

provided that the limit exists for any possible partition of $[a, b]$ and any choice of sample point x_i^* in each subinterval.

For a continuous nonnegative function f on $[a, b]$, there is an even stronger argument for defining $\lim_{\|P\| \to 0} \sum_{i=1}^{n} f(x_i^*) \Delta x_i$ to be the area A. In Section 2.6.3, the extreme value theorem showed that a continuous function on a closed interval must attain its maximum value and its minimum value at some x-value(s) in that interval. Hence, if we choose the sample points x_i^* of the partition such that $f(x_i^*)$ is the minimum value of the function on $[x_{i-1}, x_i]$ for each i, then all of the rectangles combined will be completely contained in the region below or on the curve. Hence, with these x_i^*, as we take the limit of the sum, we have

$$\lim_{\|P\| \to 0} \sum_{i=1}^{n} f(x_i^*) \Delta x_i \leqslant A.$$

On the other hand, if we instead choose the sample points x_i^* of the partition such that $f(x_i^*)$ is the maximum value of the function on $[x_{i-1}, x_i]$ for each i, then all of the rectangles combined completely contain the region below the curve. Hence, for these x_i^*, as we take the limit of the sum, we find

$$\lim_{\|P\| \to 0} \sum_{i=1}^{n} f(x_i^*) \Delta x_i \geqslant A.$$

By the squeeze theorem, it follows that $\lim_{\|P\| \to 0} \sum_{i=1}^{n} f(x_i^*) \Delta x_i = A$.

These two problems, and many others, look different at first sight, but mathematically speaking, both require evaluating a limit of this type:

$$\lim_{\|P\| \to 0} \sum_{i=1}^{n} f(x_i^*) \Delta x_i.$$

Definition 5.1.1 (Definite integral of a function f from a to b). Given any partition of $[a, b]$, let $\|P\| = \max\{\Delta x_1, \ldots, \Delta x_n\}$ be the maximum length of the subintervals of the partition. The *definite integral of a function f from a to b* is written as $\int_a^b f(x)\,dx$ and is defined as

$$\int_a^b f(x)\,dx = \lim_{\|P\| \to 0} \sum_{i=1}^{n} f(x_i^*) \Delta x_i,$$

provided the limit exists and has the same value for all possible partitions and all possible choices of the sample points x_i^*. When the limit exists, we say that $f(x)$ is *integrable* on $[a, b]$. The sum $t \sum_{i=1}^{n} f(x_i^*) \Delta x_i$ is called a *Riemann sum* and sometimes denoted by R_n.

Using ε-δ language, the precise meaning of the limit that defines a definite integral is the following: for any number $\varepsilon > 0$, there is a number $\delta > 0$ (depending on ε) such that

$$\left| \int_a^b f(x)\, dx - \sum_{i=1}^{n} f(x_i^*) \Delta x_i \right| < \varepsilon$$

for any partition with $\|P\| < \delta$ and for every possible choice of $x_i^* \in [x_{i-1}, x_i]$ in that partition.

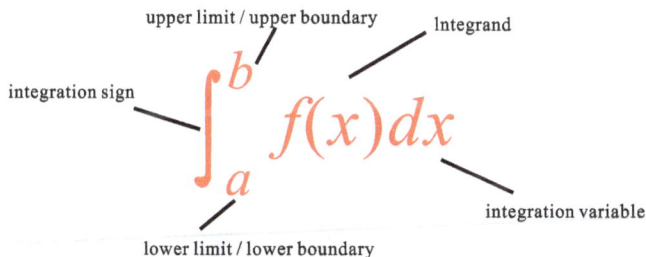

upper limit / upper boundary   Integrand

integration sign

$$\int_a^b f(x)\, dx$$

lower limit / lower boundary   integration variable

The symbol $\int$ is called an integral sign. It was chosen as an elongated S because an integral is the limit of a sum. The function $f(x)$ is called the integrand. The numbers a and b are called the limits of integration; b is the upper limit and a is the lower limit. The process of calculating the value of an integral is called integration. The definite integral $\int_a^b f(x)\, dx$ is a number and it does not depend on x, so x is a "dummy" variable. In fact, we could use any variable in place of x without changing the value of the integral, so that

$$\int_a^b f(x)\, dx = \int_a^b f(t)\, dt = \int_a^b f(u)\, du.$$

5.1.1.1 Are all functions integrable?

At first sight, it seems that a definite integral for any particular function might never exist, or at least might be impossible to compute because there are so many variables involved (choices of the partition, choices of the sample points, etc.). It is true that not all functions are integrable. In particular, if f fails to be continuous at too many points in $[a, b]$, then the integral limit may fail to exist. For example, the Dirichlet function is not integrable.

Bernhard Riemann
(1826–1866) was an influential German mathematician who made lasting contributions to analysis, number theory, and differential geometry, some of them enabling the later development of general relativity. http://en.wikipedia.org/wiki/Bernhard_Riemann

Peter G. L. Dirichlet (1805–1859) was a German mathematician who made deep contributions to number theory (including creating the field of analytic number theory) and to the theory of Fourier series and other topics in mathematical analysis; he is credited with being one of the first mathematicians to give the modern formal definition of a function. http://en.wikipedia.org/wiki/Peter_Gustav_Lejeune_Dirichlet

Example 5.1.1. Show that the Dirichlet function

$$f(x) = \begin{cases} 1, & \text{when } 0 \leqslant x \leqslant 1 \text{ and } x \text{ is rational} \\ 0, & \text{when } 0 \leqslant x \leqslant 1 \text{ and } x \text{ is irrational} \end{cases}$$

is not integrable.

Proof. For any partition of $[0,1]$, say, $0 = x_0 < x_1 < x_2 < \cdots < x_n = 1$, if we choose a rational point x_i^* in each subinterval $[x_{i-1}, x_i]$, then $f(x_i^*)$ is always 1, for $i = 1, 2, \ldots, n$, so the Riemann sum gives

$$\sum_{i=1}^{n} f(x_i^*)(x_i - x_{i-1}) = \sum_{i=1}^{n} 1 \cdot (x_i - x_{i-1}) = 1.$$

Of course the limit as n approaches infinity of this Riemann sum is 1.

However, if instead we choose an irrational point x_i^* in each subinterval $[x_{i-1}, x_i]$, then $f(x_i^*)$ is always 0 for $i = 1, 2, \ldots, n$, so the Riemann sum becomes

$$\sum_{i=1}^{n} f(x_i^*)(x_i - x_{i-1}) = \sum_{i=1}^{n} 0 \cdot (x_i - x_{i-1}) = 0.$$

Of course, the limit as n approaches infinity of this Riemann sum is zero.

Now we have two different choices of the sample points in each subinterval resulting in different limits of the corresponding Riemann sum. By the definition of the definite integral, the definite integral of this function on $[0,1]$ does not exist. Therefore, this function is not integrable. □

However, it can be proved that the definite integral always exists for several types of functions, as stated without proof in the following theorems.

Theorem 5.1.1. *A function f defined and continuous on an interval* $[a,b]$ *is integrable on* $[a,b]$.

Theorem 5.1.2. *A function f is integrable on* $[a,b]$ *if f is bounded and has a finite number of removable or jump discontinuities on* $[a,b]$.

NOTE. Under the definition of integrable functions, we can also deduce that *an integrable function on* $[a,b]$ *must be bounded*. If a function f is unbounded on $[a,b]$, then, for any partition $a = x_0 < x_1 < \cdots < x_n = b$, there are some subintervals on which f is unbounded (say, unbounded above). Given any positive number M, we can then choose sample points x_i^* such that $\sum f(x_i^*)\Delta x_i$ is larger than M. Therefore, the corresponding limit of the Riemann sum will not exist, so *an unbounded function is not integrable*.

If a function is integrable, then the limit of the Riemann sum exists for all possible partitions and for all possible choices of the sample points x_i^* in each subinterval $[x_{i-1}, x_i]$. Therefore, in order to evaluate a definite integral, we can choose a convenient partition, say, the one in which each subinterval has equal width and the sample point in each subinterval is the left endpoint, right endpoint or the midpoint of that subinterval. The corresponding Riemann sums are called the left Riemann sum (LRS), the right Riemann sum (RRS), and the midpoint Riemann sum (MRS), respectively.

Example 5.1.2. Calculate the Riemann sums, LRS, RRM, and MRS, for the function $f(x) = x^2$ on $[0,1]$ using five subintervals with equal widths.

Solution. The interval $[0,1]$ must be divided into five subintervals with equal width, so the subintervals are

$$\left[0, \frac{1}{5}\right], \quad \left[\frac{1}{5}, \frac{2}{5}\right], \quad \left[\frac{2}{5}, \frac{3}{5}\right], \quad \left[\frac{3}{5}, \frac{4}{5}\right], \quad \text{and} \quad \left[\frac{4}{5}, 1\right].$$

Clearly, $\Delta x_i = \frac{1}{5}$ for each subinterval, so:
1. if each sample point x_i^* is chosen to be the left endpoint of each subinterval, then

$$x_1^* = 0, \quad x_2^* = \frac{1}{5}, \quad x_3^* = \frac{2}{5}, \quad x_4^* = \frac{3}{5}, \quad \text{and} \quad x_5^* = \frac{4}{5}$$

and the corresponding LRS, as seen in Figure 5.1.3, is

$$\sum_{i=1}^{5} f(x_i^*)\Delta x_i$$

$$= f(0) \cdot \frac{1}{5} + f\left(\frac{1}{5}\right) \cdot \frac{1}{5} + f\left(\frac{2}{5}\right) \cdot \frac{1}{5} + f\left(\frac{3}{5}\right) \cdot \frac{1}{5} + f\left(\frac{4}{5}\right) \cdot \frac{1}{5}$$

$$= 0^2 \cdot \frac{1}{5} + \left(\frac{1}{5}\right)^2 \cdot \frac{1}{5} + \left(\frac{2}{5}\right)^2 \cdot \frac{1}{5} + \left(\frac{3}{5}\right)^2 \cdot \frac{1}{5} + \left(\frac{4}{5}\right)^2 \cdot \frac{1}{5}$$

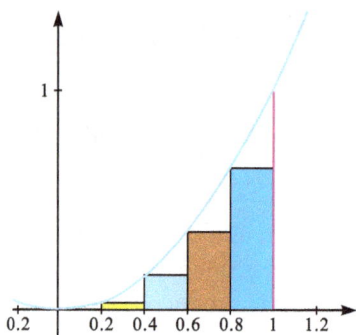

Figure 5.1.3: LRS for $y = x^2$ over $[0,1]$ with 5 subintervals.

$$= \frac{6}{25} = 0.24;$$

2. if each sample point x_i^* is chosen to be the right endpoint of each subinterval, then

$$x_1^* = \frac{1}{5}, \quad x_2^* = \frac{2}{5}, \quad x_3^* = \frac{3}{5}, \quad x_4^* = \frac{4}{5}, \quad \text{and} \quad x_5^* = 1$$

and the corresponding RRS, as seen in Figure 5.1.4, is

$$\sum_{i=1}^{5} f(x_i^*)\Delta x_i$$

$$= f\left(\frac{1}{5}\right) \cdot \frac{1}{5} + f\left(\frac{2}{5}\right) \cdot \frac{1}{5} + f\left(\frac{3}{5}\right) \cdot \frac{1}{5} + f\left(\frac{4}{5}\right) \cdot \frac{1}{5} + f(1) \cdot \frac{1}{5}$$

$$= \left(\frac{1}{5}\right)^2 \cdot \frac{1}{5} + \left(\frac{2}{5}\right)^2 \cdot \frac{1}{5} + \left(\frac{3}{5}\right)^2 \cdot \frac{1}{5} + \left(\frac{4}{5}\right)^2 \cdot \frac{1}{5} + (1)^2 \cdot \frac{1}{5}$$

$$= \frac{11}{25} = 0.44;$$

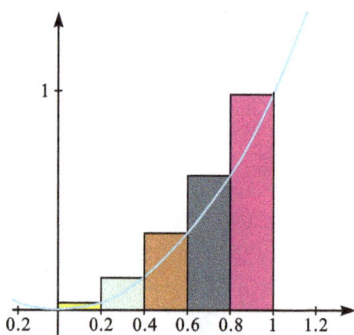

Figure 5.1.4: RRS for $y = x^2$ over $[0,1]$ with 5 subintervals.

3. if each sample point x_i^* is chosen to be the midpoint of each subinterval, then

$$x_1^* = \frac{1}{10}, \quad x_2^* = \frac{3}{10}, \quad x_3^* = \frac{5}{10}, \quad x_4^* = \frac{7}{10}, \quad \text{and} \quad x_5^* = \frac{9}{10}$$

and the corresponding MRS, as seen in Figure 5.1.5, is

$$\sum_{i=1}^{5} f(x_i^*)\Delta x_i$$

$$= f\left(\frac{1}{10}\right)\cdot\frac{1}{5} + f\left(\frac{3}{10}\right)\cdot\frac{1}{5} + f\left(\frac{5}{10}\right)\cdot\frac{1}{5} + f\left(\frac{7}{10}\right)\cdot\frac{1}{5} + f\left(\frac{9}{10}\right)\cdot\frac{1}{5}$$

$$= \left(\frac{1}{10}\right)^2\cdot\frac{1}{5} + \left(\frac{3}{10}\right)^2\cdot\frac{1}{5} + \left(\frac{5}{10}\right)^2\cdot\frac{1}{5} + \left(\frac{7}{10}\right)^2\cdot\frac{1}{5} + \left(\frac{9}{10}\right)^2\cdot\frac{1}{5}$$

$$= \frac{33}{100} = 0.33.$$

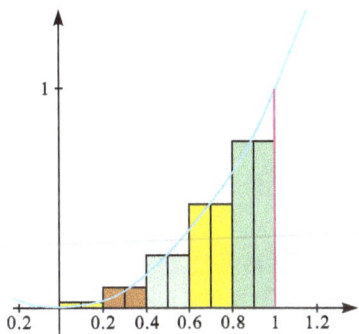

Figure 5.1.5: MRS for $y = x^2$ over [0,1] with 5 subintervals.

If we wish to use these Riemann sums to approximate the area under the curve $f(x) = x^2$ on [0,1], which of these three sums will provide the best approximation? From Figure 5.1.5, it appears that the midpoint sum is the best in this case. However, if we increase the number of subintervals, we know that the limit of all three Riemann sums must be the same. Table 5.1.1 shows some numerical values of these Riemann sums for several different numbers of subintervals.

Table 5.1.1: Selected values for LRS, MRS, and RRS.

n	LRS	MRS	RRS
5	0.24	0.33	0.44
10	0.285	0.3325	0.385
50	0.3234	0.3333	0.3434
100	0.328 35	0.333 33	0.338 35
200	0.330 84	0.333 33	0.335 84

From the table, it seems that all three Riemann sums approach $0.\overline{3}$ as n (the number of subintervals) gets large. In fact, this is true, as shown in the following example.

Example 5.1.3. Calculate the Riemann sum R_n for the integral $\int_0^1 x^2\, dx$, in which all subintervals have the same length and the smallest value of x in each subinterval is chosen as the sample point. Compute $\lim_{n\to\infty} R_n$ to find the value of this definite integral.

Solution. The function $f(x) = x^2$ is continuous on $[0,1]$, so the definite integral exists by Theorem 5.1.1. Hence, the value of the integral must be equal to the limit of the particular Riemann sum, R_n. The partition of $[0,1]$ into n subintervals by $0 = x_0 < x_1 < \cdots < x_n = 1$ has the common length of the subintervals $\Delta x = \frac{(b-a)}{n} = \frac{1-0}{n} = \frac{1}{n}$, so that $x_0 = 0$, $x_1 = \frac{1}{n}$, $x_2 = \frac{2}{n}$, and in general $x_i = \frac{i}{n}$. In the subinterval $[x_{i-1}, x_i] = [\frac{i-1}{n}, \frac{i}{n}]$, the sample point x_i^* will be $x_{i-1} = \frac{i-1}{n}$, as this is the smallest number in the subinterval. The Riemann sum R_n is therefore

$$R_n = \sum_{i=1}^{n} f(x_i^*)\Delta x = \sum_{i=1}^{n} f(x_{i-1})\Delta x = \sum_{i=1}^{n} x_{i-1}^2 \Delta x$$

$$= \sum_{i=1}^{n} \left(\frac{i-1}{n}\right)^2 \cdot \frac{1}{n} = \frac{1}{n^3} \sum_{i=1}^{n} (i-1)^2.$$

Hence,

$$R_n = \frac{1}{n^3}\left(0^2 + 1^2 + 2^2 + \cdots + (n-1)^2\right)$$

$$= \frac{1}{n^3} \frac{(n-1)\cdot n \cdot (2(n-1)+1)}{6}$$

$$= \frac{(n-1)(2n-1)}{6n^2}.$$

Thus,

$$\int_0^1 x^2\, dx = \lim_{n\to\infty} R_n$$

$$= \lim_{n\to\infty} \frac{(n-1)(2n-1)}{6n^2}$$

$$= \lim_{n\to\infty} \frac{2n^2 - 3n + 1}{6n^2} = \frac{1}{3}.$$

As seen above, the definite integral can be found by evaluating the limit of a Riemann sum. However, we were fortunate because this limit was easy to determine by using the following convenient identity:

$$1^2 + 2^2 + \cdots + n^2 = \frac{n(n+1)(2n+1)}{6}.$$

We would not be so lucky with other functions, such as $\sin x$ or $\ln x$. There will not be any identities to help us find the limit in these cases.

5.1.2 Properties of the definite integral

When we defined the definite integral $\int_a^b f(x)\, dx$, we assumed that $a < b$. However, the definition as a limit of Riemann sums is still valid even if $a > b$. If we interchange a and b, then Δx changes from $x_i - x_{i-1}$ to $x_{i-1} - x_i$. Therefore, we define

$$\int_a^b f(x)\, dx = -\int_b^a f(x)\, dx.$$

In the case where $a = b$, we have $\Delta x = 0$, so we also define

$$\int_a^a f(x)\, dx = 0.$$

We now list some basic properties of definite integrals. We assume that both f and g are integrable functions.

Properties of the definite integral

Property 1. When c is any constant, $\int_a^b c\, dx = c(b - a)$. In particular,

$$\int_a^b 1\, dx = b - a \quad \text{and} \quad \int_a^b 0\, dx = 0.$$

Figure 5.1.6 illustrates this property.

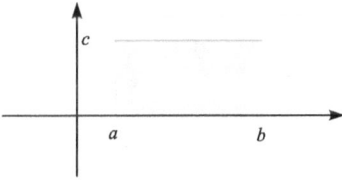

Figure 5.1.6: The definite integral of a constant function is the area of the rectangle.

Property 2. The linearity property tells us

$$\int_a^b \left[kf(x) \pm hg(x) \right] dx = k \int_a^b f(x)\, dx \pm h \int_a^b g(x)\, dx,$$

where k, h are constants. In particular,

$$\int_a^b kf(x)\, dx = k \int_a^b f(x)\, dx, \quad \text{for any constant } k.$$

Property 3. The additive property tells us $\int_a^b f(x)\, dx = \int_a^c f(x)\, dx + \int_c^b f(x)\, dx$.

Figure 5.1.7 illustrates the additive property.

Property 4. If $f(x) \geqslant 0$ for $a \leqslant x \leqslant b$, then $\int_a^b f(x)\, dx \geqslant 0$.

Property 5. If $f(x) \geqslant g(x)$ for $a \leqslant x \leqslant b$, then $\int_a^b f(x)\, dx \geqslant \int_a^b g(x)\, dx$.

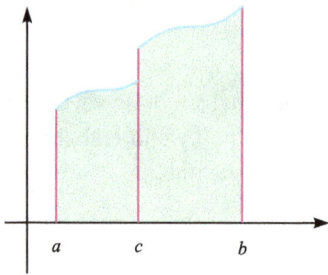

Figure 5.1.7: Additivity properpty of definite integrals.

Property 6. If $m \leqslant f(x) \leqslant M$ for $a \leqslant x \leqslant b$, then

$$m(b-a) \leqslant \int_a^b f(x)\,dx \leqslant M(b-a).$$

Figure 5.1.8 illustrates this property.

Property 7 (Mean value theorem). If $f(x)$ is continuous on $[a,b]$, then there must be a number $c \in [a,b]$ such that

$$\int_a^b f(x)\,dx = f(c)(b-a).$$

Figure 5.1.9 illustrates this property.

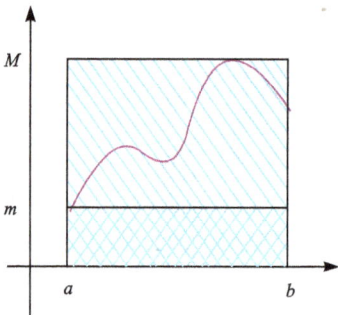

Figure 5.1.8: Estimate a definite integral.

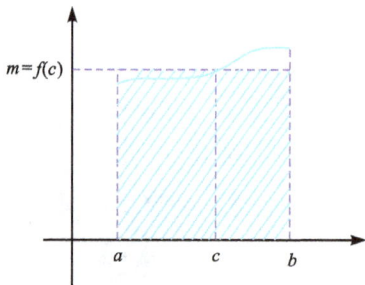

Figure 5.1.9: Mean value theorem for a definite integral.

NOTE. If $f(x) \geq 0$, the geometric interpretation of this theorem is that there is a number c in $[a, b]$ such that the area of the region between the x-axis, the curve, and the vertical lines $x = a$ and $x = b$ is equal to area of the rectangle with base length $(b - a)$ and height $f(c)$. The value of $f(c)$ is defined as the mean/average value of $f(x)$ over the interval $[a, b]$, so

$$\text{the average value of } f(x) \text{ on } [a, b] \text{ is } \quad f_{\text{ave}} = \frac{\int_a^b f(x)\, dx}{b - a}.$$

Before proving these properties, we give two examples to illustrate some of these properties.

Example 5.1.4. Find $\int_0^3 [x]\, dx$.

Solution. The greatest integer function $[x]$ is not continuous on $[0, 3]$, but it is bounded and has only three discontinuities. By Theorem 5.1.2, it is therefore integrable. Because we know that a definite integral is actually the area under a curve, we can use geometry to find the value of this definite integral. We have

$$\int_0^3 [x]\, dx = \int_0^1 [x]\, dx + \int_1^2 [x]\, dx + \int_2^3 [x]\, dx$$
$$= \int_0^1 0\, dx + \int_1^2 1\, dx + \int_2^3 2\, dx$$
$$= 0 + 1(2 - 1) + 2(3 - 2)$$
$$= 0 + 1 + 2 = 3,$$

which is the area of the shaded region, as shown in Figure 5.1.10.

Example 5.1.5. Show that

$$\frac{2}{e^2} \leq \int_0^2 e^{x - x^2}\, dx \leq 2\sqrt[4]{e}.$$

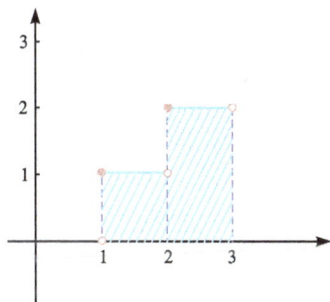

Figure 5.1.10: Area of the region below $[x]$ and above the x-axis for $0 \leq x \leq 3$.

Proof. Since $f(x) = e^{x-x^2}$ is continuous on $[0,2]$, it has extreme values on $[0,2]$ and is also integrable on $[0,2]$. To find its extrema, we take the derivative

$$f'(x) = e^{x-x^2}(x - x^2)' = e^{x-x^2}(1 - 2x).$$

The only stationary point is $x = \frac{1}{2}$ and the endpoints are $x = 0$ and $x = 2$. We compare the values of the function at these three points:

$$f(0) = e^{0-0} = 1, \quad f\left(\frac{1}{2}\right) = e^{\frac{1}{2}-(\frac{1}{2})^2} = e^{\frac{1}{4}} = \sqrt[4]{e},$$

and

$$f(2) = e^{2-2^2} = e^{-2}.$$

We conclude that the minimum value of $f(x)$ is e^{-2} and the maximum value of $f(x)$ is $\sqrt[4]{e}$. Therefore,

$$(2-0)e^{-2} \leqslant \int_0^2 e^{x-x^2}\, dx \leqslant \sqrt[4]{e}(2-0).$$

This completes the proof. Figure 5.1.11 illustrates this estimation. □

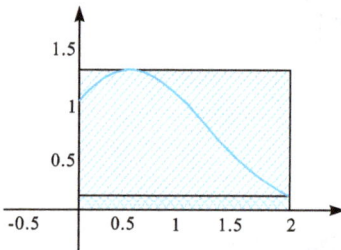

Figure 5.1.11: Estimate the integral $\int_0^2 e^{x-x^2}\, dx$.

Proofs of the properties
Proof of Property 1. Since $f(x) = c$, for any partition P of $[a,b]$ and any sample point x_i^* in each subinterval $[x_{i-1}, x_i]$, we have $f(x_i^*) = c$ and

$$\int_a^b f(x)\, dx = \lim_{\|P\| \to 0} \sum_{i=1}^n f(x_i^*)\Delta x_i = \lim_{\|P\| \to 0} \sum_{i=1}^n c\Delta x_i$$

$$= c \lim_{\|P\| \to 0} \sum_{i=1}^n \Delta x_i = c(b-a). \qquad \square$$

Proof of Property 2. For any partition P, $a = x_0 < x_1 < \cdots < x_n = b$, of $[a,b]$ and any sample point x_i^* in each subinterval $[x_{i-1}, x_i]$, we have

$$\int_a^b [kf(x) \pm hg(x)]\, dx$$

$$= \lim_{\|P\|\to 0} \sum_{i=1}^{n} [kf(x_i^*) \pm hg(x_i^*)] \Delta x_i$$

$$= \lim_{\|P\|\to 0} \sum_{i=1}^{n} kf(x_i^*) \Delta x_i \pm \lim_{\|P\|\to 0} \sum_{i=1}^{n} hg(x_i^*) \Delta x_i$$

$$= k \lim_{\|P\|\to 0} \sum_{i=1}^{n} f(x_i^*) \Delta x_i \pm h \lim_{\|P\|\to 0} \sum_{i=1}^{n} g(x_i^*) \Delta x_i$$

$$= k \int_a^b f(x)\, dx \pm h \int_a^b g(x)\, dx.$$

The above limits exist because both $f(x)$ and $g(x)$ are integrable.   □

Proof of Property 3. Since $f(x)$ is integrable on $[a,b]$, the definite integral $\int_a^b f(x)\, dx$ exists and does not depend on the partition of $[a,b]$. If $a < c < b$, we fix the point c and subdivide $[a,c]$ and $[c,b]$ in any way. Then

$$a = x_0 < x_1 < \cdots < x_m = c < x_{m+1} < \cdots < x_n = b$$

and the Riemann sum

$$\sum_{i=1}^{n} f(x_i^*) \Delta x_i = \sum_{i=1}^{m} f(x_i^*) \Delta x_i + \sum_{i=m+1}^{n} f(x_i^*) \Delta x_i$$

is one of infinitely many ways of subdividing the interval $[a,b]$. The limit of this Riemann sum, of course, exists as $\|P\| \to 0$. This gives

$$\int_a^b f(x)\, dx = \int_a^c f(x)\, dx + \int_c^b f(x)\, dx.$$

If c is not in (a,b), the additive property still holds. For example, if $a < b < c$, then

$$\int_a^c f(x)\, dx = \int_a^b f(x)\, dx + \int_b^c f(x)\, dx$$

$$= \int_a^b f(x)\, dx - \int_c^b f(x)\, dx,$$

so we still have

$$\int_a^b f(x)\, dx = \int_a^c f(x)\, dx + \int_c^b f(x)\, dx.$$ □

Proof of Property 4. For each point x_i^* in each subinterval of any partition P, we have $f(x_i^*) \geq 0$ and $\sum_{i=1}^{n} f(x_i^*) \Delta x_i \geq 0$, so

$$\int_a^b f(x)\, dx = \lim_{\|P\|\to 0} \sum_{i=1}^{n} f(x_i^*) \Delta x_i \geq 0.$$ □

Proof of Property 5. This follows from Property 4, since $f(x) - g(x) \geq 0$ means that

$$0 \leq \int_a^b f(x) - g(x)\, dx.$$

By the linearity property,

$$0 \leq \int_a^b f(x) - g(x)\, dx = \int_a^b f(x)\, dx - \int_a^b g(x)\, dx.$$

This implies

$$\int_a^b f(x)\, dx \geq \int_a^b g(x)\, dx. \qquad \square$$

Proof of Property 6. Since $m \leq f(x) \leq M$, by Property 5, we have

$$\int_a^b m\, dx \leq \int_a^b f(x)\, dx \leq \int_a^b M\, dx.$$

By Property 1, we have $\int_a^b m\, dx = m(b-a)$ and $\int_a^b M\, dx = M(b-a)$, so

$$m(b-a) \leq \int_a^b f(x)\, dx \leq M(b-a). \qquad \square$$

NOTE. This property says that, even when we do not know the exact value of the definite integral of $f(x)$ on $[a, b]$, we could estimate the value by bounding the function as we did in Example 5.1.5.

Proof of Property 7. Since $f(x)$ is continuous on $[a, b]$, it obtains its maximum value M and minimum value m on $[a, b]$. By Property 3, we have

$$m(b-a) \leq \int_a^b f(x)\, dx \leq M(b-a).$$

Dividing this inequality by $(b - a)$, we obtain

$$m \leq \frac{\int_a^b f(x)\, dx}{b - a} \leq M.$$

Hence, the number $\frac{\int_a^b f(x)\, dx}{b-a}$ is between the maximum value and the minimum value of $f(x)$ on $[a, b]$. Then, by the intermediate value theorem, we know there must be a number $c \in [a, b]$ such that

$$f(c) = \frac{\int_a^b f(x)\, dx}{b - a}. \qquad \square$$

5.1.3 Interpreting $\int_a^b f(x)\,dx$ in terms of area

As discussed previously, we know that, if $f(x) \geqslant 0$, the definite integral $\int_a^b f(x)\,dx$, if it exists, is the area under the curve $f(x)$ above the x-axis and between the two lines $x = a$ and $x = b$. But what happens if $f(x)$ is negative on $[a, b]$? By the definition of the definite integral, it is not hard to see that $f(x_i^*) < 0$ and in the Riemann sum, $f(x_i^*)\Delta x_i$ is equal to the length of the base times the negative height of the rectangle, so the value of the definite integral is the negative value of the area of the region that is below the x-axis, above the graph of $f(x)$, and between the lines $x = a$ and $x = b$.

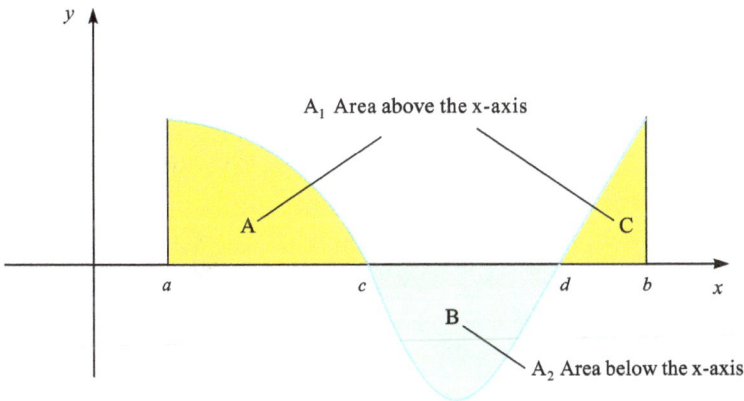

Figure 5.1.12: Geometric interpretation of the definite integral.

When f takes negative and positive values as in Figure 5.1.12, there is a more complicated interpretation of the definite integral in terms of areas. By the additive property of the definite integral,

$$\int_a^b f(x)\,dx = \int_a^c f(x)\,dx + \int_c^d f(x)\,dx + \int_d^b f(x)\,dx$$
$$= \text{Area } A - \text{Area } B + \text{Area } C$$
$$= (\text{area above the } x\text{-axis}) - (\text{area below the } x\text{-axis})$$

In another words, the integral $\int_a^b f(x)\,dx$ is the sum of the areas that lie above the x-axis and under the graph of $f(x)$ minus the sum of those areas that lie below the x-axis and the above the graph of $f(x)$. A definite integral can thus be interpreted as a net area, that is, a difference of areas. We have

$$\int_a^b f(x)\,dx = A_1 - A_2 = (\text{area above } x\text{-axis}) - (\text{area below the } x\text{-axis}),$$

where A_1 is the sum of the area of the regions above the x-axis and below the graph of f and A_2 is the sum of the area of the regions below the x-axis, above the graph of f, between the two lines $x = a$ and $x = b$.

Example 5.1.6. Evaluate the following definite integrals by interpreting each in terms of areas:

$$\text{(a)} \int_0^2 (x+1)\,dx, \quad \text{(b)} \int_0^2 (2x-1)\,dx, \quad \text{(c)} \int_0^\pi \cos x\,dx, \quad \text{(d)} \int_{-1}^1 \frac{|x|}{x}\,dx.$$

Solution. (a) Since $f(x) = x + 1 \geqslant 0$, we can interpret this integral as the area under the curve $y = x + 1$ from 0 to 2, as shown in Figure 5.1.13 (a). The graph of f for $0 \leqslant x \leqslant 2$ is a trapezoid. Hence, the area is

$$\int_0^2 (x+1)\,dx = \frac{3+1}{2} \cdot 2 = 4.$$

(b) The graph of $y = 2x - 1$ is the straight line shown in Figure 5.1.13 (b). We compute the definite integral as the difference between the areas of two triangles, namely Area C − Area B in Figure 5.1.13 (b). We have

$$\int_0^2 (2x-1)\,dx = \text{Area C} - \text{Area B} = \frac{1}{2}\left(\frac{3}{2} \times 3\right) - \frac{1}{2}\left(\frac{1}{2} \times 1\right) = 2.$$

(c) The graph of $\cos x$ for $0 \leqslant x \leqslant \pi$ is shown in Figure 5.1.13 (c). By symmetry, $\int_0^\pi \cos x\,dx = 0$.

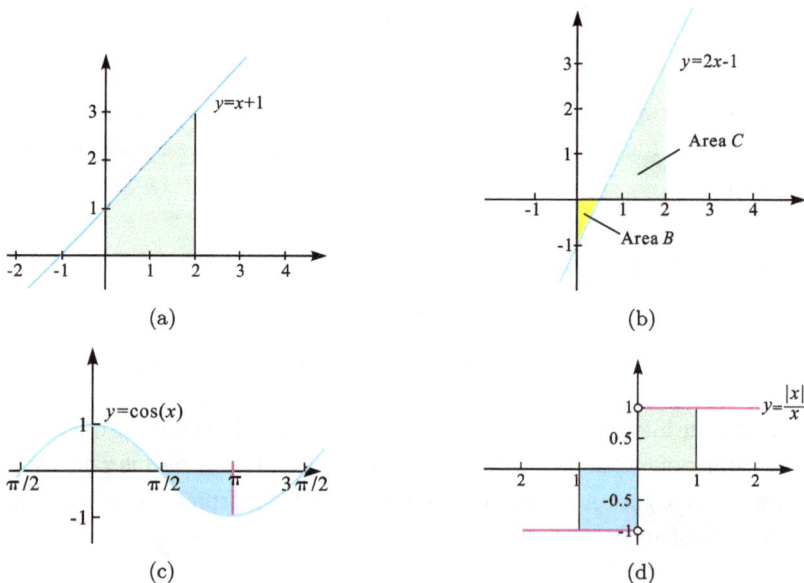

Figure 5.1.13: Graphs for Example 5.1.6.

(*d*) The function $\frac{|x|}{x}$ has a jump discontinuity at $x = 0$, but the graph determines two rectangles, one below the x-axis and one above. Using the idea of net area, we have

$$\int_{-1}^{1} \frac{|x|}{x}\, dx = 1 - 1 = 0.$$

In fact, for any integrable odd function $f(x)$, the definite integral $\int_{-a}^{a} f(x)\, dx = 0$.

5.1.4 Interpreting $\int_{a}^{b} v(t)\, dt$ as a distance or displacement

In a similar way, if $f(t) = v(t) \geqslant 0$ is the velocity of a particle moving along a straight line, then $\int_{a}^{b} v(t)\, dt$ is the distance traveled by the particle during time interval $[a,b]$. If $v(t) \leqslant 0$, the particle is moving in a negative direction and $\int_{a}^{b} v(t)\, dt$ is the negative of the distance traveled by the particle during this time interval. If $v(t)$ has both positive and negative values on $[a,b]$, this means that the particle moves sometimes to the right (positive direction) and sometimes to the left (negative direction). The definite integral $\int_{a}^{b} v(t)\, dt$ is the distance traveled by the particle in the positive direction minus the distance traveled in the negative direction. The total distance traveled is given by

$$\int_{a}^{b} |v(t)|\, dt.$$

Suppose a particle is moving along the x-axis and we let $x(t)$ represent its position at time t. Since $x'(t) = v(t)$, that is, the derivative of the position/displacement function is the velocity. During the time interval $[a,b]$, the particle moves along the x-axis from an initial position $x(a)$ to an ending position of $x(b)$. The change in a position of the particle, $x(b) - x(a)$, is exactly equal to the distance that the particle moves in a positive direction minus the distance that the particle moves in a negative direction. That is,

$$\int_{a}^{b} v(t)\, dt = \int_{a}^{b} \frac{dx(t)}{dt}\, dt = x(b) - x(a).$$

In this context, we interpret the definite integral $\int_{a}^{b} v(t)\, dt$ as $x(b) - x(a)$, the net change in a position, where x is an antiderivative of v. This result anticipates the FTC, given in the next section.

5.2 The fundamental theorem of calculus

As seen in the previous section, the FTC does confirm us that there exists a nice connection between definite integrals and antiderivatives. The first part of the FTC deals with a function $F(x)$ for $x \in [a,b]$ defined by an equation of the form

$$F(x) = \int_{a}^{x} f(t)\, dt,$$

where f is a continuous function on $[a,b]$. Observe that $F(x)$ depends only on x, the upper limit on the integral sign, and has no connection with the variable t since that is not present in the value of $\int_a^x f(t)\,dt$. If x is a fixed number, then the integral $\int_a^x f(t)\,dt$ is a definite number. If we let x vary, then the value of the integral $\int_a^x f(t)\,dt$ also varies and defines a function of x denoted by $F(x)$. If $f(t)$ happens to be a positive function, then $F(x)$ can be interpreted as the area under the graph of f from a to x, where x can vary from a to b, as in Figure 5.2.1 (a), so it is sometimes called the area function. If $f(x) \not\geqslant 0$ for all $x \in [a,b]$, then $F(x)$ requires a more complicated interpretation in terms of areas, as illustrated in Figure 5.2.1 (b).

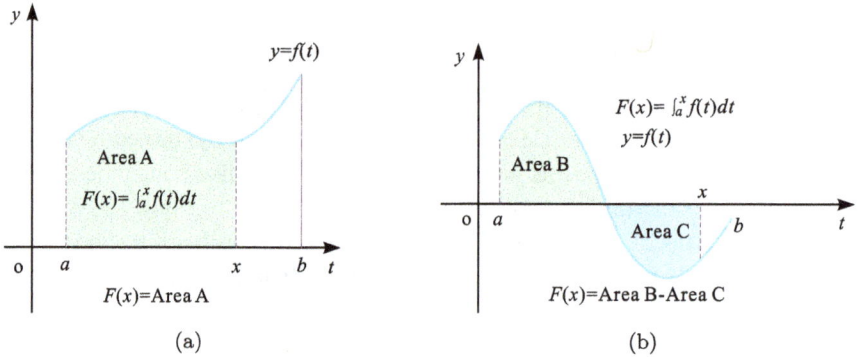

Figure 5.2.1: The area function.

Theorem 5.2.1 (Fundamental theorem of calculus, Part I). *Suppose that f is continuous on $[a,b]$. Define $F : [a,b] \to R$ by*

$$F(x) = \int_a^x f(t)\,dt, \quad a \leqslant x \leqslant b.$$

Then F is continuous, differentiable on (a,b) and $F'(x) = f(x)$ for $x \in (a,b)$. That is, F is an antiderivative of f.

Proof. By definition, the derivative of F at x is

$$F'(x) = \lim_{h \to 0} \frac{F(x+h) - F(x)}{h}.$$

Notice that

$$F(x+h) - F(x) = \int_a^{x+h} f(t)\,dt - \int_a^x f(t)\,dt$$

$$= \int_a^x f(t)\,dt + \int_x^{x+h} f(t)\,dt - \int_a^x f(t)\,dt$$

$$= \int_x^{x+h} f(t)\, dt.$$

Since f is continuous on $[x, x+h]$, or on $[x+h, x]$ if $h < 0$, the extreme value theorem tells us that there are numbers c and d between x and $x + h$ such that $f(c) = m$ and $f(d) = M$, where m and M are the absolute minimum and maximum values of f on the closed interval between x and $x + h$. Assuming that $h > 0$ (a similar proof can be given for the theorem when $h < 0$), the properties of definite integrals from Section 5.1.2 establish that

$$m(x + h - x) \leqslant \int_x^{x+h} f(t)\, dt \leqslant M(x + h - x).$$

Simplifying these inequalities gives

$$m \cdot h \leqslant \int_x^{x+h} f(t)\, dt \leqslant M \cdot h,$$

$$m \leqslant \frac{1}{h} \int_x^{x+h} f(t)\, dt \leqslant M,$$

$$f(c) \leqslant \frac{1}{h}[F(x+h) - F(x)] \leqslant f(d).$$

Then

$$\lim_{h \to 0} f(c) \leqslant \lim_{h \to 0} \frac{F(x+h) - F(x)}{h} \leqslant \lim_{h \to 0} f(d)$$

and

$$\lim_{h \to 0} f(c) \leqslant F'(x) \leqslant \lim_{h \to 0} f(d).$$

Since f is continuous and $c, d \in [x, x+h]$, it follows that, when $h \to 0$, both $c, d \to x$ and both $f(c), f(d) \to f(x)$. Hence, by the squeeze theorem, we have

$$f(x) \leqslant F'(x) \leqslant f(x).$$

Thus $F'(x) = f(x)$, or

$$\frac{d}{dx} \int_a^x f(t)\, dt = f(x). \qquad \square$$

NOTES. 1. The theorem tells us that, for any continuous function f on $[a, b]$, we can define a function F that is an antiderivative of f on (a, b). In particular, it says that every continuous function has an antiderivative defined by the equation

$$F(x) = \int_a^x f(t)\, dt, \quad a \leqslant x \leqslant b.$$

2. In the formula $\int_a^x f(t)\,dt$, there are two variables, namely x and t. The variable t is the dummy variable of integration and this means we can replace it by any other letter we like, without changing the meaning or value of the integral. Well, almost any other variable. There is one variable that is not allowed, namely x. The reason is simply that x has been already used as the independent variable for the function F, so we cannot use it to replace t.

3. The theorem can also be written as

$$\frac{d}{dx}\int_a^x f(t)\,dt = f(x).$$

4. Functions defined by definite integrals often appear strange at first sight. However, there are many examples in mathematics and physics of functions defined by integrals. For example, in the theory of optics we find the Fresnel function

$$S(x) = \int_0^x \sin\left(\frac{\pi t^2}{2}\right) dt.$$

The graph of $S(x)$ is shown in Figure 5.2.2.

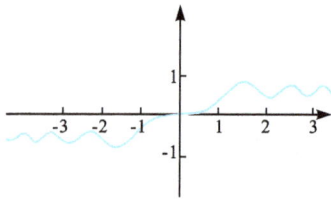

Figure 5.2.2: Graph of the Fresnel function.

In mathematics we have the gamma function

$$\Gamma(x) = \int_0^\infty t^{x-1} e^{-t}\,dt, \quad x > 0.$$

Both are defined by integrals. Figure 5.2.3 shows the graph of $\Gamma(x)$.

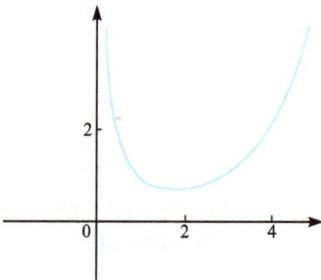

Figure 5.2.3: Graph of the Gamma function.

Example 5.2.1. Find the derivative of the function

$$F(x) = \int_2^x \sin(t^3 + 1)\, dt.$$

Solution. Since the integrand function

$$f(t) = \sin(t^3 + 1)$$

is continuous everywhere, it follows from the FTC that the function $F(x)$ is differentiable everywhere and

$$F'(x) = f(x) = \sin(x^3 + 1).$$

Example 5.2.2. Find the derivative of the function

$$G(x) = \int_0^x \sqrt{t^4 + t^2 + 1}\, dt.$$

Solution. Since the integrand function

$$g(t) = \sqrt{t^4 + t^2 + 1}$$

is continuous everywhere, it follows from the FTC that the function $G(x)$ is differentiable everywhere and

$$G'(x) = g(x) = \sqrt{x^4 + x^2 + 1}.$$

Example 5.2.3. Find $\frac{d}{dx} \int_a^{x^2} \sin t^2\, dt$.

Solution. The function $\sin t^2$ is continuous everywhere, so we apply the FTC. However, we have to be careful to use the chain rule in conjunction with the FTC. Let $u = x^2$. Then

$$\begin{aligned}
\frac{d}{dx} \int_a^{x^2} \sin t^2\, dt &= \frac{d}{dx} \int_a^u \sin t^2\, dt \\
&= \frac{d}{du}\left(\int_a^u \sin t^2\, dt \right) \frac{du}{dx} \\
&= \sin(u^2) \cdot \frac{du}{dx} \\
&= \sin(x^4) \cdot 2x.
\end{aligned}$$

Example 5.2.4. Find $\frac{d}{dx} \int_{-x^2}^{2x} e^{-t^2}\, dt$.

Solution. In addition to the chain rule, we need the additive property of the definite integral, because the lower limit of integration is not a constant in this case. Let $u = -x^2$ and $v = 2x$. Then

$$\frac{d}{dx} \int_{-x^2}^{2x} e^{-t^2}\, dt = \frac{d}{dx} \left(\int_{-x^2}^{0} e^{-t^2}\, dt + \int_{0}^{2x} e^{-t^2}\, dt \right)$$

$$= \frac{d}{dx} \left(\int_{u}^{0} e^{-t^2}\, dt + \int_{0}^{v} e^{-t^2}\, dt \right)$$

$$= \frac{d}{dx} \left(-\int_{0}^{u} e^{-t^2}\, dt + \int_{0}^{v} e^{-t^2}\, dt \right)$$

$$= -\frac{d}{dx} \int_{0}^{u} e^{-t^2}\, dt + \frac{d}{dx} \int_{0}^{v} e^{-t^2}\, dt$$

$$= -\frac{d}{du} \int_{0}^{u} e^{-t^2}\, dt \cdot \frac{du}{dx} + \frac{d}{dv} \int_{0}^{v} e^{-t^2}\, dt \cdot \frac{dv}{dx}$$

$$= -e^{-u^2} \frac{du}{dx} + e^{-v^2} \frac{dv}{dx}$$

$$= -e^{-x^4}(-2x) + e^{-(2x)^2} \cdot 2$$

$$= 2e^{-4x^2} + 2xe^{-x^4}.$$

NOTE. In general, if f is integrable and ϕ and ψ are differentiable, then

$$\frac{d}{dx} \int_{\phi(x)}^{\psi(x)} f(t)\, dt = f(\psi(x))\psi'(x) - f(\phi(x))\phi'(x).$$

The second part of the FTC shows that, if we know an antiderivative F of a function f, the value of a definite integral of f on an interval can be computed simply by evaluating the antiderivative F at the two endpoints of the interval. This is also called the *Newton–Leibnitz theorem*.

Theorem 5.2.2 (Fundamental theorem of calculus, Part II). *Suppose that $f(x)$ is continuous on $[a,b]$ and $F(x)$ is any antiderivative of $f(x)$ on $[a,b]$. Then*

$$\int_{a}^{b} f(x)\, dx = F(b) - F(a).$$

Proof. Since $\int_{a}^{x} f(t)\, dt$ is an antiderivative of $f(x)$ by the FTC, Part I and since $F(x)$ is also an antiderivative of $f(x)$, there must be a constant C such that $F(x) = \int_{a}^{x} f(t)\, dt + C$ (see Theorem 4.2.4, Chapter 4). Therefore,

$$F(a) = \int_{a}^{a} f(t)\, dt + C = C \quad \text{and}$$

$$F(b) = \int_{a}^{b} f(t)\, dt + C = \int_{a}^{b} f(x)\, dx + F(a).$$

Then

$$\int_{a}^{b} f(x)\, dx = F(b) - F(a). \tag{5.1}$$

$\square$

NOTES. 1. Sometimes we write $F(b) - F(a)$ as $F(x)|_{x=a}^{x=b}$ for the sake of convenience.

2. We know that $F'(x)$ represents the rate of change of $y = F(x)$ with respect to x and $F(b) - F(a)$ is the net change in y when x changes from a to b. Although y could, for instance, increase, decrease, and then increase again, $F(b) - F(a)$ represents the *net change* in y over $[a, b]$, so we reformulate the above theorem as follows. The definite integral of the rate of change of a function $F(x)$ is equal to its net change over the interval $[a, b]$, or, in symbols,

$$\int_a^b F'(x)\, dx = F(b) - F(a) = F(x)|_a^b.$$

As seen in Theorem 4.2.4, to find an antiderivative of a basic function $f(x)$, we could use the basic derivative formulas. For example, since

$$(-\cos x)' = \sin x, \quad (e^x)' = e^x, \quad \left(\frac{x^{n+1}}{n+1}\right)' = x^n,$$

$-\cos x$, e^x, and $\frac{x^{n+1}}{n+1}$ are an antiderivative of $\sin x$, e^x, and x^n, respectively.

Example 5.2.5. Find the area of the region bounded by $y = x^2$, $y = 0$, $x = 0$, and $x = 2$.

Solution. The area of the region is the definite integral $\int_0^2 x^2\, dx$. Since $\frac{1}{3}x^3$ is an antiderivative of x^2, the FTC, Part II gives

$$\int_a^b x^2\, dx = \frac{1}{3}x^3\Big|_{x=0}^{x=2}$$
$$= \frac{1}{3}(2^3 - 0^3) = \frac{8}{3}.$$

Figure 5.2.4 shows the region.

Example 5.2.6. Evaluate the integral $\int_1^3 e^x\, dx$.

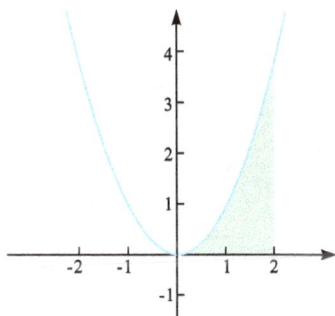

Figure 5.2.4: Region bounded by $y = x^2$, $y = 0$, $x = 0$, and $x = 2$.

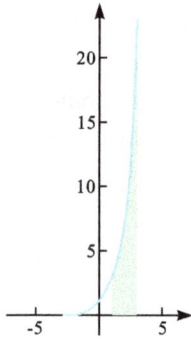

Figure 5.2.5: Region bounded by $y = e^x$, $y = 0$, $x = 1$, and $x = 3$.

Solution. The function $f(x) = e^x$ is continuous everywhere and we know that an antiderivative of f is $F(x) = e^x$. Hence, the FTC, Part II gives

$$\int_1^3 e^x \, dx = F(3) - F(1) = e^3 - e.$$

Figure 5.2.5 shows the region.

NOTE. The FTC, Part II says we can use any antiderivative F of f, so we may as well use the simplest one, namely $F(x) = e^x$ (if we use one with a nonzero constant C, $e^x + C$, the constant C would cancel out in the calculation of $F(3) - F(1)$).

Example 5.2.7. Find the area of the region bounded by $y = \sin x$, $y = 0$, $x = 0$, and $x = \pi$.

Solution. An antiderivative of $y = \sin x$ is $-\cos x$, since $(-\cos x)' = \sin x$ and $\sin x \geqslant 0$ on $[0, \pi]$. The area is given by the definite integral

$$\int_0^\pi \sin x \, dx = -\cos x|_0^\pi$$
$$= -(\cos \pi - \cos 0)$$
$$= -(-1 - 1)$$
$$= 2.$$

Figure 5.2.6 shows this region.

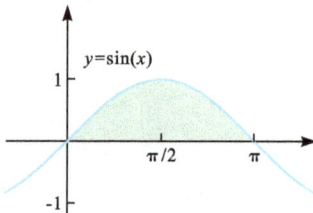

Figure 5.2.6: Region bounded by $y = \sin x$, $y = 0$, $x = 0$, and $x = \pi$.

Example 5.2.8. Find the area between the two curves $y = \sqrt{x}$ and $y = x^2$.

Solution. The intersections of the two curves are $(0,0)$ and $(1,1)$, as seen in Figure 5.2.7. The curve $y = \sqrt{x}$ is above the curve $y = x^2$, between $x = 0$ and $x = 1$. The area A between the curves is given by

$$A = \int_0^1 \sqrt{x}\, dx - \int_0^1 x^2\, dx$$

$$= \left(\frac{2}{3}x^{\frac{3}{2}}\right)\Big|_0^1 - \left(\frac{1}{3}x^3\right)\Big|_0^1$$

$$= \frac{2}{3} - \frac{1}{3} = \frac{1}{3}.$$

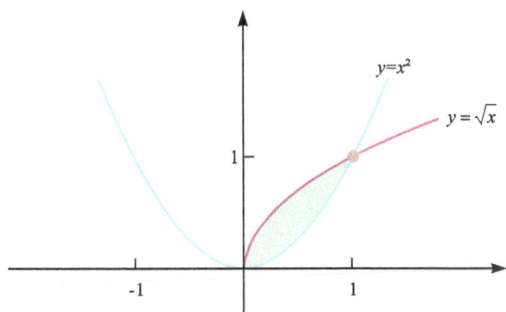

Figure 5.2.7: Region bounded by $y = x^2$ and $y = \sqrt{x}$.

5.3 Numerical integration

The integrals used to calculate the length of the orbit of Mars or the angular position of a simple pendulum are among the many integrals that cannot be evaluated by substitution, integration by parts, or indeed any known integration technique. Some other examples of integrals that cannot be evaluated exactly include

$$\int_a^b e^{-x^2}\, dx, \quad \int_a^b \sin x^2\, dx, \quad \text{and} \quad \int_1^5 \frac{e^x}{x}\, dx.$$

In each of these cases, the integrands are continuous functions over the stated intervals, so we know the definite integrals must exist and have some finite value. However, there is no closed-form antiderivative for any of these functions, so we cannot compute the definite integrals exactly. We therefore use numerical methods to approximate the values of such definite integrals. Two such numerical methods are the trapezoidal rule and Simpson's rule.

5.3.1 Trapezoidal rule

The trapezoidal rule approximates the definite integral $\int_a^b f(x)\,dx$ by partitioning $[a,b]$ into n equal subintervals of length $h = \frac{(b-a)}{n}$ by the points $a = x_0 < x_1 < \cdots < x_n = b$ (so $x_1 = x_0 + h$, $x_2 = x_1 + h$, and in general $x_k = x_0 + kh$).

The approximation given by this rule is the definite integral of a new function that is a straight line between any two consecutive points of the partition, $(x_{i-1}, f(x_{i-1}))$ and $(x_i, f(x_i))$, as seen in Figure 5.3.1. The value of this approximation is given in the next theorem.

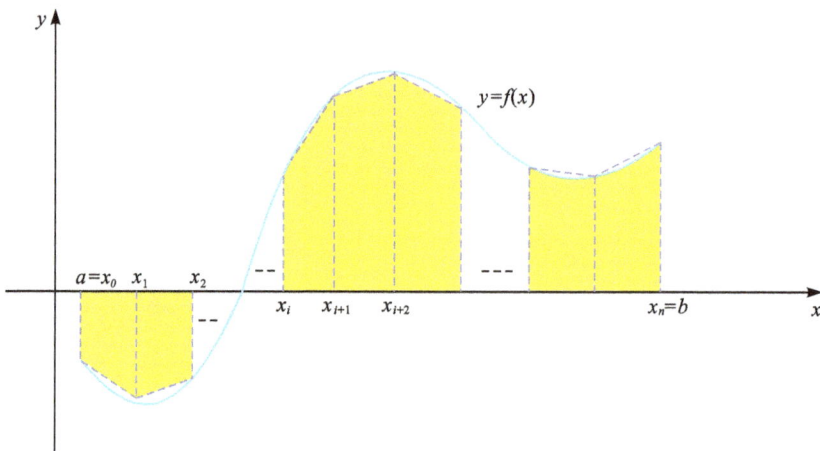

Figure 5.3.1: Trapezoidal approximation.

Theorem 5.3.1 (The trapezoidal rule is also called the trapezium rule). *Let* $a = x_0 < x_1 < \cdots < x_n = b$ *be a partition of* $[a,b]$ *into n equal subintervals of length* $h = \frac{(b-a)}{n}$ *and let*

$$y_0 = f(x_0), \quad y_1 = f(x_1), \quad \ldots, \quad y_n = f(x_n).$$

If f is continuous on $[a,b]$, then an approximate value of the integral $\int_a^b f(x)$ is given by

$$\int_a^b f(x) \approx T_n = \frac{1}{2}h(y_0 + 2(y_1 + y_2 + \cdots + y_{n-1}) + y_n). \tag{5.2}$$

The error $|\int_a^b f(x) - T_n| \to 0$ *as* $n \to \infty$.

Example 5.3.1. Compute the trapezoidal rule approximation, T_6, with $n = 6$ subdivisions for

$$\int_1^4 \sqrt{x}\,dx.$$

Use the known exact value of this integral, $\int_1^4 \sqrt{x}\,dx = \frac{2}{3}x^{\frac{3}{2}}|_{x=1}^{x=4} = \frac{14}{3}$, to calculate an upper bound on the error $|14/3 - T_6|$.

Solution. For $n = 6$, we have $h = \frac{(4-1)}{6} = \frac{1}{2}$. The subdivision of $[1, 4]$ is

$$x_0 = 1, \quad x_1 = \frac{3}{2}, \quad x_2 = 2, \quad x_3 = \frac{5}{2}, \quad x_4 = 3, \quad x_5 = \frac{7}{2}, \quad x_6 = 4, \quad \text{so}$$

$$y_0 = 1, \quad y_1 = \sqrt{\frac{3}{2}}, \quad y_2 = \sqrt{2}, \quad y_3 = \sqrt{\frac{5}{2}}, \quad y_4 = \sqrt{3}, \quad y_5 = \sqrt{\frac{7}{2}}, \quad y_6 = 2.$$

By the trapezoidal rule,

$$\int_1^4 \sqrt{x}\,dx \approx R_6 = \frac{1}{2}h(y_0 + 2(y_1 + y_2 + y_3 + y_4 + y_5) + y_6)$$

$$= \frac{1}{2}\cdot\frac{1}{2}\left(1 + 2\left(\sqrt{\frac{3}{2}} + \sqrt{2} + \sqrt{\frac{5}{2}} + \sqrt{3} + \sqrt{\frac{7}{2}}\right) + 2\right)$$

$$\approx 4.661\,488.$$

The difference between the exact value and the approximation R_6 is $|\frac{14}{3} - 4.614\,88| < 0.0052$.

5.3.2 Simpson's rule

If $[a, b]$ is partitioned into an even number of n equal subintervals of length $h = \frac{b-a}{n}$ by the points $a = x_0 < x_1 < \cdots < x_n = b$, then Simpson's rule for approximating $\int_a^b f(x)\,dx$ is based on approximating f on two consecutive intervals $[x_{i-1}, x_i]$ and $[x_i, x_{i+1}]$ by a quadratic function of the form $y = Ax^2 + Bx + C$, whose graph is a parabola. The values of A, B, and C are determined by requiring the quadratic to go through the corresponding three points of the graph, $(x_{i-1}, f(x_{i-1}))$, $(x_i, f(x_i))$, and $(x_{i+1}, f(x_{i+1}))$. Most functions can be more closely fit by parabolas than straight lines, so we expect Simpson's rule to be more accurate than the trapezoidal rule. The details of this approximation are given in Theorem 5.3.2.

Theorem 5.3.2 (Simpson's rule). *For an even number n, let $a = x_0 < x_1 < \cdots < x_n = b$ be a subdivision of $[a, b]$ into equal subintervals with width $h = \frac{b-a}{n}$ and let*

$$y_0 = f(x_0), y_1 = f(x_1), \ldots, y_n = f(x_n).$$

If f is continuous on $[a, b]$, then an approximate value of the integral $\int_a^b f(x)$ is given by

$$\int_a^b f(x)\,dx \approx S_n$$

$$= \frac{1}{3}h(y_0 + 4(y_1 + y_3 + \cdots + y_{n-1}) + 2(y_2 + y_4 + \cdots + y_{n-2}) + y_n).$$

The difference $|\int_a^b f(x) - S_n| \to 0$ as $n \to \infty$.

Example 5.3.2. Use Simpson's rule with $n = 6$ to approximate the integral

$$\int_0^{\sqrt{2\pi}} \sin x^2 \, dx.$$

Solution. Note that $n = 6$ is even, as required, and $h = (\sqrt{2\pi} - 0)/6$. The subdivision of $[0, \sqrt{2\pi}]$ is

$$x_0 = 0, \quad x_1 = \frac{\sqrt{2\pi}}{6}, \quad x_2 = \frac{2\sqrt{2\pi}}{6}, \quad x_3 = \frac{3\sqrt{2\pi}}{6},$$

$$x_4 = \frac{4\sqrt{2\pi}}{6}, \quad x_5 = \frac{5\sqrt{2\pi}}{6}, \quad x_6 = \frac{6\sqrt{2\pi}}{6}.$$

Thus,

$$S_6 = \frac{1}{3}h[y_0 + 4(y_1 + y_3 + y_5) + 2(y_2 + y_4) + y_6]$$

$$= \frac{\sqrt{2\pi}}{18} \left\{ \begin{array}{l} \sin 0^2 + 4\left(\sin\left(\frac{\sqrt{2\pi}}{6}\right)^2 + \sin\left(\frac{3\sqrt{2\pi}}{6}\right)^2 + \sin\left(\frac{5\sqrt{2\pi}}{6}\right)^2\right) \\ + 2\left(\sin\left(\frac{2\sqrt{2\pi}}{6}\right)^2 + \sin\left(\frac{4\sqrt{2\pi}}{6}\right)^2\right) + \sin\left(\frac{6\sqrt{2\pi}}{6}\right)^2 \end{array} \right\}$$

$$\approx 0.404\,6.$$

NOTE. A more exact approximation to the integral is $\int_0^{\sqrt{2\pi}} \sin x^2 \, dx \approx 0.430\,41$, so this particular Simpson's rule approximation is not very accurate. The reason for this inaccuracy is that the curve makes a tight turn between $x = 1.65$ and $x = 2.5$ that is not approximated well by parabolas, as seen in Figure 5.3.2 (the curve $y = \sin(x^2)$ is the dashed line and the approximating curves are solid lines). The remedy for this is to significantly increase n, the number of subdivisions of $[0, \sqrt{2\pi}]$.

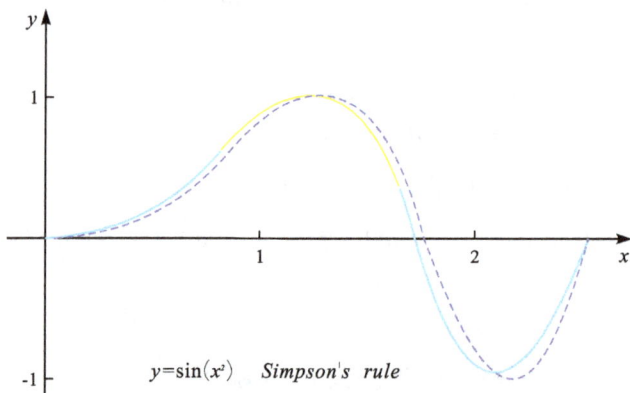

Figure 5.3.2: Simpson's rule to approximate $\int_0^{\sqrt{2\pi}} \sin x^2 \, dx$.

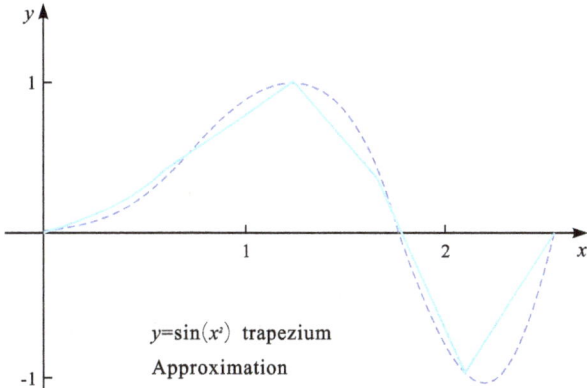

Figure 5.3.3: Trapezoidal approximation to $\int_0^{\sqrt{2\pi}} \sin(x^2)\, dx$.

NOTE. The trapezoidal rule gives a much worse approximation of 0.509 16. Figure 5.3.3 shows why this happens. The straight line approximations of the trapezoidal rule are very poor approximations to the curve for $x \geqslant 1$.

5.4 Exercises

1. Express each of the following limits of a Riemann sum as a definite integral, assuming each c_k is chosen from the kth subinterval of a regular partition of the indicated interval into n subintervals of equal length Δx:

   (a) $\lim_{\Delta x \to 0} \sum_{k=1}^{n} (\cos c_k) \Delta x$, $[0,3]$;   (b) $\lim_{\Delta x \to 0} \sum_{k=1}^{n} \frac{e^{2c_k}}{c_k^2} \Delta x$, $[1,4]$;

   (c) $\lim_{\Delta x \to 0} \sum_{k=1}^{n} 4 \frac{\ln(1+x_k^2)}{x_k} \Delta x$, $[2,6]$;   (d) $\lim_{n \to \infty} \sum_{k=1}^{n} (\sin^2 c_k - 2c_k) \frac{b-a}{n}$, $[a,b]$;

   (e) $\lim_{\Delta x \to 0} \sum_{k=1}^{n} \frac{x_k}{x_k^2+4} \Delta x$, $[1,3]$.

2. Let P be a partition of $[2,7]$. Express $\lim_{\|P\| \to 0} \sum_{i=1}^{n} (5(x_i^*)^3 - 4x_i^*) \Delta x_i$ as a definite integral.

3. Express the following limits as definite integrals:

   (a) $\lim_{n \to \infty} (\frac{\sqrt{1}}{n^{3/2}} + \frac{\sqrt{2}}{n^{3/2}} + \cdots + \frac{\sqrt{n}}{n^{3/2}})$;   (b) $\lim_{n \to \infty} \sum_{i=1}^{n} \frac{e^{\frac{i}{n}}}{n}$;

   (c) $\lim_{n \to \infty} \frac{1^p + 2^p + \cdots + n^p}{n^{p+1}}$, $(p > 0)$.

4. (*Group activity) Give an argument based on Riemann sums to explain why the following functions are or are not integrable on $[-1,1]$:

   (a) $f(x) = \begin{cases} \frac{1}{x^2}, & x \neq 0 \\ 0, & x = 0; \end{cases}$ (b) $f(x) = \begin{cases} \sin \frac{1}{x}, & x \neq 0 \\ 0, & x = 0; \end{cases}$

   (c) $g(x) = \begin{cases} x^2, & x \geqslant 0 \\ 1, & x < 0. \end{cases}$

5. Calculate the LRS, RRS, and MRS for the following functions using four subintervals of equal width:

   (a) $f(x) = \cos x$, $[0,\pi]$;   (b) $f(x) = \frac{1}{x}$, $[1,9]$;   (c) $f(x) = 2x - x^2$, $[-1,3]$.

6. Assume $\int_a^b f(x)\,dx = 4$ and $\int_a^b g(x)\,dx = -2$. Find:

   (a) $\int_a^b 5f(x)\,dx;$ (b) $\int_a^a 4f(u) - 7g^2(u)\,du;$

   (c) $\int_b^a f(t) - \frac{g(t)}{3}\,dt;$ (d) $\int_a^b 3f(s) - \frac{g(s)}{\sqrt{2}}\,ds.$

7. Use the properties of definite integrals to verify the following inequalities without evaluating the integrals:

   (a) $\int_0^4 (x^2 - 4x + 4)\,dx \geq 0;$ (b) $\int_0^1 \sqrt{1 + x^2}\,dx \leq \int_0^1 \sqrt{1 + x}\,dx;$

   (c) $2 \leq \int_{-1}^1 \sqrt{1 + x^2}\,dx \leq 2\sqrt{2}.$

8. If $f(x)$ is differentiable everywhere, $f(3) = 3$, and $\int_0^1 f(x)\,dx = 3$, show that there is a number $\xi \in (0, 3)$ such that $f'(\xi) = 0$.

9. (a) Assume $f(x)$ is continuous and differentiable on $[a, b]$ and $\frac{1}{b-a}\int_a^b f(x)\,dx = f(b)$. Prove that there is at least one point $\varepsilon \in (a, b)$ such that $f'(\varepsilon) = 0$.

   (b) Give a counterexample for a function $f(x)$ that is not differentiable.

10. Evaluate each of the following limits:

    (a) $\lim_{n \to \infty} \int_n^{n+p} \frac{\sin x}{x}\,dx,\ p > 0;$

    (b) $\lim_{h \to 0} \frac{\int_x^{x+h} f(t)\,dt}{h}$, where f is continuous;

    (c) $\lim_{n \to \infty} \int_0^{\frac{\pi}{4}} \sin^n x\,dx;$

    (d) $\lim_{x \to +\infty} \int_x^{x+2} t \sin \frac{2}{t} f(t)\,dt$, where $f(x)$ is differentiable everywhere and $\lim_{x \to +\infty} f(x) = 1$.

11. If $f(x)$ is integrable on $[a, b]$, prove that $|\int_a^b f(x)\,dx| \leq \int_a^b |f(x)|\,dx$.

12. (**Extended mean value theorem of integrals**) If both $f(x)$ and $g(x)$ are continuous on $[a, b]$ and $g(x)$ does not change sign over $[a, b]$, show that there is number $\xi \in [a, b]$ such that

$$\int_a^b f(x)g(x)\,dx = f(\xi) \int_a^b g(x)\,dx.$$

13. The mean value of n numbers $f(x_1), f(x_2), \dots, f(x_n)$ is defined by

$$\frac{f(x_1) + f(x_2) + \cdots + f(x_n)}{n}.$$

If $f(x)$ is continuous on $[a, b]$ and $x_1, x_2, \dots, x_n$ are n distinct numbers chosen from each subinterval of a partition of $[a, b]$ with n subintervals of equal width, show that

$$\lim_{n \to \infty} \frac{f(x_1) + f(x_2) + \cdots + f(x_n)}{n} = \frac{\int_a^b f(x)\,dx}{b - a}.$$

14. Evaluate the following definite integrals by interpreting them as areas:

    (a) $\int_{-1}^1 (2 - |x|)\,dx;$ (b) $\int_2^4 (\frac{t}{2} + 3);$ (c) $\int_{-1}^3 [u]\,du;$

    (d) $\int_0^{2\pi} (\cos 2\theta)\,d\theta;$ (e) $\int_2^{-1} 3\,dx;$ (f) $\int_{-1}^2 |2x - 3|\,dx;$

    (g) $\int_{-1}^1 \sqrt{1 - x^2}\,dx;$ (h) $\int_{-5}^5 (x - \sqrt{25 - x^2})\,dx.$

15. The graph of $f(t)$ consisting of line segments and a half circle is given below.

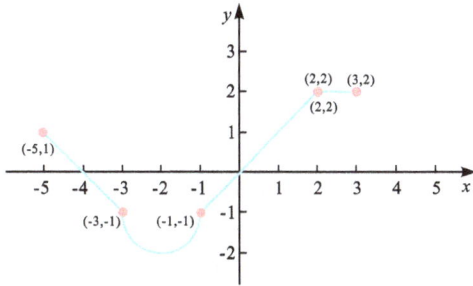

Question 15

Assume $H(x) = \int_1^x f(t)\,dt$. Then:
(a) find $H(-2)$, $H(0)$, $H(1)$, $H(3)$, and $H'(-3)$;
(b) find the extreme values of $H(x)$ on $[-5,3]$;
(c) find any inflection points of $H(x)$.

16. Find the derivative of each of the following functions:
   (a) $f(x) = \int_{-1}^x \sqrt{1+t^2}\,dt$; (b) $g(t) = \int_2^{t^2} (\sin u^2)\,du$;
   (c) $g(y) = \int_{\sqrt{y}}^3 (s^2 + \ln s)\,ds$; (d) $k(u) = \int_{\sin u}^{\cos u} e^{-\pi t^2}\,dt$;
   (e) $f(x) = \int_0^x (x-1)\sqrt{1+t^4}\,dt$.

17. Find each of the following limits:
   (a) $\lim_{x\to 0} \frac{\int_0^x \cos t^2\,dt}{x}$; (b) $\lim_{x\to 0} \frac{x^2}{\int_{\cos x}^1 e^{-t^2}\,dt}$; (c) $\lim_{n\to\infty} \sum_{i=0}^n \frac{i^3}{n^4}$;
   (d) $\lim_{n\to\infty} \frac{1}{n}\sum_{i=0}^n \cos(\frac{i}{n})$; (e) $\lim_{n\to\infty} \int_0^1 \frac{x^n}{1+x}\,dx$.

18. If $f(x)$ is continuous on $[0,1]$, $0 < f'(x) < 1$ for all $x \in (0,1)$, and $f(0) = 0$, then show that $(\int_0^1 f(x)\,dx)^2 > \int_0^1 f^3(x)\,dx$. [Hint: consider $F(t) = (\int_0^t f(x)\,dx)^2 - \int_0^t f^3(x)\,dx$.]

19. If $f(x)$ is continuous on $[0,1]$ and $a = \int_0^1 f(x)\,dx$, then, by considering $\int_0^1 (f(x) - a)^2\,dx$ or by any other method, show that $\int_0^1 f^2(x)\,dx \geq (\int_0^1 f(x)\,dx)^2$. More generally, show that

$$\left(\int_a^b f(x)g(x)\,dx\right)^2 \leq \int_a^b f^2(x)\,dx \int_a^b g^2(x)\,dx$$

for any integrable functions $f(x)$ and $g(x)$ on $[a,b]$.

20. Use the FTC, Part II to evaluate each of the following definite integrals:
   (a) $\int_0^\pi \sin x\,dx$; (b) $\int_{-1}^1 \frac{1}{1+x^2}\,dx$; (c) $\int_1^e \frac{1}{x}\,dx$;
   (d) $\int_1^2 (x^3 - \sqrt{x})\,dx$; (e) $\int_0^{\pi/4} \sec^2 x\,dx$; (f) $\int_3^9 8\,dx$;
   (g) $\int_0^{1/2} \frac{1}{\sqrt{1-x^2}}\,dx$; (h) $\int_0^1 e^x\,dx$; (i) $\int_0^\pi (e^x + 2\cos x)\,dx$;
   (j) $\int_1^2 \frac{(x+1)^2}{x}\,dx$; (k) $\int_1^x t^2 - \frac{1}{t}\,dt$; (l) $\int_a^x (\cos\theta - 2\sin\theta)\,d\theta$.

21. Find a function f and a positive number a such that $6 + \int_a^x \frac{f(t)}{t^2}\,dt = 2\sqrt{x}$.

22. Assume the function f is differentiable on $[0,1]$ and $f(0) = 0$.
    (a) Show that $8 \int_0^{\frac{1}{2}} f(x)\,dx \leq \max_{0 \leq x \leq 0.5} |f'(x)|$.
    (b) If $f(1) = f(0) = 0$, prove that $|\int_0^1 f(x)\,dx| \leq \frac{1}{4} \max_{0 \leq x \leq 1} |f'(x)|$.

23. **The Gauss error function** is a special nonelementary function that occurs in probability and statistics. It is defined as $\mathrm{erf}(x) = \frac{2}{\sqrt{\pi}} \int_0^x e^{-t^2}\,dt$. The error function also occurs in the solutions of the heat equation when boundary conditions are given by the Heaviside step function.
    (a) Show that $\int_a^b e^{-t^2}\,dt = \frac{\sqrt{\pi}}{2}(\mathrm{erf}(b) - \mathrm{erf}(a))$.
    (b) Show that the function $y = e^{x^2}\,\mathrm{erf}(x)$ satisfies the equation $y' - 2xy = \frac{2}{\sqrt{\pi}}$.

24. If $\int_1^3 f(x)\,dx + x^2 = f(x)$, find $\int_1^3 f(x)\,dx$.

25. Find the average value of the following functions on the indicated interval:
    (a) $y = \cos x$, $[0, \pi]$;   (b) $y = 3x^2 + 2e^x$, $[0,3]$;
    (c) $y = \sec x \tan x$, $[0, \frac{\pi}{3}]$;   (d) $y = \frac{1}{x}$, $[e, e^2]$.

26. Are the following statements true or false? Explain.
    (a) $\int_{-1}^3 \frac{1}{x^2}\,dx = \frac{x^{-1}}{-1}\big|_{-1}^3 = -\frac{1}{3} - 1 = -\frac{4}{3}$;
    (b) $\int_{\frac{\pi}{3}}^{\pi} \sec\theta \tan\theta\,d\theta = \sec\theta\big|_{\frac{\pi}{3}}^{\pi} = \frac{1}{\cos\pi} - \frac{1}{\cos\frac{\pi}{3}} = -3$.

27. Find the area of the shaded regions.

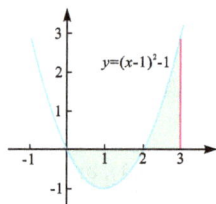

(a)

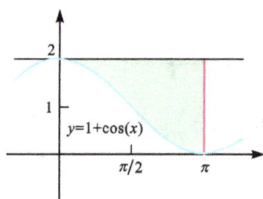

(b)

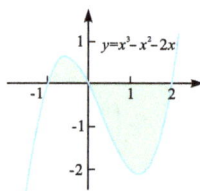

(c)

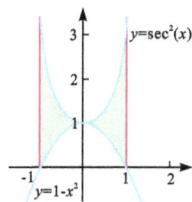

(d)

28. A parabola $y = ax^2 + bx + c$ passes through the three points (x_1, y_1), (x_2, y_2), and (x_3, y_3).
    (a) Find, in terms of x_i, y_i, $i = 1, 2, 3$, the coefficients a, b, and c.
    (b) Show that $\frac{ax^3}{3} + \frac{bx^2}{2} + cx$ is an antiderivative of y.

(c) Find, in terms of x_i, y_i, $i = 1,2,3$, the area of the region bounded by the parabola, the x-axis, and the two lines $x = x_1$ and $x = x_3$.

29. Use the trapezoidal rule to find an estimate of the following integrals with n subintervals of equal length:

(a) $\int_0^{\pi/2} \cos x \, dx$, $n = 5$;   (b) $\int_0^3 \sqrt{x+1} \, dx$, $n = 10$;

(c) $\int_0^1 (\cos x^2) \, dx$, $n = 4$.

30. Use Simpson's rule to find an approximation of the following integrals with n subintervals of equal length:

(a) $\int_{-1}^1 e^{-x^2} \, dx$, $n = 10$;   (b) $\int_{-1}^2 x\sqrt{2+x^3} \, dx$, $n = 10$;

(c) $\int_0^1 (\cos x^2) \, dx$, $n = 4$.

31. Approximate the following integrals by using a Taylor polynomial of degree 3 for a suitable function:

(a) $\int_0^1 e^{-x^2} \, dx$;   (b) $\int_0^1 \frac{6}{\sqrt{4-x^2}} \, dx$;

(c) $\int_0^1 \frac{8}{x^2+1} \, dx$, $n = 10$;   (d) $\int_0^1 (\cos x^2) \, dx$.

32. (**Acceleration and velocity**) The graph of the acceleration $a(t)$ of a car measured in ft/sec^2 is shown below. Use (a) the trapezoidal rule and (b) Simpson's rule with $n = 6$ to estimate the increase in the velocity of the car during a six-second time interval.

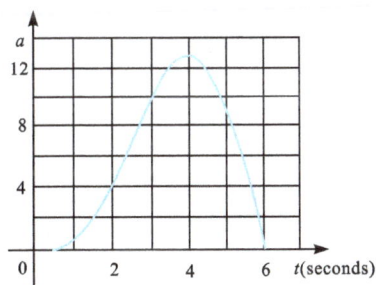

Question 32

Time	0	5	10	15	20	25	30
Velocity $v(t)$ ft/s	0	48	85	108	125	138	141

33. (**Particle motion**) The table above shows selected values of the velocity of a particle moving along a straight line.

(a) Set up an integral that represents the distance traveled by the particle during the first 30 seconds.

(b) Use the trapezoidal rule with six subintervals of equal length to estimate how far the particle moved during the first 30 seconds.

34. (**Optimization**) A certain type of machine depreciates at a continuous rate $r = r(t)$, where t is the time measured in months since the last overhaul. A fixed cost M is

incurred for each overhaul. The owner wants to determine the optimal time T (in months) between overhauls.

(a) Set up an integral expression that represents the loss in value of the machine over the period of time t since the last overhaul.

(b) Let C be given by $C(t) = \frac{1}{t}[M + \int_0^t r(s)\, ds]$. What does C represent in this situation? Why would the owner want to minimize C?

(c) Show that C has a minimum value at the times $t = T$ where $C(T) = r(T)$.

6 Techniques for integration and improper integrals

In this chapter, you will learn about:
- *indefinite integrals;*
- *integration by substitution;*
- *integration by parts;*
- *partial fractions;*
- *substitution in definite integrals;*
- *integrations by parts in definite integrals;*
- *improper integrals.*

Any function can be differentiated, using the rules of differentiation, provided it has an equation created from the basic mathematical functions. However, there is no set of rules for finding antiderivatives of a function and it can be proved that some very simple functions do not have an antiderivative function given by a formula using only basic functions. Consequently, we now present several methods that help us to find antiderivative functions, such as *integration by substitution*, *integration by parts*, and the *partial fractions method*. Finally, in the last section, we will discuss improper integrals.

6.1 Indefinite integrals

6.1.1 Definition of indefinite integrals and basic antiderivatives

The fundamental theorem of calculus (FTC) reveals the connection between the definite integral of a function on an interval and the antiderivatives of the function on that interval. To evaluate a definite integral of a function on an interval, the only thing we need to do is to find an antiderivative of the function. We now focus on some techniques to find antiderivatives of functions. First, we recall the definition of an antiderivative.

Definition 6.1.1. A function F is an *antiderivative* of f if $F'(x) = f(x)$ for all points x in the domain of f.

Finding an antiderivative of a function on an interval may be very difficult. However, we know from the FTC, Part I that a continuous function must have antiderivatives, since $\int_a^x f(t)\,dt$ is an antiderivative of f. Therefore, we have the following theorem.

Theorem 6.1.1. *A continuous function f, defined on an interval $\mathbf{I}$, has antiderivatives. All of its antiderivatives are given by $F(x) + C$, where C is an arbitrary constant and $F(x)$ is an antiderivative of f.*

https://doi.org/10.1515/9783110527780-006

Definition 6.1.2. The collection of all the antiderivatives of a function f, if they exist, is called the *indefinite integral* of f and is denoted by $\int f(x)\,dx$, where $\int f(x)\,dx = F(x) + C$ and $F(x)$ is an antiderivative of f.

The indefinite integral of a function f is a family of curves, called *integral curves*. The difference between any two integral curves is a constant, so these integral curves are "parallel" to each other.

Example 6.1.1. For the function $f(x) = x^2$, we know that the indefinite integral of $f(x)$ is $\int x^2\,dx = \frac{x^3}{3} + C$. By assigning specific values to the constant C, we obtain a family of functions whose graphs are vertical translations of one another, some of which, for $C = -2, -1, 0, 1, 2, 5$, are shown in Figure 6.1.1.

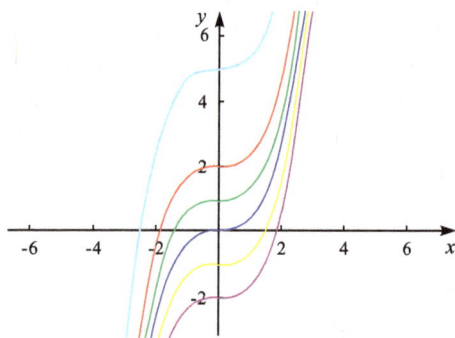

Figure 6.1.1: Some integral curves of $y' = x^2$.

From the definition of the antiderivative and the indefinite integral, it is easy to see the following theorem holds.

Theorem 6.1.2. *If $f(x)$ has an antiderivative for $x \in I$, then, for $x \in I$:*
(a) $\frac{d}{dx}[\int f(x)\,dx] = f(x);$ *(b)* $d[\int f(x)\,dx] = f(x)\,dx;$
(c) $\int F'(x)\,dx = F(x) + C;$ *(d)* $\int dF(x) = F(x) + C.$

By the definition, we know an indefinite integral is the reverse process of differentiation, so from the differentiation rules for basic functions, we have the following indefinite integral rules for basic functions.

Basic antiderivatives
1. When k is a constant, $\int k\,dx = kx + C$;
2. $\int \frac{1}{x}\,dx = \ln|x| + C$;

3. $\int x^n \, dx = \frac{1}{n+1} x^{n+1} + C, \, n \neq -1;$

4. $\int \cos x \, dx = \sin x + C;$

5. $\int \sin x \, dx = -\cos x + C;$

6. $\int \sec^2 x \, dx = \tan x + C;$

7. $\int \csc^2 x \, dx = -\cot x + C;$

8. $\int \tan x \sec x \, dx = \sec x + C;$

9. $\int \cot x \csc x \, dx = -\csc x + C;$

10. $\int \frac{dx}{1+x^2} = \arctan x + C;$

11. $\int \frac{1}{\sqrt{1-x^2}} \, dx = \arcsin x + C;$

12. $\int a^x \, dx = \frac{a^x}{\ln a} + C$ (provided $a > 0, \, a \neq 1$);

13. $\int e^x \, dx = e^x + C$, when a is a constant.

Check for yourself that each of the indefinite integrals above is correct by differentiating the right-hand side. For example, to show that $\int \frac{dx}{1+x^2} = \arctan x + C$, we use the known result that $\frac{d}{dx}(\arctan x) = \frac{1}{1+x^2}$.

NOTE. The notation $\int \frac{dx}{\sqrt{a^2-x^2}}$ means the same as $\int \frac{1}{\sqrt{a^2-x^2}} \, dx.$

Example 6.1.2. Find $\int \frac{1}{\sqrt[3]{x}} \, dx.$

Solution. We have

$$\int \frac{dx}{\sqrt[3]{x}} = \int \frac{dx}{x^{\frac{1}{3}}} = \int x^{-\frac{1}{3}} \, dx$$

$$= \frac{x^{-\frac{1}{3}+1}}{-\frac{1}{3}+1} + C = \frac{3}{2} x^{\frac{2}{3}} + C.$$

NOTE. The indefinite integral function $\frac{3}{2} x^{\frac{2}{3}}$ is defined for all x, but the integrand $\frac{1}{\sqrt[3]{x}}$ is not defined at $x = 0$. Nevertheless, we generally state that $\frac{3}{2} x^{\frac{2}{3}} + C$ is the indefinite integral of $\frac{1}{\sqrt[3]{x}}$ for all x, since the value, or nonexistence of a value, of the integrand at a single x-value (or even a finite number of x-values) has no effect on the indefinite integral function and its properties.

Theorem 6.1.3 (Linearity property of indefinite integral). *The integral*

$$\int (kf(x) \pm hg(x)) \, dx = k \int f(x) \, dx \pm h \int g(x) \, dx,$$

where k and h are any constants.

Proof. By the linearity property of differentiation, we have

$$\frac{d}{dx}\left(k\int f(x)\,dx \pm h\int g(x)\,dx\right)$$

$$= \frac{d}{dx}\left(k\int f(x)\,dx\right) \pm \frac{d}{dx}\left(h\int g(x)\,dx\right)$$

$$= k\frac{d}{dx}\left(\int f(x)\,dx\right) \pm h\frac{d}{dx}\left(\int g(x)\,dx\right)$$

$$= kf(x) \pm hg(x),$$

so $\int kf(x) \pm hg(x)\,dx = k\int f(x)\,dx \pm h\int g(x)\,dx + C$.

However, this constant C could be merged with any of the indefinite integrals in the theorem, since the sum or difference of two arbitrary constants is another arbitrary constant, so we have

$$k\int f(x)\,dx \pm h\int g(x)\,dx + C = k\int f(x)\,dx \pm h\int g(x)\,dx. \qquad \square$$

Example 6.1.3. Find $\int (\cos x + 2\sin x)\,dx$.

Solution. Since $\int \cos x\,dx = \sin x + C_1$ and $\int \sin x\,dx = -\cos x + C_2$, we have

$$\int \cos x + 2\sin x\,dx = \int \cos x\,dx + 2\int \sin x\,dx$$

$$= \sin x + C_1 + 2(-\cos x) + C_2$$

$$= \sin x - 2\cos x + C.$$

NOTE. Each of the indefinite integrals $\int \cos x\,dx$ and $2\int \sin x\,dx$ produces its own arbitrary constant. However, we can simply combine these two constants into a single arbitrary constant C.

Example 6.1.4. Find $\int (2 - \sec^2 x)\,dx$.

Solution. We have

$$\int (2 - \sec^2 x)\,dx = \int 2\,dx - \int \sec^2 x\,dx$$

$$= 2x - \tan x + C.$$

Example 6.1.5. Find $\int \frac{(x+1)^2}{x}\,dx$.

Solution. We have

$$\int \frac{(x+1)^2}{x}\,dx = \int \left(\frac{x^2 + 2x + 1}{x}\right)dx$$

$$= \int \left(x + 2 + \frac{1}{x}\right)dx$$

$$= \int x\,dx + \int 2\,dx + \int \frac{1}{x}\,dx$$

$$= \frac{x^2}{2} + 2x + \ln|x| + C.$$

Example 6.1.6. Find all functions g such that

$$g'(x) = e^x + 2x^3 - \sqrt{x} + 2.$$

Solution. The indefinite integral produces the general antiderivative of g, so it must give the most general form for g. Hence,

$$g(x) = \int \left(e^x + 2x^3 - \sqrt{x} + 2\right) dx.$$

Using the formulas in the list of basic antiderivatives, we obtain

$$g(x) = e^x + \frac{1}{2}x^4 - \frac{2}{3}x^{3/2} + 2x + C.$$

Example 6.1.7. Suppose

$$f(x) = \begin{cases} \sin x, & \text{when } x > 0 \\ \sqrt[3]{x}, & \text{when } x \leqslant 0. \end{cases}$$

Find $\int f(x)\,dx$.

Solution. We integrate $f(x)$ separately on the intervals $(-\infty, 0)$ and $(0, \infty)$, obtaining

$$\int f(x)\,dx = \begin{cases} -\cos x + C_1, & \text{when } x > 0 \\ \frac{3}{4}x^{\frac{4}{3}} + C_2, & \text{when } x \leqslant 0. \end{cases}$$

Since $\int f(x)\,dx$ must be continuous at $x = 0$, we have

$$-\cos 0 + C_1 = C_2 \quad \Longrightarrow \quad C_2 = C_1 - 1.$$

Thus,

$$\int f(x)\,dx = \begin{cases} -\cos x + C_1, & \text{when } x > 0 \\ \frac{3}{4}x^{\frac{4}{3}} + C_1 - 1, & \text{when } x \leqslant 0. \end{cases}$$

6.1.2 Differential equations

In applications of calculus, it is very common to have a situation as in the following example, where it is required to find a function, based on the knowledge about its derivatives. An equation that involves the derivatives of a known function is called a *differential equation*. The general solution of a differential equation always involves an arbitrary constant (or constants). However, there may be some extra conditions given that will determine the constants and therefore uniquely specify the solution, as in the next example.

Example 6.1.8. A particle starts from the origin and moves along the x-axis with velocity $v(t) = 2t + \sin t$. Find the position of the particle after 2 seconds.

Solution. If the position function is $x(t)$, then

$$x(t) = \int v(t)\,dt = \int (2t + \sin t)\,dt = t^2 - \cos t + C.$$

When $t = 0$, $x = 0$, which implies $0 = 0^2 - \cos 0 + C$, so $C = 1$. Therefore,

$$x(t) = t^2 - \cos t + 1.$$

When $t = 2$,

$$x(2) = 2^2 - \cos 2 + 1 \approx 5.416.$$

Thus, after 2 seconds, the particle is approximately at the point $x = 5.416$.

Example 6.1.9. Find f if $f'(x) = \frac{5}{x^2} + \frac{2}{x} - 3x + 1$ and $f(1) = 2$.

Solution. The general antiderivative of f' is f given by

$$f(x) = \int \left(\frac{5}{x^2} + \frac{2}{x} - 3x + 1 \right) dx$$

$$= -\frac{5}{x} + 2\ln|x| - \frac{3}{2}x^2 + x + C.$$

To determine C, we use the fact that $f(1) = 2$. We have

$$f(1) = -5 + 0 - \frac{3}{2} + 1 + C = 2$$

$$\implies C = \frac{15}{2}.$$

Therefore, the particular solution we want is

$$f(x) = -\frac{5}{x} + 2\ln|x| - \frac{3}{2}x^2 + x + \frac{15}{2}.$$

Example 6.1.10. Find f if $f''(x) = 12x^2 - 6x + 2$, $f(0) = 2$, and $f(1) = 6$.

Solution. The general antiderivative of $f''(x) = 12x^2 - 6x + 2$ is computed with arbitrary constant A as follows:

$$f'(x) = \int (12x^2 - 6x + 2)\, dx$$
$$f'(x) = 12\frac{x^3}{3} - 6\frac{x^2}{2} + 2x + A$$
$$= 4x^3 - 3x^2 + 2x + A.$$

Using the rules of antiderivatives once more, we find (with a second arbitrary constant B)

$$f(x) = 4\frac{x^4}{4} - 3\frac{x^3}{3} + 2\frac{x^2}{2} + Ax + B$$
$$= x^4 - x^3 + x^2 + Ax + B.$$

To determine A and B, we use the given conditions that $f(0) = 2$ and $f(1) = 6$. Since $f(0) = 0 + B = 2$, we have $B = 2$. Since

$$f(1) = 1 - 1 + 1 + A + 2 = 6,$$

we have $A = 3$. Therefore, the required function is

$$f(x) = x^4 - x^3 + x^2 + 3x + 2.$$

Example 6.1.11. Solve the differential equation $y' = \frac{x}{y}$.

Solution. Note that $y' = \frac{dy}{dx}$ and

$$\frac{dy}{dx} = \frac{x}{y}.$$

If we put all the terms involving x together and all the terms involving y together, then we have

$$y\, dy = x\, dx.$$

We then integrate with respect to y and with respect to x, respectively, to obtain

$$\int y\, dy = \int x\, dx.$$

This gives

$$\frac{y^2}{2} = \frac{x^2}{2} + C.$$

This is the general solution to the differential equation, even though it is given in an implicit form.

The technique for solving the differential equation shown above is called *separation of variables*. The type of differential equation is called a *separable differential equation*.

Example 6.1.12. The radioactive decay of Sm-151 can be modeled by the differential equation $\frac{dy}{dt} = -0.0077y$, where t is measured in years. This means a substance consisting of Sm-151 loses 0.77% of its mass each year. Find the half-life of Sm-151.

Solution. Since

$$\frac{dy}{dt} = -0.0077y,$$

$$\frac{1}{y} dy = -0.0077\, dt,$$

$$\int \frac{1}{y} dy = \int -0.0077\, dt,$$

$$\ln y = -0.0077t + C_1,$$

$$y = e^{C_1} e^{-0.0077t}$$

$$= Ce^{-0.0077t}.$$

Suppose $t = 0$, $y = y_0$. This means

$$y_0 = Ce^0 \rightarrow C = y_0,$$

so

$$y = y_0 e^{-0.0077t}.$$

When $y = \frac{y_0}{2}$, we have

$$\frac{y_0}{2} = y_0 e^{-0.0077t},$$

$$\ln \frac{1}{2} = -0.0077t,$$

$$t = \frac{\ln \frac{1}{2}}{-0.0077} \approx 90.02 \text{ years},$$

so after approximately 90 years, the substance will have lost half of its mass.

NOTE. If a substance's decay is modeled by $\frac{dy}{dt} = -ky$, then its half-life is given by

$$\text{half-life} = \frac{\ln 2}{k}.$$

6.1.3 Substitution in indefinite integrals

Often we will not know how to compute an indefinite integral because it is not in the list of basic antiderivatives from Section 6.1.1 and we do not recognize it as being the derivative of another function. Integration by substitution is a method for changing such an integral into another form that we may know how to compute. It is based on the observation that, if we know an antiderivative formula

$$\int f(x)\,dx = F(x) + C,$$

we also know the more general formula

$$\int f(g(x))g'(x)\,dx = F(g(x)) + C, \tag{6.1}$$

where g is any differentiable function.

This follows from the chain rule

$$\frac{d}{dx}F(g(x)) = F'(g(x))g'(x) = f(g(x))g'(x).$$

That is, the composite function $F(g(x))$ is an antiderivative of the function $f(g(x))g'(x)$.

If we substitute $u = g(x)$ in equation (6.1), then we can write the equation as

$$\int f(u)\frac{du}{dx}\,dx = F(u) + C.$$

However, F is an antiderivative of f, so we know

$$\int f(u)\,du = F(u) + C.$$

Hence, it follows that the substitution method is equivalent to the following equation:

$$\int f(u)\frac{du}{dx}\,dx = \int f(u)\,du. \tag{6.2}$$

In other words, we can use differential methods in an integral, allowing us to replace $\frac{du}{dx}\,dx$ by du.

In order to use this method, called *integration by substitution*, to compute an integral $\int h(x)\,dx$, we first identify in $h(x)$ a possible substitution function $u = g(x)$ and we differentiate this to give the differential form $du = g'(x)\,dx$. If we have chosen $g(x)$ well, we will be able to use the substitutions, $u = g(x)$ and $du = g'(x)\,dx$, to rewrite the integral totally in terms of the new variable u. That is, for some new function f we will have transformed the integral

$$\int h(x)\,dx = \int f(u)\,du.$$

If our choice of g was a good one, we may be able to find an antiderivative F of f and write

$$\int h(x)\,dx = \int f(u)\,du = F(u)|_{u=g(x)} + C = F(g(x)) + C.$$

Example 6.1.13. Find $\int 2\cos 2x\,dx$.

Solution. Since we know how to integrate $\cos x$, set $u = 2x$, $du = 2\,dx$, so that

$$\int 2\cos 2x\,dx = \int \cos 2x \cdot (2x)'\,dx \overset{u=2x}{=} \int \cos u\,\frac{du}{dx}\,dx$$

$$= \int \cos u\,du = \sin u + C$$

$$= \sin 2x + C.$$

NOTE. You might be able to see this result immediately, without the substitution, if you notice that $(\sin 2x)' = 2\cos 2x$.

Example 6.1.14. Find $\int e^{2x}\,dx$.

Solution.

$$\int e^{2x}\,dx = \frac{1}{2}\int e^{2x}\,d(2x) \overset{u=2x}{=} \frac{1}{2}\int e^u\,du$$

$$= \frac{1}{2}e^u + C = \frac{1}{2}e^{2x} + C.$$

Example 6.1.15. Find $\int 2xe^{x^2}\,dx$.

Solution. The substitution $u = x^2$ is suggested since $u' = 2x$. We have

$$\int 2xe^{x^2}\,dx = \int e^{x^2}(x^2)'\,dx \overset{u=x^2}{=} \int e^u\,\frac{du}{dx}\,dx$$

$$= \int e^u\,du = e^u + C = e^{x^2} + C.$$

Example 6.1.16. Evaluate the integral

$$\int \frac{(\ln x)^2}{x}\,dx.$$

Solution. Since

$$\int (\ln x)^2\frac{1}{x}\,dx = \int (\ln x)^2(\ln x)'\,dx,$$

the substitution $u = \ln x$ is suggested. Calculating $du = (\ln x)'\,dx$ and substituting gives

$$\int (\ln x)^2\frac{1}{x}\,dx = \int u^2\,du = \frac{1}{3}u^3\Big|_{u=\ln x} + C = \frac{1}{3}(\ln x)^3 + C.$$

NOTE. Verify for yourself that $\frac{1}{3}(\ln x)^3$ is an antiderivative of $(\ln x)^2(1/x)$.

Example 6.1.17. Find $\int (2x+1)(x^2+x+5)^{17}\,dx$.

Solution. Let $u = x^2 + x + 5$. Then $\frac{du}{dx} = 2x + 1$, which gives $du = (2x+1)\,dx$. Hence, the integral can be written as

$$\int (x^2+x+5)^{17}(2x+1)\,dx = \int u^{17}\,du$$
$$= \frac{1}{18}u^{18} + C$$
$$= \frac{1}{18}(x^2+x+5)^{18} + C.$$

Example 6.1.18. Find $\int x\sin(x^2+1)\,dx$.

Solution. Let $u = x^2 + 1$. Then we have $du = (x^2+1)'\,dx$, so

$$\int x\sin(x^2+1)\,dx = \frac{1}{2}\int \sin(x^2+1)(x^2+1)'\,dx$$
$$= \frac{1}{2}\int \sin u\,du$$
$$= -\frac{1}{2}\cos u + C$$
$$= -\frac{1}{2}\cos(x^2+1) + C.$$

NOTE. If you are familiar with this technique, then the u-substitution does not need to be specified.

Example 6.1.19. Find $\int \sin^2 x\cos x\,dx$.

Solution. We have

$$\int \sin^2 x\cos x\,dx = \int \sin^2 x\,d(\sin x) = \frac{1}{3}\sin^3 x + C.$$

Example 6.1.20. Find $\int te^{t^2}\,dt$.

Solution. We have

$$\int te^{t^2}\,dt = \frac{1}{2}\int e^{t^2}\,d(t^2) = \frac{1}{2}e^{t^2} + C.$$

Example 6.1.21. Find $\int \tan x\,dx$.

Solution. We have

$$\int \tan x\,dx = \int \frac{\sin x}{\cos x}\,dx = -\int \frac{(\cos x)'}{\cos x}\,dx = -\ln|\cos x| + C.$$

Example 6.1.22. Find $\int \csc x \, dx$.

Solution. This problem is considerably more difficult and we use trigonometric formulas to transform it until we find a form suitable for substitution. We have

$$\int \csc x \, dx = \int \frac{dx}{\sin x} = \int \frac{dx}{2 \sin \frac{x}{2} \cos \frac{x}{2}}$$

$$= \int \frac{dx}{2 \tan \frac{x}{2} \cos^2 \frac{x}{2}}$$

$$= \int \frac{\sec^2 \frac{x}{2}}{2 \tan \frac{x}{2}} \, dx = \int \frac{\sec^2 \frac{x}{2}}{\tan \frac{x}{2}} \, d\frac{x}{2}$$

$$= \int \frac{d(\tan \frac{x}{2})}{\tan \frac{x}{2}}$$

$$= \ln \left| \tan \frac{x}{2} \right| + C.$$

NOTE. This integral is most often written in the form

$$\int \csc x \, dx = \ln |\csc x - \cot x| + C,$$

but this is the same result since

$$\csc x - \cot x = \frac{1}{\sin x} - \frac{\cos x}{\sin x} = \frac{1 - \cos x}{\sin x}$$

$$= \frac{1 - (1 - 2 \sin^2 \frac{x}{2})}{2 \sin \frac{x}{2} \cos \frac{x}{2}}$$

$$= \frac{\sin \frac{x}{2}}{\cos \frac{x}{2}} = \tan \frac{x}{2}.$$

Example 6.1.23. Find $\int \sec x \, dx$.

Solution. We have

$$\int \sec x \, dx = \int \frac{1}{\cos x} \, dx$$

$$= \int \frac{1}{\sin(x + \frac{\pi}{2})} \, d\left(x + \frac{\pi}{2}\right)$$

$$= \int \csc\left(x + \frac{\pi}{2}\right) d\left(x + \frac{\pi}{2}\right)$$

$$= \ln \left| \csc\left(x + \frac{\pi}{2}\right) - \cot\left(x + \frac{\pi}{2}\right) \right| + C$$

$$= \ln |\sec x + \tan x| + C.$$

Example 6.1.24. Find $\int \frac{1}{a^2 + x^2} \, dx$, $a > 0$.

Solution. We have

$$\int \frac{1}{a^2(1+(\frac{x}{a})^2)}\,dx = \frac{1}{a^2}\int \frac{1}{1+(\frac{x}{a})^2}\,dx$$

$$= \frac{1}{a}\int \frac{1}{1+(\frac{x}{a})^2}\,d\left(\frac{x}{a}\right)$$

$$= \frac{1}{a}\arctan\frac{x}{a} + C.$$

Example 6.1.25. Find $\int \frac{1}{\sqrt{a^2-x^2}}\,dx$, $a > 0$.

Solution. We have

$$\int \frac{1}{\sqrt{a^2 - x^2}}\,dx = \frac{1}{a}\int \frac{1}{\sqrt{1-(\frac{x}{a})^2}}\,dx$$

$$= \int \frac{1}{\sqrt{1-(\frac{x}{a})^2}}\,d\left(\frac{x}{a}\right)$$

$$= \arcsin\frac{x}{a} + C.$$

Example 6.1.26. Find $\int \frac{1}{\sqrt{1-x-x^2}}\,dx$.

Solution. We have

$$\int \frac{1}{\sqrt{1 - x - x^2}}\,dx = \int \frac{1}{\sqrt{\frac{5}{4} - (x + \frac{1}{2})^2}}\,dx$$

$$= \int \frac{1}{\sqrt{(\frac{\sqrt{5}}{2})^2 - (x + \frac{1}{2})^2}}\,d\left(x + \frac{1}{2}\right)$$

$$= \arcsin\frac{x + \frac{1}{2}}{\frac{\sqrt{5}}{2}} + C.$$

6.1.4 Further results using integration by substitution

Integration by substitution may be used to solve other types of problems than those shown thus far. If the substitution $u = g(x)$ is instead written in the reverse form $x = h(u)$, for some function h, then $dx = h'(u)\,du$ and this can be used to transform an integral in the following way:

$$\int f(x)\,dx = \int f(h(u))h'(u)\,du.$$

This does not directly help us to compute the integral because the right-hand side only has an immediate solution, $F(h(u)) + C$, if we know F (an antiderivative of f), but if we knew this F, then we would not need to make the substitution. However, the new integral $\int f(h(u))h'(u)\,du$ might be in a form that we know how to compute by some other method, as we see in the following examples.

Example 6.1.27. Find $\int \frac{dx}{1+\sqrt{x}}$.

Solution. Let $\sqrt{x} = t$. Then $x = t^2$ and $dx = d(t^2) = 2t\,dt$, so

$$\int \frac{dx}{1+\sqrt{x}} = \int \frac{2t\,dt}{1+t} = 2\int \frac{t}{1+t}\,dt$$

$$= 2\int \left(1 - \frac{1}{1+t}\right)dt = 2\int 1\,dt - 2\int \frac{1}{1+t}\,dt$$

$$= 2t - 2\ln|1+t| + C$$

$$= 2\sqrt{x} - 2\ln(1+\sqrt{x}) + C.$$

Example 6.1.28. Find $\int \sqrt{1-x^2}\,dx$.

Solution. Since $\sqrt{1-\sin^2 x} = \sqrt{\cos^2 x} = |\cos x|$, we may be able to simplify this integral by using the substitution $x = \sin t$, with $-\frac{\pi}{2} \leqslant t \leqslant \frac{\pi}{2}$, to ensure $\cos t \geqslant 0$. Then $dx = \cos t\,dt$, so

$$\int \sqrt{1-x^2}\,dx = \int \cos t \cos t\,dt = \int \cos^2 t\,dt$$

$$= \frac{1}{2}\int (1 + \cos 2t)\,dt$$

$$= \frac{1}{2}t + \frac{1}{4}\sin 2t + C.$$

Since $x = \sin t$, with $-\frac{\pi}{2} \leqslant t \leqslant \frac{\pi}{2}$, we have $t = \arcsin x$ and $\sin 2t = 2\sin t \cos t = 2\sin t\sqrt{1-\sin^2 t} = 2x\sqrt{1-x^2}$.

Substituting these results in the equation above we obtain

$$\int \sqrt{1-x^2}\,dx = \frac{1}{2}\arcsin x + \frac{1}{2}x\sqrt{1-x^2} + C.$$

Example 6.1.29. Find $\int \frac{1}{x\sqrt{x^2-1}}\,dx$.

Solution. Let $t = \frac{1}{x}$. Then $x = \frac{1}{t}$, giving $dx = -\frac{1}{t^2}\,dt$, so

$$\int \frac{1}{x\sqrt{x^2-1}}\,dx = \int \frac{-\frac{1}{t^2}\,dt}{\frac{1}{t}\sqrt{\frac{1}{t^2}-1}} = -\int \frac{dt}{\sqrt{1-t^2}}$$

$$= -\arcsin t + C$$

$$= -\arcsin \frac{1}{x} + C.$$

NOTE. In this example, we can also use the substitution $x = \sec t$ for $-\frac{\pi}{2} < t < \frac{\pi}{2}$, so that $\sqrt{x^2 - 1} = \sqrt{\sec^2 t - 1} = \tan t$ and $dx = (\sec t \tan t)\, dt$. Thus,

$$\int \frac{1}{x\sqrt{x^2 - 1}}\, dx = \int \frac{1}{\sec t \times \tan t} \times \sec t \times \tan t\, dt$$

$$= \int 1\, dt = t + C = \arccos \frac{1}{x} + C.$$

Example 6.1.30. Evaluate $\int \frac{\sqrt{4-x^2}}{x^2}\, dx$.

Solution. Let $x = 2\sin\theta$, where $-\pi/2 \leqslant \theta \leqslant \pi/2$. Then $dx = 2\cos\theta\, d\theta$ and

$$\sqrt{4 - x^2} = \sqrt{4 - 4\sin^2\theta} = 2|\cos\theta| = 2\cos\theta.$$

Since $\cos\theta \geqslant 0$ when $-\pi/2 \leqslant \theta \leqslant \pi/2$, we have

$$\int \frac{\sqrt{4 - x^2}}{x^2}\, dx = \int \frac{2\cos\theta}{4\sin^2\theta} 2\cos\theta\, d\theta$$

$$= \int \frac{\cos^2\theta}{\sin^2\theta}\, d\theta = \int \cot^2\theta\, d\theta$$

$$= \int (\csc^2\theta - 1)\, d\theta$$

$$= -\cot\theta - \theta + C.$$

We must now return to the original variable x. This can be done by using trigonometric identities to express $\cot\theta$ in terms of $\sin\theta = x/2$. Alternatively, we can draw a diagram, as in Figure 6.1.2, so we simply read the value of $\cot\theta$ to obtain

$$\cot\theta = \frac{\sqrt{4 - x^2}}{x}.$$

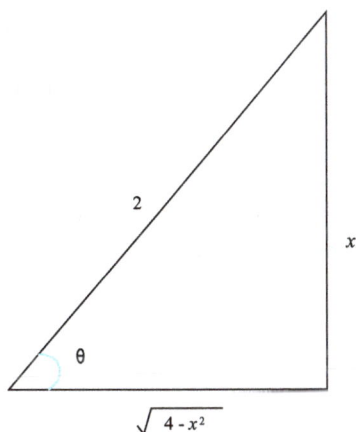

Figure 6.1.2: Trigonometric function in a right triangle.

Since $\sin\theta = x/2$, we have $\theta = \arcsin(\frac{x}{2})$, so

$$\int \frac{\sqrt{4-x^2}}{x^2}\,dx = -\frac{\sqrt{4-x^2}}{x} - \arcsin\left(\frac{x}{2}\right) + C.$$

NOTE. Although $\theta > 0$ in Figure 6.1.2, check for yourself that the expression for $\cot\theta$ is still valid when $\theta < 0$.

6.1.5 Integration by parts

Let $u(x)$ and $v(x)$ be functions of x. Recall the formula for the derivative of the product of two functions $u(x)v(x)$, which is

$$(uv)' = uv' + u'v, \quad \text{or}$$
$$\frac{d(uv)}{dx} = u\frac{dv}{dx} + v\frac{du}{dx}.$$

Integrating both sides gives

$$\int \frac{d(uv)}{dx}\,dx = \int u\frac{dv}{dx}\,dx + \int \frac{du}{dx}v\,dx,$$
$$uv = \int u\frac{dv}{dx}\,dx + \int \frac{du}{dx}v\,dx.$$

Thus we have obtained a very useful formula, called *integration by parts*. There are several forms of this formula.

Integration by parts formula
Integration by parts makes use of the following integration by parts formula:

$$\int \frac{du}{dx}v\,dx = uv - \int u\frac{dv}{dx}\,dx \quad \text{or} \tag{6.3}$$

$$\int u'v\,dx = uv - \int uv'\,dx \quad \text{or} \tag{6.4}$$

$$\int u\,dv = uv - \int v\,du. \tag{6.5}$$

Like the substitution method, the integration by parts formula allows us to change an integral that we cannot compute into an expression involving a new integral that we may know how to compute.

Example 6.1.31. Find $\int \ln x\,dx$.

Solution. Since

$$\int \ln x\,dx = \int (x)' \ln x\,dx,$$

let $u = x$, $v = \ln x$, so that the integration by parts formula

$$\int u'v\,dx = uv - \int uv'\,dx$$

gives

$$\int \ln x\,dx = \int (x)' \ln x\,dx$$
$$= x\ln x - \int x \cdot (\ln x)'\,du$$
$$= x\ln x - \int x \cdot \frac{1}{x}\,dx$$
$$= x\ln x - \int 1\,dx$$
$$= x\ln x - x + C.$$

Example 6.1.32. Find $\int x^2 \ln x\,dx$.

Solution. Choose $u = \frac{x^3}{3}$, $v = \ln x$, so that $u' = x^2$. Then we use integration by parts to obtain

$$\int x^2 \ln x\,dx = \int \left(\frac{x^3}{3}\right)' \ln x\,dx$$
$$= \frac{x^3}{3} \ln x - \int \frac{x^3}{3}(\ln x)'\,dx$$
$$= \frac{x^3}{3} \ln x - \int \frac{x^3}{3} \times \frac{1}{x}\,dx$$
$$= \frac{x^3}{3} \ln x - \frac{x^3}{9} + C.$$

Example 6.1.33. Find $\int x^2 e^x\,dx$.

Solution. Choose $u = e^x$, $v = x^2$, so that $u' = e^x$. Integrating by parts gives

$$\int x^2 e^x\,dx = \int (e^x)' x^2\,dx = x^2 e^x - \int e^x (x^2)'\,dx$$
$$= x^2 e^x - 2\int xe^x\,dx.$$

Now we use integration by parts again to obtain

$$\int x^2 e^x\,dx = x^2 e^x - 2\int x(e^x)'\,dx$$
$$= x^2 e^x - 2\left(xe^x - \int (x)'e^x\,dx\right)$$
$$= x^2 e^x - 2xe^x + 2\int e^x\,dx$$
$$= x^2 e^x - 2xe^x + 2e^x + C.$$

Example 6.1.34. Find $\int x \cos x \, dx$.

Solution. Choose $u = x$, $v = \sin x$, so that $\int x \cos x \, dx = \int x \, d(\sin x)$. Then

$$\int x \cos x \, dx = \int x \, d(\sin x) = x \sin x - \int \sin x \, dx = x \sin x + \cos x + C.$$

In this example, we might have tried writing $\int x \cos x \, dx = \int \cos x \, d(\frac{x^2}{2})$, so that $u = \cos x$, $v = \frac{x^2}{2}$. This time, integrating by parts gives

$$\int x \cos x \, dx = \int \cos x \, d\left(\frac{x^2}{2}\right) = \frac{x^2}{2} \cos x - \int \frac{x^2}{2} d(\cos x)$$

$$= \frac{x^2}{2} \cos x + \int \frac{x^2}{2} \sin x \, dx.$$

However, the integral on the right-hand side is even more difficult than the original integral, so this was a bad choice for $u(x)$ and $v(x)$.

In general, when applying integration by parts to an integral, $\int f(x) \, dx$, we try to write $f(x)$ as a product, $f(x) = g(x)h(x)$, such that $u = g(x)$ has a simple derivative and $dv = h(x) \, dx$ has a simple antiderivative.

Example 6.1.35. Find $\int \arctan(\sqrt{x}) \, dx$.

Solution. Let $u = \arctan(\sqrt{x})$ and $dv = dx$. Then $v = x$ and, by the chain rule of differentiation, $du = \frac{1}{1+(\sqrt{x})^2} \cdot \frac{d}{dx}(\sqrt{x}) = \frac{1}{(1+x)2\sqrt{x}}$. Integration by parts gives

$$\int \arctan(\sqrt{x}) \, dx = x \arctan(\sqrt{x}) - \int \frac{x}{(1+x)2\sqrt{x}} \, dx$$

$$= x \arctan \sqrt{x} - \frac{1}{2} \int \frac{\sqrt{x}}{1+x} \, dx.$$

Now use the substitution $\sqrt{x} = t$, so that $x = t^2$ and $dx = 2t \, dt$, giving

$$\int \frac{\sqrt{x}}{1+x} \, dx = \int \frac{t}{1+t^2} 2t \, dt$$

$$= 2 \int \frac{t^2 + 1 - 1}{1+t^2} \, dt$$

$$= 2 \int \left(1 - \frac{1}{1+t^2}\right) \, dt$$

$$= 2(t - \arctan t) + C$$

$$= 2(\sqrt{x} - \arctan(\sqrt{x})) + C.$$

Hence, the complete integral is

$$\int \arctan(\sqrt{x}) \, dx = x \arctan(\sqrt{x}) - \sqrt{x} + \arctan(\sqrt{x}) + C$$

$$= (x+1) \arctan \sqrt{x} - \sqrt{x} + C.$$

In the next example, we see that integration by parts must sometimes be applied repeatedly and may eventually produce the original integral on the right-hand side, but in such a way that we can find its value.

Example 6.1.36. Evaluate $I = \int e^x \cos x\, dx$.

Solution. Choose $u = \cos x$, $dv = e^x\, dx$, so that

$$v = e^x, \quad du = -\sin x\, dx.$$

Then

$$\int e^x \cos x\, dx = \int (e^x)' \cos x\, dx = e^x \cos x - \int e^x (\cos x)'\, dx$$

$$= e^x \cos x + \int e^x \sin x\, dx.$$

We apply integration by parts a second time to obtain

$$\int e^x \cos x\, dx = e^x \cos x + \int e^x \sin x\, dx = e^x \cos x + \int (e^x)' \sin x\, dx$$

$$= e^x \cos x + e^x \sin x - \int e^x (\sin x)'\, dx$$

$$= e^x \cos x + e^x \sin x - \int e^x \cos x\, dx.$$

Observe that the original integral appears on the right-hand side. Hence, combining the two values of I on the left-hand side gives the required solution for the integral I as follows:

$$I = \int e^x \cos x\, dx = \frac{1}{2} e^x (\sin x + \cos x) + C.$$

Using integration by parts, we can obtain many useful reduction formulas.

Example 6.1.37. If $I_n = \int \sin^n x\, dx$, then prove the reduction formula

$$I_n = \frac{n-1}{n} I_{n-2} - \frac{\sin^{n-1} x \cos x}{n} \quad \text{for all integers } n \geqslant 2.$$

Proof. We have

$$I_n = \int \sin^n x\, dx = \int \sin^{n-1} x \cdot \sin x\, dx$$

$$= \int \sin^{n-1} x \cdot (-\cos x)'\, dx$$

$$= -\sin^{n-1} x \cos x - \int (\sin^{n-1} x)'(-\cos x)\, dx$$

$$= -\sin^{n-1} x \cos x + \int (n-1) \sin^{n-2} x \cos^2 x\, dx$$

$$= -\sin^{n-1} x \cos x + (n-1) \int \sin^{n-2} x(1 - \sin^2 x)\, dx$$

$$= -\sin^{n-1} x \cos x + (n-1) \int (\sin^{n-2} x - \sin^n x)\, dx$$

$$= -\sin^{n-1} x \cos x + (n-1)(I_{n-2} - I_n).$$

Solving for I_n, we have

$$I_n = \frac{n-1}{n} I_{n-2} - \frac{\sin^{n-1} x \cos x}{n}. \qquad \square$$

Example 6.1.38 (Tabular method). Find $\int x^3 e^x\, dx$.

Solution. As in example 6.1.33, we would expect that we will need integration by parts three times. However, we can use a shortcut by the tabular method shown below.

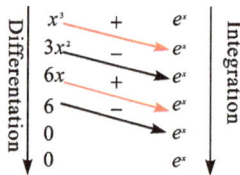

The indefinite integral is therefore given by

$$\int x^3 e^x\, dx = x^3 e^x - 3x^2 e^x + 6xe^x - 6e^x + C.$$

(*Group activity) Why does the tabular method work? When does it not work?

6.1.6 Partial fractions in integration

Unlike either integration by substitution or integration by parts, which offer hope but not certainty of success, using partial fractions always enables us to compute integrals of the form

$$\int \frac{p(x)}{q(x)}\, dx,$$

where p and q are polynomials, provided $q(x)$ can be factored into polynomial factors of degree 1 or degree 2.

Ratios of the form $R(x) = p(x)/q(x)$ where p and q are polynomial functions are called *rational functions*. The technique of integrating a rational function R is to decompose R using partial fractions into a sum of simpler fractions, each of which has a known antiderivative. The *partial fractions* decomposition procedure, applied to $\frac{p(x)}{q(x)}$, has the following three main steps, which are all algebraic.

1. If the degree of $p(x)$ is greater than or equal to the degree of $q(x)$, then *divide* the polynomial $p(x)$ by $q(x)$ (using polynomial long division) to obtain a quotient (a polynomial) and a *remainder term* that is another rational function, $\frac{r(x)}{q(x)}$, with denominator $q(x)$ but where the degree of $q(x)$ is now greater than the degree of its numerator $r(x)$.

2. *Factor* $q(x)$ into linear factors, irreducible quadratic factors (irreducible means that it cannot be factored into real linear factors), or a combination of both.

3. *Decompose* the remainder term into a sum of fractions, each with one of the factors of $q(x)$ as denominator. The numerator of each fraction will be a constant if the denominator is linear and, if the denominator is quadratic, it will be a linear polynomial of the form $Ax + B$. The use of partial fractions when a factor of $q(x)$ is repeated (appears more than once) is more complicated and is described in the more formal description of the partial fractions procedure following the first example. The values of the constants $A, B, C, \dots$ must be determined using techniques shown in the examples below.

Example 6.1.39. Compute

$$\int \frac{x + 2}{2x^3 - x^2 + 2x - 1}\, dx.$$

Solution. Because the degree of the numerator of the rational function is less than the degree of the denominator, we skip the division step. The denominator polynomial factors as

$$2x^3 - x^2 + 2x - 1 = (2x - 1)(x^2 + 1).$$

The factor $(2x - 1)$ corresponds to the real zero $1/2$ and the factor $x^2 + 1$ corresponds to the complex zeros i and $-i$. Thus, the standard partial fraction decomposition has three constants A, B, C and it reads

$$\frac{x + 2}{(2x - 1)(x^2 + 1)} = \frac{Ax + B}{x^2 + 1} + \frac{C}{2x - 1}.$$

The values of A, B, and C must be determined. Clearing fractions by multiplying both sides by $(x^2 + 1)(2x - 1)$ gives the identity

$$x + 2 = (Ax + B)(2x - 1) + C(x^2 + 1).$$

We generate three equations in the unknowns A, B, and C, by replacing x by three different numbers. We use the real zero $x = 1/2$ for one value and $x = 0$ and $x = 1$ for the others (the complex zeros i and $-i$ can also be used). Substituting these values, we obtain

$$x = \frac{1}{2} \text{ gives } \frac{5}{2} = C\frac{5}{4} \implies C = 2.$$
$$x = 0 \text{ gives } 2 = B(-1) + C(1) = -B + 2 \implies B = 0.$$
$$x = 1 \text{ gives } 3 = (A + B)(1) + C(2) = A + 4 \implies A = -1.$$

Hence, the partial fractions expansion is

$$\frac{x + 2}{(2x - 1)(x^2 + 1)} = \frac{-x}{x^2 + 1} + \frac{2}{2x - 1}.$$

Therefore,

$$\int \frac{x + 2}{(2x - 1)(x^2 + 1)}\, dx = \int \frac{-x}{x^2 + 1}\, dx + \int \frac{2}{2x - 1}\, dx$$
$$= -\frac{1}{2} \ln(x^2 + 1) + \ln\left|x - \frac{1}{2}\right| + C.$$

NOTES. 1. Check for yourself that the partial fractions expansion is correct by combining the partial fractions back into a single fraction.
2. All polynomials with real coefficients can, in theory, be factored, as in this example, into a product of real linear or quadratic factors. However, if the degree of the polynomial is greater than 4, then it is often not easy to find these factors.
3. Complex number zeros of a real polynomial always occur as complex conjugate pairs. The product of the linear factors corresponding to a complex conjugate pair is always a real quadratic factor, like $x^2 + 1$ above, and it is called an *irreducible factor*. Irreducible quadratics are not factored into linear (complex) factors in the standard partial fractions decomposition, but it is possible to work with partial fractions using complex number factors.
4. The method used previously to determine the coefficients when performing the partial fraction expansion of a rational function is called the *Heaviside cover-up method*, named after Oliver Heaviside.

Oliver Heaviside (1850–1925) was a self-taught English electrical engineer, mathematician, and physicist who adapted complex numbers to the study of electrical circuits, invented mathematical techniques for the solution of differential equations (later found to be equivalent to Laplace transforms), reformulated Maxwell's field equations in terms of electric and magnetic forces and energy flux, and independently co-formulated vector analysis. Although at odds with the scientific establishment for most of his life, Heaviside changed the face of telecommunications, mathematics, and science for years to come. http://en.wikipedia.org/wiki/Oliver_Heaviside

General description of partial fractions
A rational function $R(x) = p(x)/q(x)$ where the polynomial $p(x)$ has a degree greater than or equal to the degree of the polynomial $q(x)$ can be rewritten by using the poly-

nomial form of long division as

$$R(x) = \frac{p(x)}{q(x)} = Q(x) + \frac{r(x)}{q(x)},$$

where $Q(x)$ and $r(x)$ are polynomials and the degree of $r(x)$ is less than the degree of $q(x)$. If the degree of $p(x)$ is less than the degree of $q(x)$, then $Q(x)$ is identical to 0 in this formula and $r(x) = p(x)$. Integration gives

$$\int R(x)\, dx = \int Q(x)\, dx + \int \frac{r(x)}{q(x)}\, dx,$$

where the integral $\int Q(x)\, dx$ is easy to compute. Hence, we focus on the second integral.

From advanced algebra it is known that the polynomial $q(x)$ with real coefficients can, in theory, be factored as the product of a constant c (the coefficient of the highest power of x in $q(x)$), real linear factors of the form $(x - r)^m$, and irreducible real quadratic factors of the form $(x^2 + Ux + V)^n$ ($m \geqslant 1$ and $n \geqslant 1$ are integers). If these factors of $q(x)$ are found, then $\frac{r(x)}{q(x)}$ can be decomposed into a sum of terms of the following types. Each linear factor $(x - r)^m$ contributes a group of m terms with constants $A_1, A_2, \ldots, A_m$, giving

$$\frac{A_1}{(x-r)^1} + \frac{A_2}{(x-r)^2} + \cdots + \frac{A_m}{(x-r)^m}. \tag{6.6}$$

Each irreducible quadratic factor $(x^2 + Ux + V)^n$ contributes a group of n terms with constants $B_1, B_2, \ldots, B_n, C_1, C_2, \ldots, C_n$, giving

$$\frac{B_1 x + C_1}{(x^2 + Ux + V)^1} + \frac{B_2 x + C_2}{(x^2 + Ux + V)^2} + \cdots + \frac{B_n x + C_n}{(x^2 + Ux + V)^n}. \tag{6.7}$$

The sum of all such groups of terms from all factors of $q(x)$, multiplied by $\frac{1}{c}$, is called the *standard decomposition* of $r(x)/q(x)$. The individual terms in such sums are called *partial fractions*. The values of the real constants $A_1, B_1, C_1, \ldots$ can be determined by methods given in the examples.

NOTE. In practice, the constant c is usually incorporated, by multiplication, into one or more of the factors of $q(x)$.

Any individual fraction in the standard decomposition can be integrated, using one of the six integrals for partial fractions given below. It is assumed that the quadratic $x^2 + Ux + V$ can be written as $(x - a)^2 + b^2$, after completing the square (it could not be $(x - a)^2 - b^2$ since that can be factored). A numerator, $Ax + B$, of one of these terms is rewritten as $Ax + B = A(x - a) + (B + Aa)$ in order to apply these formulas. We have

$$\int \frac{dx}{x - r} = \ln |x - r| + C, \tag{6.8}$$

$$\int \frac{dx}{(x-r)^k} = \frac{(x-r)^{-k+1}}{-k+1} + C, \tag{6.9}$$

$$\int \frac{x-a}{(x-a)^2 + b^2} \, dx = \frac{1}{2} \ln((x-a)^2 + b^2) + C, \tag{6.10}$$

$$\int \frac{dx}{(x-a)^2 + b^2} = \frac{1}{b} \arctan \frac{x-a}{b} + C. \tag{6.11}$$

For $k > 1$, we have

$$\int \frac{x-a}{((x-a)^2 + b^2)^k} \, dx = \frac{-1}{2(k-1)((x-a)^2 + b^2)^{k-1}} + C \tag{6.12}$$

and

$$\int \frac{dx}{((x-a)^2 + b^2)^k} = \frac{x-a}{2b^2(k-1)((x-a)^2 + b^2)^{k-1}}$$
$$+ \frac{2k-3}{2b^2(k-1)} \int \frac{dx}{((x-a)^2 + b^2)^{k-1}}. \tag{6.13}$$

NOTE. The last of the integrals above is in an iterative format because it gives a formula for the integral that involves the same integral but with one less power in the denominator. Consequently, it would have to be applied repeatedly until the integral on the right is reduced to the form of equation (6.11) (that is, when the power in the denominator is one).

Example 6.1.40. Evaluate $\int \frac{x^2+1}{x^2-1} \, dx$.

Solution. Since both polynomials have degree two, we first perform the long division, obtaining

$$\frac{x^2 + 1}{x^2 - 1} = 1 + \frac{2}{x^2 - 1}.$$

Integrating and applying partial fractions to the remainder fraction, we obtain

$$\int \frac{x^2 + 1}{x^2 - 1} \, dx = \int 1 + \frac{2}{(x-1)(x+1)} \, dx = \int 1 + \frac{1}{x-1} - \frac{1}{x+1} \, dx$$
$$= x + \ln|x-1| - \ln|x+1| + C.$$

Example 6.1.41. Write out the form of the partial fraction decomposition of the function $f(x)$, without evaluating the constants, to obtain

$$f(x) = \frac{x^4 + x^3 + 1}{x(x-1)(x^2 + 3x + 3)(x^2 + 1)^3}.$$

Solution. We must use 10 constants, $A, B, C, D, E, F, G, H, I, J$, and the required partial fraction expansion is

$$f(x) = \frac{A}{x} + \frac{B}{x-1} + \frac{Cx+D}{(x^2 + 3x + 3)} + \frac{Ex+F}{x^2 + 1} + \frac{Gx+H}{(x^2 + 1)^2} + \frac{Ix+J}{(x^2 + 1)^3}.$$

Example 6.1.42. Find $\int \frac{x+3}{6x^2+x-2} \, dx$.

Solution. The standard partial fraction decomposition of the integrand takes the form

$$\frac{x+3}{6x^2+x-2} = \frac{x+3}{6(x^2+\frac{1}{6}x-\frac{1}{3})} = \frac{x+3}{6(x+\frac{2}{3})(x-\frac{1}{2})}$$

$$= \frac{1}{6}\left(\frac{A}{x+\frac{2}{3}} + \frac{B}{x-\frac{1}{2}}\right).$$

Generally, however, the multiplying constant would be combined with the linear factors, giving the decomposition as

$$\frac{x+3}{(3x+2)(2x-1)} = \frac{C}{3x+2} + \frac{D}{2x-1}.$$

To determine the values of C and D, we multiply both sides of this equation by the product of the denominators $(3x+2)(2x-1)$, obtaining

$$x+3 = C(2x-1) + D(3x+2).$$

Using $x = -\frac{2}{3}$ (so that $3x+2 = 0$) gives

$$\frac{7}{3} = C \cdot \left(-\frac{7}{3}\right) \implies C = -1$$

and using $x = \frac{1}{2}$ (so that $2x-1 = 0$) gives $\frac{7}{2} = D \cdot \frac{7}{2} \implies D = 1$, so

$$\int \frac{x+3}{6x^2+x-2} \, dx = \int \frac{-1}{3x+2} + \frac{1}{2x-1} \, dx$$

$$= -\frac{1}{3}\ln|3x+2| + \frac{1}{2}\ln|2x-1| + C.$$

Example 6.1.43. Find $\int \frac{1}{x(x-1)^2} \, dx$.

Solution. The partial fraction decomposition has the form

$$\frac{1}{x(x-1)^2} = \frac{A}{x} + \frac{B}{x-1} + \frac{C}{(x-1)^2}.$$

To determine the values of A, B, and C, we multiply both sides of this equation by a multiple of the denominators $x(1-x)^2$, obtaining the identity

$$1 = A(x-1)^2 + B(x-1)x + Cx.$$

Now combining like terms, we have

$$1 = (A+B)x^2 + (-2A-B+C)x + A.$$

Both sides must be exactly the same function, so this time, instead of substituting values for x, we just equate coefficients of like powers of x on both sides:

$$A + B = 0,$$
$$-2A - B + C = 0,$$
$$A = 1.$$

Solving the system of three linear equations, we obtain $A = 1$, $B = -1$, and $C = 1$, so

$$\int \frac{1}{x(1-x)^2}\, dx = \int \frac{1}{x} - \frac{1}{x-1} + \frac{1}{(x-1)^2}\, dx$$
$$= \ln|x| - \ln|x-1| - \frac{1}{x-1} + C.$$

Example 6.1.44. Find $\int \frac{1}{(1+x)(1+x^2)}\, dx$.

Solution. The form of the partial fraction decomposition is

$$\frac{1}{(1+x)(1+x^2)} = \frac{A}{1+x} + \frac{Bx+C}{1+x^2}.$$

Multiplying by $(1+x)(1+x^2)$, we have

$$1 = A(1+x^2) + (Bx+C)(1+x).$$

Combining like terms, we obtain

$$1 = (A+B)x^2 + (B+C)x + A + C.$$

If we equate the coefficients, we obtain the system of equations

$$A+B = 0, \quad B+C = 0, \quad A+C = 1,$$

which has the solution

$$A = \frac{1}{2}, \quad B = -\frac{1}{2}, \quad \text{and} \quad C = \frac{1}{2}.$$

Thus,

$$\int \frac{1}{(1+x)(1+x^2)}\, dx$$
$$= \int \frac{1}{2(x+1)} - \frac{1}{2}\frac{x-1}{x^2+1}\, dx$$
$$= \frac{1}{2}\ln|x+1| - \frac{1}{4}\int \frac{2x}{x^2+1}\, dx + \frac{1}{2}\int \frac{1}{1+x^2}\, dx$$
$$= \frac{1}{2}\ln|x+1| - \frac{1}{4}\ln(x^2+1) + \frac{1}{2}\arctan x + C.$$

Example 6.1.45. Find $\int \frac{1}{a^2-x^2}\, dx$, $a > 0$.

Solution. Start by writing in partial fractions as follows:

$$\frac{1}{a^2 - x^2} = \frac{A}{a+x} + \frac{B}{a-x}.$$

Solving for A and B, we have $A = B = \frac{1}{2a}$ and

$$
\begin{aligned}
\int \frac{1}{a^2 - x^2}\, dx &= \frac{1}{2a} \int \frac{1}{a+x} + \frac{1}{a-x}\, dx \\
&= \frac{1}{2a}(\ln|a+x| - \ln|a-x|) + C \\
&= \frac{1}{2a} \ln\left|\frac{a+x}{a-x}\right| + C.
\end{aligned}
$$

Example 6.1.46. Find $\int \frac{x-2}{x^2+2x+3}\, dx$.

Solution. The denominator does not factor into real linear factors, so partial fractions cannot be used. However, the integral is easily converted into a standard integral given in equation (6.10). Using this integral, we obtain

$$
\begin{aligned}
&\int \frac{x-2}{x^2 + 2x + 3}\, dx \\
&= \int \frac{x+1-3}{(x+1)^2 + 2}\, dx \\
&= \int \frac{x+1}{(x+1)^2 + 2}\, dx - 3 \int \frac{1}{(x+1)^2 + 2}\, dx \\
&= \frac{1}{2} \ln((x+1)^2 + 2) - \frac{3}{\sqrt{2}} \arctan \frac{x+1}{\sqrt{2}} + C.
\end{aligned}
$$

Alternatively, we could solve the integral directly using the substitution $t = x + 1$, $dt = dx$. This gives

$$
\begin{aligned}
&\int \frac{x-2}{x^2 + 2x + 3}\, dx \\
&= \int \frac{t-3}{t^2 + 2}\, dt \\
&= \int \frac{t}{t^2 + 2}\, dt - 3 \int \frac{1}{t^2 + 2}\, dt \\
&= \frac{1}{2} \ln(t^2 + 2) - \frac{3}{\sqrt{2}} \arctan \frac{t}{\sqrt{2}} + C \\
&= \frac{1}{2} \ln(x^2 + 2x + 3) - \frac{3}{\sqrt{2}} \arctan \frac{x+1}{\sqrt{2}} + C.
\end{aligned}
$$

Example 6.1.47. Find $\int \frac{\ln x}{(1-x)^2}\, dx$.

Solution. Choose $u = \ln x$, $dv = \frac{1}{(1-x)^2}\,dx$, so that we can take $v = \int \frac{1}{(1-x)^2}\,dx = \frac{1}{1-x}$ and $du = \frac{1}{x}\,dx$. Then

$$\int \frac{\ln x}{(1-x)^2}\,dx = \int u\,dv = uv - \int v\,du$$

$$= \frac{\ln x}{1-x} - \int \frac{1}{x(1-x)}\,dx$$

$$= \frac{\ln x}{1-x} - \int \frac{(1-x)+x}{x(1-x)}\,dx$$

$$= \frac{\ln x}{1-x} - \int \left(\frac{1}{x} + \frac{1}{1-x}\right)dx$$

$$= \frac{\ln x}{1-x} - \ln|x| + \ln|1-x| + C$$

$$= \frac{\ln x}{1-x} + \ln \frac{|1-x|}{|x|} + C.$$

6.1.7 Rationalizing substitutions

Some nonrational functions can be changed into rational functions by means of appropriate substitutions. In that case, we say that the function has been *rationalized* and we can then apply the partial fractions method to compute the integral.

Example 6.1.48. Evaluate $\int \frac{1}{x-\sqrt{x+2}}\,dx$.

Solution. Substitute $u = \sqrt{x+2}$. Then $x = u^2 - 2$, giving $dx = 2u\,du$. Thus

$$\int \frac{1}{x - \sqrt{x+2}}\,dx = \int \frac{2u}{(u^2-2)-u}\,du.$$

The integrand is now a rational function and we integrate using partial fractions as follows:

$$\int \frac{2u}{(u^2-2)-u}\,du = \int \frac{2u}{(u-2)(u+1)}\,du$$

$$= \int \frac{4}{3(u-2)} + \frac{2}{3(u+1)}\,du$$

$$= \frac{4}{3}\ln|u-2| + \frac{2}{3}\ln|u+1| + C$$

$$= \frac{4}{3}\ln|\sqrt{x+2}-2| + \frac{2}{3}\ln|\sqrt{x+2}+1| + C.$$

Example 6.1.49. Find $\int \frac{1}{2+\sin x}\,dx$.

Solution. Since

$$\sin x = \frac{2\tan\frac{x}{2}}{1 + \tan^2\frac{x}{2}},$$

we substitute $u = \tan\frac{x}{2}$, so that $x = 2\arctan u$ and $dx = \frac{2}{1+u^2}\,du$. This gives

$$\int \frac{1}{2 + \sin x}\,dx = \int \frac{1}{2 + \frac{2u}{1+u^2}} \cdot \frac{2}{1+u^2}\,du$$

$$= \int \frac{1}{u^2 + u + 1}\,du.$$

The denominator does not factor into real factors, so we cannot use partial fractions. However, by completing the square we can turn this into an integral from the standard set (see equation (6.11)). We have

$$\int \frac{1}{2 + \sin x}\,dx = d\int \frac{1}{(u + \frac{1}{2})^2 + (\frac{\sqrt{3}}{2})^2}\,du$$

$$= \frac{2}{\sqrt{3}}\arctan\left(\frac{2}{\sqrt{3}}\left(u + \frac{1}{2}\right)\right) + C$$

$$= \frac{2}{\sqrt{3}}\arctan\left(\frac{2}{\sqrt{3}}\tan\frac{x}{2} + \frac{1}{\sqrt{3}}\right) + C.$$

6.2 Substitution in definite integrals

To evaluate a definite integral, we can use the same substitutions method that we saw in Section 6.1.2. However, now we must either convert to the original variable in order to evaluate at the limits, or, if we wish, we can transform the limits using the same substitution.

Example 6.2.1. Evaluate $\int_0^4 \frac{1}{1+\sqrt{x}}\,dx$.

Solution. Let $t = \sqrt{x}$, so $x = t^2$ and $dx = 2t\,dt$. Transforming the limits of integration, we find that, when $x = 0$, $t = 0$ and, when $x = 4$, $t = 2$. Hence, substituting gives

$$\int_0^4 \frac{dx}{1 + \sqrt{x}} = \int_0^2 \frac{2t}{1+t}\,dt = 2\int_0^2 \frac{t}{1+t}\,dt = 2\int_0^2 \frac{t + 1 - 1}{1+t}\,dt$$

$$= 2\int_0^2\left(1 - \frac{1}{1+t}\right)dt$$

$$= 2(t - \ln|1 + t|)|_0^2$$

$$= 2\{(2 - \ln 3) - (0 - \ln 1)\}$$

$$= 2(2 - \ln 3).$$

Example 6.2.2. Evaluate $\int_0^{\frac{1}{2}} \frac{x^2}{\sqrt{1-x^2}}\, dx$.

Solution. Let $x = \sin t$ (restrict t to $[0, \frac{\pi}{2}]$ so that $\sin t$ is an increasing function and $\sin t \geqslant 0$ and $\cos t \geqslant 0$). Then $dx = d(\sin t) = \cos t\, dt$. When $x = 0$, $\sin t = 0 \Longrightarrow t = 0$. When $x = \frac{1}{2}$, $\sin t = \frac{1}{2} \Longrightarrow t = \frac{\pi}{6}$. By substitution we obtain

$$\int_0^{\frac{1}{2}} \frac{x^2\, dx}{\sqrt{1-x^2}} = \int_0^{\frac{\pi}{6}} \frac{(\sin^2 t)\cos t\, dt}{\sqrt{1 - \sin^2 t}} = \int_0^{\frac{\pi}{6}} \sin^2 t\, dt$$

$$= \int_0^{\frac{\pi}{6}} \frac{1 - \cos 2t}{2}\, dt = \left(\frac{t}{2} - \frac{1}{4}\sin 2t\right)\Big|_0^{\frac{\pi}{6}}$$

$$= \left(\frac{\pi}{12} - \frac{1}{4}\sin\frac{\pi}{3}\right) - \left(\frac{0}{2} - \frac{1}{4}\sin 0\right)$$

$$= \frac{\pi}{12} - \frac{\sqrt{3}}{8}.$$

Example 6.2.3. Show that the area of the circle with radius r is πr^2.

Proof. Since $x^2 + y^2 = r^2$ is a circle centered at the origin with radius r, $y = \sqrt{r^2 - x^2}$, $-r \leqslant x \leqslant r$ is the upper half of the circle. The area of the circle is four times the area of the quarter circle in the first quadrant, as seen in Figure 6.2.1. That is,

$$\text{area} = 4 \times \int_0^r \sqrt{r^2 - x^2}\, dx.$$

Let $x = r\sin t$, $0 \leqslant t \leqslant \frac{\pi}{2}$. Then $dx = r\cos t\, dt$. When $x = 0$, $t = 0$ and when $x = r$, $t = \frac{\pi}{2}$, so

$$\int_0^r \sqrt{r^2 - x^2}\, dx = \int_0^{\frac{\pi}{2}} \sqrt{r^2 - r^2\sin^2 t}\,(r\cos t)\, dt$$

$$= \int_0^{\frac{\pi}{2}} (r\cos t)(r\cos t)\, dt = r^2 \int_0^{\frac{\pi}{2}} \cos^2 t\, dt$$

$$= \frac{r^2}{2} \int_0^{\frac{\pi}{2}} (1 + \cos 2t)\, dt$$

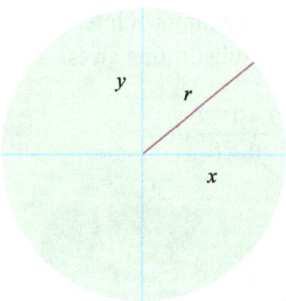

Figure 6.2.1: The area of a circle with radius r is πr^2.

$$= \frac{r^2}{2}\left[t + \frac{1}{2}\sin 2t\right]_0^{\frac{\pi}{2}}$$

$$= \frac{r^2}{2}\left(\frac{\pi}{2} + \frac{1}{2}\sin \pi - 0\right)$$

$$= \frac{\pi r^2}{4}.$$

Therefore, the area of the circle with radius r is $4 \times \frac{\pi r^2}{4} = \pi r^2$. □

Example 6.2.4. Find the area of the ellipse defined parametrically by

$$\begin{cases} x = a\cos t \\ y = b\sin t, \end{cases} \quad \text{where } 0 \leqslant t \leqslant 2\pi.$$

Solution. The area of the ellipse, A, is four times the area A_1, which is below the curve and above the x-axis in the first quadrant, as shown in Figure 6.2.2. Observe that, if $x = a\cos t$, $dx = -a\sin t \, dt$. When $x = 0$, $t = \frac{\pi}{2}$ and when $x = a$, $t = 0$. Thus,

$$A = 4A_1 = 4\int_0^a y \, dx = 4\int_{\frac{\pi}{2}}^0 b\sin t(-a\sin t) \, dt$$

$$= 4ab\int_0^{\frac{\pi}{2}} \sin^2 t \, dt = 2ab\int_0^{\frac{\pi}{2}} (1 - \cos 2t) \, dt$$

$$= 2ab\left(t - \frac{\sin 2t}{2}\right)\Big|_0^{\frac{\pi}{2}} = 2ab\left(\frac{\pi}{2} - \frac{\sin \pi}{2} - 0\right)$$

$$= \pi ab.$$

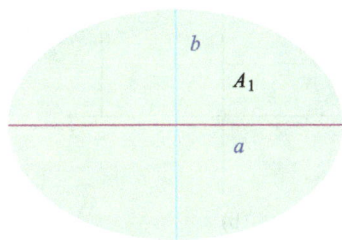

Figure 6.2.2: Area of the ellipse $\frac{x^2}{a^2} + \frac{y^2}{b^2} = 1$ is πab.

Example 6.2.5. If f is continuous, find $\frac{d}{dx}\int_0^x f(x - t) \, dt$.

Solution. Let $x - t = u$ and $t = x - u$, so $\frac{dt}{du} = -1$. When $t = 0$, $u = x$ and when $t = x$, $u = 0$. Then

$$\int_0^x f(x - t) \, dt = \int_x^0 f(u)(-du) = -\int_x^0 f(u) \, du = \int_0^x f(u) \, du,$$

so

$$\frac{d}{dx}\int_0^x f(x-t)\,dt = \frac{d}{dx}\int_0^x f(u)\,du = f(x).$$

Example 6.2.6. Prove that, for any constant a,

$$\int_{-a}^a f(x)\,dx = \begin{cases} 2\int_0^a f(x)\,dx, & \text{if } f(x) \text{ is an even function} \\ 0, & \text{if } f(x) \text{ is an odd function.} \end{cases}$$

Proof. We know that $\int_{-a}^a f(x)\,dx = \int_{-a}^0 f(x)\,dx + \int_0^a f(x)\,dx$. For the integral $\int_{-a}^0 f(x)\,dx$, we substitute $t = -x$ so that $dx = -dt$ and we obtain

$$\int_{-a}^0 f(x)\,dx = -\int_a^0 f(-t)\,dt = \int_0^a f(-t)\,dt = \int_0^a f(-x)\,dx.$$

Hence,

$$\int_{-a}^a f(x)\,dx = \int_{-a}^0 f(x)\,dx + \int_0^a f(x)\,dx$$
$$= \int_0^a [f(-x) + f(x)]\,dx.$$

Figure 6.2.3 illustrates the definite integral of an even/odd function.

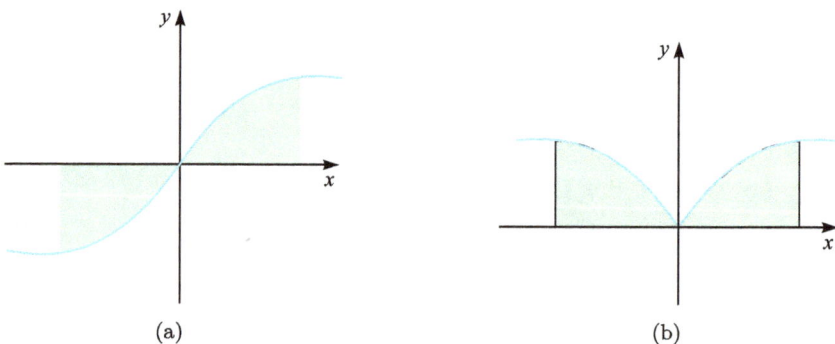

(a)   (b)

Figure 6.2.3: Integral of odd/even function over $[-a, a]$.

If $f(x)$ is an odd function, then $f(-x) = -f(x)$ and the integrand is $f(-x) + f(x) = 0$, so $\int_{-a}^a f(x)\,dx = 0$.

If $f(x)$ is an even function, then $f(-x) = f(x)$ and the integrand is $f(-x) + f(x) = 2f(x)$, so $\int_{-a}^a f(x)\,dx = 2\int_0^a f(x)\,dx$. □

Example 6.2.7. Evaluate $\int_{-1}^1 (\frac{x}{1+x^2+\cos x} + \cos x)\,dx$.

Solution. Since $\frac{x}{1+x^2+\cos x}$ is odd and $\cos x$ is even,

$$\int_{-1}^{1}\left(\frac{x}{1 + x^2 + \cos x} + \cos x\right)dx$$

$$= \int_{-1}^{1}\frac{x}{1 + x^2 + \cos x}\,dx + \int_{-1}^{1}\cos x\,dx$$

$$= 0 + 2\int_{0}^{1}\cos x\,dx$$

$$= 2\sin x|_{0}^{1} = 2\sin 1.$$

6.3 Integration by parts in definite integrals

The integration by parts formula used for indefinite integration, introduced in Section 6.1.5, can be used for definite integrals with minor changes (putting in the limits of integration). If $u = u(x)$ and $v = v(x)$ are differentiable on $[a, b]$, then the modified formula is

$$\int_{a}^{b}u\,dv = uv|_{x=a}^{x=b} - \int_{a}^{b}v\,du.$$

Example 6.3.1. Evaluate $\int_{0}^{\frac{\pi}{2}}x\cos x\,dx$.

Solution. Use integration by parts with $u = x$, $dv = \cos x\,dx$ so that $du = dx$ and let $v = \sin x$. Then

$$\int_{0}^{\pi/2}x\cos x\,dx = \int_{0}^{\pi/2}x(\sin x)'\,dx$$

$$= x\sin x|_{x=0}^{x=\pi/2} - \int_{0}^{\frac{\pi}{2}}(x)'\sin x\,dx$$

$$= \frac{\pi}{2} - (-\cos x)|_{0}^{\pi/2}$$

$$= \frac{\pi}{2} - 1.$$

Example 6.3.2. Find $\int_{0}^{\frac{1}{2}}\arcsin x\,dx$.

Solution. Use integration by parts with $u = \arcsin x$, $dv = dx$ so that $du = \frac{1}{\sqrt{1-x^2}}dx$ and let $v = x$. Then

$$\int_{0}^{\frac{1}{2}}\arcsin x\,dx = [x\arcsin x]_{0}^{\frac{1}{2}} - \int_{0}^{\frac{1}{2}}\frac{x}{\sqrt{1-x^2}}\,dx$$

$$= \frac{1}{2}\frac{\pi}{6} + [\sqrt{1-x^2}]_{0}^{\frac{1}{2}}$$

$$= \frac{\pi}{12} + \frac{\sqrt{3}}{2} - 1.$$

Example 6.3.3. Given $I_n = \int_0^{\pi/2} \sin^n x \, dx$, prove the reduction formula

$$I_n = \frac{n-1}{n} I_{n-2} \quad \text{for all integers } n \geq 3.$$

Then evaluate $I_7 = \int_0^{\pi/2} \sin^7 x \, dx$.

Solution. Similar to Example 6.1.37, we have

$$
\begin{aligned}
I_n &= \int_0^{\pi/2} \sin^n x \, dx = \int_0^{\pi/2} \sin^{n-1} x \cdot (\sin x) \, dx \\
&= \int_0^{\pi/2} \sin^{n-1} x \cdot (-\cos x)' \, dx \\
&= -\cos x \sin^{n-1} x \big|_0^{\pi/2} + \int_0^{\pi/2} (\sin^{n-1} x)' \cos x \, dx \\
&= (n-1) \int_0^{\pi/2} (\sin^{n-2} x \cdot \cos x \cdot \cos x) \, dx \\
&= (n-1) \int_0^{\pi/2} \sin^{n-2} x \cdot (1 - \sin^2 x) \, dx \\
&= (n-1) \int_0^{\pi/2} (\sin^{n-2} x - \sin^n x) \, dx = (n-1)(I_{n-2} - I_n),
\end{aligned}
$$

so

$$I_n = \frac{n-1}{n} I_{n-2}.$$

Since

$$I_1 = \int_0^{\pi/2} \sin x \, dx = -\cos x \big|_0^{\pi/2} = -(0-1) = 1,$$

we find

$$
\begin{aligned}
I_3 &= \frac{3-1}{3} I_1 = \frac{2}{3} \times 1 = \frac{2}{3}, \\
I_5 &= \frac{5-1}{5} I_3 = \frac{4}{5} \times \frac{2}{3} = \frac{8}{15}, \\
I_7 &= \frac{7-1}{7} I_5 = \frac{6}{7} \times \frac{8}{15} = \frac{16}{35}.
\end{aligned}
$$

6.4 Improper integrals

6.4.1 Improper integrals of the first kind

An integral $\int_a^b f(x) \, dx$ is a proper integral if f is continuous or piecewise continuous and the range of integration $[a, b]$ is finite. These are integrals we have been studying. If the interval of integration is unbounded, the integral is an *improper integral of the first kind*.

Definition 6.4.1. If f is continuous on the infinite interval $[a, \infty)$, the *improper integral of the first kind* $\int_a^{+\infty} f(x)\, dx$ is defined to be

$$\int_a^{+\infty} f(x)\, dx = \lim_{b \to +\infty} \int_a^b f(x)\, dx.$$

When this limit $L = \lim_{b \to +\infty} \int_a^b f(x)\, dx$ exists, we write $\int_a^{+\infty} f(x)\, dx = L$ and say that the improper integral is *convergent* (or we say it *converges to L*). Otherwise we say that the improper integral is *divergent* (or it *diverges*). Similarly, we define the improper integral $\int_{-\infty}^b f(x)\, dx$ as

$$\int_{-\infty}^b f(x)\, dx = \lim_{a \to -\infty} \int_a^b f(x)\, dx.$$

The improper integral $\int_{-\infty}^{\infty} f(x)\, dx$ is defined, for any c, to be

$$\int_{-\infty}^{\infty} f(x)\, dx = \int_c^{\infty} f(x)\, dx + \int_{-\infty}^c f(x)\, dx.$$

We say that $\int_{-\infty}^{\infty} f(x)\, dx$ is convergent if and only if both $\int_c^{\infty} f(x)\, dx$ and $\int_{-\infty}^c f(x)\, dx$ are convergent.

Example 6.4.1. Determine whether the integral $\int_1^{\infty} \frac{1}{x}\, dx$ is convergent or divergent.

Solution. According to the definition, we have

$$\int_1^{\infty} \frac{1}{x}\, dx = \lim_{b \to \infty} \int_1^b \frac{1}{x}\, dx = \lim_{b \to \infty} \ln |x| \big|_1^b$$
$$= \lim_{b \to \infty} (\ln b - \ln 1) = \infty.$$

The limit does not exist as a finite number, so the improper integral is divergent (diverges to infinity). Figure 6.4.1 shows the region below $y = \frac{1}{x}$ and to the right of $x = 1$.

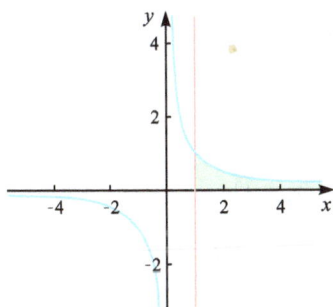

Figure 6.4.1: The area of the shaded region is undefined (diverges to infinity).

Example 6.4.2. Does the improper integral

$$\int_{-\infty}^{\infty} \frac{1}{1+x^2}\, dx$$

converge or diverge? If it converges, calculate its value.

Solution. We have

$$\int_{-\infty}^{\infty} \frac{1}{1+x^2}\, dx = \int_{-\infty}^{0} \frac{1}{1+x^2}\, dx + \int_{0}^{\infty} \frac{1}{1+x^2}\, dx$$

$$= \lim_{a\to-\infty} \int_{a}^{0} \frac{1}{1+x^2}\, dx + \lim_{b\to\infty} \int_{0}^{b} \frac{1}{1+x^2}\, dx$$

$$= \lim_{a\to-\infty} \arctan x\big|_{a}^{0} + \lim_{b\to\infty} \arctan x\big|_{0}^{b}$$

$$= \lim_{a\to-\infty} (-\arctan a) + \lim_{b\to\infty} (\arctan b)$$

$$= -\left(-\frac{\pi}{2}\right) + \frac{\pi}{2} = \pi,$$

so the integral converges and its value is π. Figure 6.4.2 illustrates this improper integral.

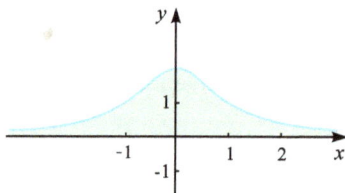

Figure 6.4.2: The area of the shaded region is π.

Example 6.4.3. Show that the improper integral $\int_{a}^{+\infty} \frac{dx}{x^p}\,(a>0)$ converges for all $p>1$ and diverges when $p\leqslant 1$.

Solution. Case 1: when $p=1$, we have

$$\int_{a}^{+\infty} \frac{dx}{x} = \lim_{b\to+\infty} \int_{a}^{b} \frac{dx}{x} = \lim_{b\to+\infty} \ln x\big|_{a}^{b} = \lim_{b\to+\infty} (\ln b - \ln a) = +\infty.$$

Case 2: when $p\neq 1$, we have

$$\int_{a}^{+\infty} \frac{dx}{x^p} = \lim_{b\to+\infty} \int_{a}^{b} \frac{dx}{x^p} = \lim_{b\to+\infty} \frac{1}{(-p+1)x^{p-1}}\bigg|_{a}^{b}$$

$$= \lim_{b\to+\infty} \left(\frac{1}{(-p+1)b^{p-1}} - \frac{1}{(-p+1)a^{p-1}} \right)$$

$$= \begin{cases} +\infty, & \text{when } p<1 \\ \frac{1}{(p-1)a^{p-1}}, & \text{when } p>1. \end{cases}$$

Therefore, we conclude that, when $p \leqslant 1$, the improper integral $\int_a^{+\infty} \frac{1}{x^p} \, dx \, (a > 0)$ diverges and, when $p > 1$, it converges.

Theorem 6.4.1 (Comparison theorem). *Assume that f and g are continuous on $[a, +\infty)$ and, for all sufficiently large x, $0 \leqslant f(x) \leqslant g(x)$. For improper integrals $\int_a^{+\infty} f(x) \, dx$ and $\int_a^{+\infty} g(x) \, dx$ of the first kind, we have:*

1. *if $\int_a^{+\infty} g(x) \, dx$ converges, then so does $\int_a^{+\infty} f(x) \, dx$;*
2. *if $\int_a^{+\infty} f(x) \, dx$ diverges, then so does $\int_a^{+\infty} g(x) \, dx$.*

Proof. Part 1: let $t \in (a, +\infty)$. Because $f(x) \geqslant 0$, $F(t) = \int_a^t f(x) \, dt$ is increasing on $(a, +\infty)$ and we have

$$0 \leqslant \int_a^t f(x) \, dt \leqslant \int_a^t g(x) \, dx \leqslant \int_a^{+\infty} g(x) \, dx.$$

However, if $\int_a^{+\infty} g(x) \, dx$ converges, say, $\int_a^{+\infty} g(x) \, dx = L$, then $F(t)$ is increasing and bounded above on $(a, +\infty)$, so the limit $\lim_{t \to +\infty} F(t)$ must exist. That is,

$$\lim_{t \to +\infty} F(t) = \lim_{t \to +\infty} \int_a^t f(x) \, dx \quad \text{exists,}$$

so $\int_a^{+\infty} f(x) \, dx$ converges.

Part 2 of the theorem is the converse negative proposition of Part 1, so it is true. ☐

Example 6.4.4. Use the comparison test to determine whether the following integrals converge or diverge:

$$\text{(a)} \int_1^{+\infty} \frac{1}{x^3 + 2x + 3} \, dx, \quad \text{(b)} \int_1^{+\infty} \frac{x^{3/2}}{1 + x^2} \, dx.$$

Solution. (a) When $x > 1$, $x^3 + 2x + 3 > x^3$, so

$$\int_1^{+\infty} \frac{1}{x^3 + 2x + 3} \, dx < \int_1^{+\infty} \frac{1}{x^3} \, dx.$$

By Example 6.4.3, we know that $\int_1^{+\infty} \frac{1}{x^3} \, dx$ converges, so

$$\int_1^{+\infty} \frac{1}{x^3 + 2x + 3} \, dx$$

also converges.

(b) When $x > 1$,

$$\frac{x^{3/2}}{1 + x^2} = \frac{1}{x^{-3/2} + x^{1/2}} = \frac{1}{\frac{1}{\sqrt[3]{x^2}} + \sqrt{x}} > \frac{1}{\sqrt{x} + \sqrt{x}} = \frac{1}{2\sqrt{x}} > 0.$$

Since $\int_1^{+\infty} \frac{1}{x^{\frac{1}{2}}} \, dx$ diverges, by the comparison test, $\int_1^{+\infty} \frac{x^{3/2}}{1+x^2} \, dx$ also diverges.

6.4.2 Improper integrals of the second kind

An integral is an improper integral of the second kind if the interval of integration is bounded but the integrand is not bounded on this interval.

Definition 6.4.2. If f is continuous on the interval $(a, b]$ and $\lim_{x \to a^+} f(x) = \pm\infty$ or the limit does not exist, then $\int_a^b f(x)\,dx$ is an *improper integral of the second kind*. It is defined to be

$$\int_a^b f(x)\,dx = \lim_{\varepsilon \to a^+} \int_\varepsilon^b f(x)\,dx.$$

When the limit $\lim_{\varepsilon \to a^+} \int_\varepsilon^b f(x)\,dx = L$ exists, we write $\int_a^b f(x)\,dx = L$ and say that the improper integral is *convergent* (or we say it *converges to L*). Otherwise we say that the improper integral is *divergent* (or we say it *diverges*).

Similarly, we define the divergence and convergence of the improper integral $\int_a^b f(x)\,dx$ for which $\lim_{x \to b^-} f(x) = \pm\infty$ as

$$\int_a^b f(x)\,dx = \lim_{\varepsilon \to b^-} \int_a^\varepsilon f(x)\,dx.$$

The improper integral $\int_a^b f(x)\,dx$ (for which $\lim_{x \to c} f(x) = \pm\infty$ for some $c \in (a, b)$) is defined to be the sum of two improper integrals of the second kind. We write

$$\int_a^b f(x)\,dx = \int_a^c f(x)\,dx + \int_c^b f(x)\,dx.$$

The improper integral $\int_a^b f(x)\,dx$ is convergent exactly when both $\int_a^c f(x)\,dx$ and $\int_c^b f(x)\,dx$ are convergent.

Example 6.4.5. Find $\int_2^5 \frac{1}{\sqrt{x-2}}\,dx$.

Solution. We note first that the given integral is improper because $\lim_{x \to 2^+} \frac{1}{\sqrt{x-2}} = +\infty$. The definition gives

$$
\begin{aligned}
\int_2^5 \frac{dx}{\sqrt{x-2}} &= \lim_{t \to 2^+} \int_t^5 \frac{dx}{\sqrt{x-2}} \\
&= \lim_{t \to 2^+} [2\sqrt{x-2}]_t^5 \\
&= \lim_{t \to 2^+} 2(\sqrt{3} - \sqrt{t-2}) \\
&= 2\sqrt{3}.
\end{aligned}
$$

Thus the given improper integral is convergent and has the value $2\sqrt{3}$. Figure 6.4.3 shows the region below the curve $y = \sqrt{x-2}$ and above the x-axis, and between $x = 2$ and $x = 5$.

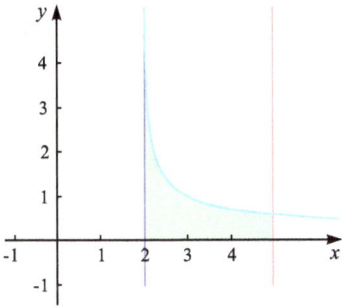

Figure 6.4.3: The area of the shaded region is $2\sqrt{3}$.

Example 6.4.6. Investigate the convergence of the improper integral $\int_{-1}^{1} \frac{1}{x^2}\,dx$.

Solution. The integrand $\frac{1}{x^2}$ has a discontinuity at $x = 0$ and $\lim_{x \to 0} \frac{1}{x^2} = +\infty$. Thus,

$$\int_{-1}^{1} \frac{1}{x^2}\,dx = \int_{-1}^{0} \frac{1}{x^2}\,dx + \int_{0}^{1} \frac{1}{x^2}\,dx$$

$$= \lim_{\varepsilon_1 \to 0^-} \int_{-1}^{\varepsilon_1} \frac{1}{x^2}\,dx + \lim_{\varepsilon_2 \to 0^+} \int_{\varepsilon_2}^{1} \frac{1}{x^2}\,dx$$

$$= \lim_{\varepsilon_1 \to 0^-} -\frac{1}{x}\Big|_{-1}^{\varepsilon_1} + \lim_{\varepsilon_2 \to 0^+} -\frac{1}{x}\Big|_{\varepsilon_2}^{1}$$

$$= \lim_{\varepsilon_1 \to 0^-} \left(\frac{-1}{\varepsilon_1} - 1\right) + \lim_{\varepsilon_2 \to 0^+} \left(-1 + \frac{1}{\varepsilon_2}\right).$$

Since

$$\lim_{\varepsilon_1 \to 0^-} \left(\frac{-1}{\varepsilon_1} - 1\right) = +\infty \quad \text{and} \quad \lim_{\varepsilon_2 \to 0^+} \left(-1 + \frac{1}{\varepsilon_2}\right) = +\infty,$$

we conclude that $\int_{-1}^{0} \frac{1}{x^2}\,dx$ and $\int_{0}^{1} \frac{1}{x^2}\,dx$ both diverge to ∞. Therefore, $\int_{-1}^{1} \frac{1}{x^2}\,dx$ also diverges to ∞. The improper integral is illustrated in Figure 6.4.4.

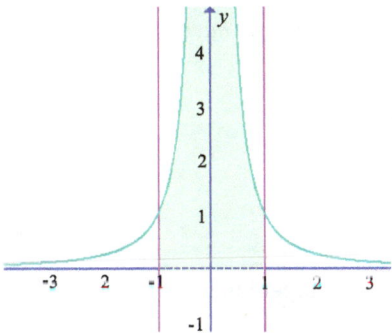

Figure 6.4.4: Graph of $y = 1/x^2$. The area of the shaded region is undefined (diverges to $+\infty$).

Example 6.4.7. Show that, for $b > a$, $\int_a^b \frac{dx}{(x-a)^p}$ diverges when $p \geq 1$ and converges when $p < 1$.

Solution. If $p > 0$, then the function $\frac{1}{(x-a)^p}$ has a discontinuity at $a = 0$ and $\lim_{x \to a^+} \frac{1}{(x-a)^p} = \infty$, but if $p \leq 0$, the integrand has no discontinuities on $[a,b]$.

Case 1: when $p = 1$, we have

$$\int_a^b \frac{1}{(x-a)^p} \, dx = \lim_{\varepsilon \to a^+} \int_\varepsilon^b \frac{1}{x-a} \, dx = \lim_{\varepsilon \to a^+} \ln|x-a| \Big|_\varepsilon^b$$
$$= \lim_{\varepsilon \to a^+} (\ln|b-a| - \ln|\varepsilon - a|) = +\infty.$$

Case 2: when $p \neq 1$, we have

$$\int_a^b \frac{1}{(x-a)^p} \, dx = \lim_{\varepsilon \to a^+} \int_\varepsilon^b \frac{1}{(x-a)^p} \, dx = \lim_{\varepsilon \to a^+} \frac{1}{1-p}(x-a)^{1-p} \Big|_\varepsilon^b$$
$$= \frac{1}{1-p} \lim_{\varepsilon \to a^+} ((b-a)^{1-p} - (\varepsilon - a)^{1-p})$$
$$= \begin{cases} \frac{1}{1-p}(b-a)^{1-p}, & \text{when } p < 1 \\ +\infty, & \text{when } p > 1. \end{cases}$$

Thus, $\int_a^b \frac{1}{(x-a)^p} \, dx$ diverges to $+\infty$ when $p \geq 1$ and converges when $p < 1$.

NOTE. Similarly, it can be proved that $\int_a^b \frac{1}{(b-x)^p} \, dx$ converges for $p < 1$ and diverges for $p \geq 1$.

We now prove the following theorem, which offers a way to determine whether or not some improper integrals converge.

Theorem 6.4.2 (Comparison theorem). *Suppose that f and g are continuous on (a,b) with $f(x) \geq g(x) \geq 0$ for all x near a. For improper integrals*

$$\int_a^b f(x) \, dx \text{ and } \int_a^b g(x) \, dx \text{ of the second kind, we have:}$$

1. *if $\int_a^b f(x) \, dx$ converges, then so does $\int_a^b g(x) \, dx$;*
2. *if $\int_a^b g(x) \, dx$ diverges, then so does $\int_a^b f(x) \, dx$.*

Proof. Similar to Theorem 6.4.1. □

Example 6.4.8. Determine the convergence of the improper integral

$$\int_1^3 \frac{1}{\ln x} \, dx.$$

Solution. Since $\ln(1 + x) < x$ for $x > 0$,

$$\ln x = \ln(1 + x - 1) < x - 1 \quad \text{for } x > 1.$$

Then

$$\frac{1}{\ln x} > \frac{1}{x - 1}, \quad \text{for } x > 1.$$

We know $\int_1^3 \frac{1}{(x-1)}\,dx$ diverges from Example 6.4.7. Therefore, $\int_1^3 \frac{1}{\ln x}\,dx$ also diverges. Figure 6.4.5 shows the region between $x = 1$ and $x = 3$, below the curve $y = \frac{1}{\ln x}$, and above the x-axis.

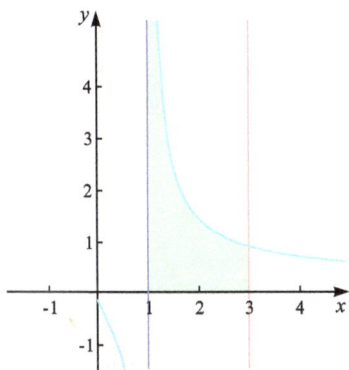

Figure 6.4.5: The area of the region is undefined (diverges to infinity).

Example 6.4.9. Determine the convergence of the improper integral

$$\int_0^1 \frac{1}{\sqrt{(1 - x^2)(1 - k^2 x^2)}}\,dx, \quad \text{where } 0 < k < 1 \text{ is a constant.}$$

Solution. For $0 < x < 1$ and $0 < k < 1$,

$$\frac{1}{\sqrt{(1 - x^2)(1 - k^2 x^2)}} = \frac{1}{\sqrt{(1 + x)(1 - x)(1 - k^2 x^2)}}$$

$$\leqslant \frac{1}{\sqrt{(1 - x)(1 - k^2 x^2)}}$$

$$\leqslant \frac{1}{\sqrt{1 - k^2}} \frac{1}{\sqrt{1 - x}},$$

but

$$\int_0^1 \frac{1}{\sqrt{1 - x}}\,dx = \lim_{b \to 1^-} \int_0^b \frac{1}{\sqrt{1 - x}}\,dx = \lim_{b \to 1^-} (-2\sqrt{1 - x})|_0^b = 2,$$

so $\int_0^1 \frac{1}{\sqrt{1-x}}\,dx$ converges. Therefore, $\int_0^1 \frac{1}{\sqrt{(1-x^2)(1-k^2x^2)}}\,dx$ converges.

6.5 Exercises

1. Evaluate each of the following integrals:

    (a) $\int f'(x)\,dx$; (b) $\frac{d}{dt}(\int f(t)\,dt)$; (c) $\int d(f(x))$; (d) $\int f'(\theta)\,d\theta$.

2. Find each of the following indefinite integrals:

    (a) $\int 4x^2\,dx$;

    (b) $\int x\sqrt{x}\,dx$;

    (c) $\int \sqrt{x}\sqrt{x}\,dx$;

    (d) $\int \frac{\sqrt{x}+2\sqrt[3]{x^2}+1}{\sqrt{x}}\,dx$;

    (e) $\int \frac{(x^2+1)^2}{x^3}\,dx$;

    (f) $\int (2-\sec^2 x)\,dx$;

    (g) $\int (\cos t + 2\sin t)\,dt$;

    (h) $\int \frac{\sin x}{3} - \frac{1}{\cos^2 x} + \frac{6}{1+x^2}\,dx$;

    (i) $\int (e^x + 2^x)\,dx$;

    (j) $\int \frac{3}{\sqrt{1-x^2}}\,dx$;

    (k) $\int \frac{1}{\sin^2 \frac{x}{2}\cos^2 \frac{x}{2}}\,dx$;

    (l) $\int \sec x(\sec x - \tan x)\,dx$.

3. Find $\int f(x)\,dx$ if $f(x) = \begin{cases} \sin x, & x < 0 \\ x^2, & x \geq 0. \end{cases}$

4. Find $f(x)$ if $f'(x) = e^x + 2(1+x^2)^{-1}$ and $f(0) = -2$.

5. Solve the following differential equations:

    (a) $\frac{dy}{dx} = 2x - 1$; (b) $r'' - 6t = 0$, $r(0) = 0$, and $r'(0) = 1$;

    (c) $y'''(x) = 1$; (d) $x\,dx - y\,dy = 0$;

    (e) $y' + xy = 0$; (f) $\frac{dP}{dt} = -kt$, $P(0) = 1000$.

6. The radioactive decay rate of the isotope C-14 in a certain substance is modeled by $\frac{dC}{dt} = -0.000\,121\,6C$, where t is measured in years and C in grams. If the substance initially contains 20 g of C-14, then find (a) how much is left after 10 000 years and (b) the time point when 10 g of the C-14 remains.

7. Use a suitable substitution to find the general antiderivative of each of the following functions:

    (a) $\int (x+1)^{20}\,dx$;

    (b) $\int \frac{dx}{(2x-3)^5}$;

    (c) $\int x^2(\sqrt{1+x^3})\,dx$;

    (d) $\int \frac{x}{(1+x^2)^2}\,dx$;

    (e) $\int \cos 2x + e^{-x}\,dx$;

    (f) $\int \sqrt{1+2x}\,dx$;

    (g) $\int (1-\cos 2x)\sin 2x\,dx$;

    (h) $\int \tan\frac{t}{2}\sec^2\frac{t}{2}\,dt$;

    (i) $\int xe^{-x^2}\,dx$;

    (j) $\int \frac{\ln^2 x}{x}\,dx$;

    (k) $\int \frac{dx}{e^x+e^{-x}}$;

    (l) $\int \sin^4 x\cos x\,dx$;

    (m) $\int \frac{\sin x}{\sqrt{\cos^3 x}}\,dx$;

    (n) $\int \frac{\arctan x}{1+x^2}\,dx$;

    (o) $\int \frac{x^3}{1+x^2}\,dx$;

    (p) $\int \frac{dx}{1-\cos x}$;

    (q) $\int \frac{dx}{(\arcsin x)^2\sqrt{1-x^2}}$;

    (r) $\int x^2\sqrt[3]{1-x}\,dx$;

    (s) $\int \frac{x^2}{(1-x)^{100}}\,dx$;

    (t) $\int \cos^3 x\sqrt{\sin x}\,dx$;

    (u) $\int \frac{e^{2x}}{\sqrt[4]{e^x+1}}\,dx$;

    (v) $\int \frac{dx}{\sqrt{1+e^x}}$;

    (w) $\int \frac{\ln x}{x\sqrt{1+\ln x}}\,dx$;

    (x) $\int \frac{\cos\sqrt{x}}{\sqrt{x}}\,dx$;

    (y) $\int \frac{dx}{x^4(1+x^2)}$;

    (z) $\int \frac{1}{1+\cos x}\,dx$;

    (aa) $\int \sqrt{\tan u}\sec^2 u\,du$;

    (bb) $\int \tan(4t+2)\,dt$;

    (cc) $\int \frac{\ln^6 y}{y}\,dy$;

    (dd) $\int \frac{1}{\sin^2 3x}\,dx$;

    (ee) $\int \frac{6\cos z}{(1+\sin z)^2}\,dz$;

    (ff) $\int \frac{x}{x^2+1}\,dx$;

    (gg) $\int \frac{1}{\theta\ln\theta\ln\ln\theta}\,d\theta$;

    (hh) $\int \tan 2x\,dx$;

    (ii) $\int \frac{1}{\sin 2x} - 3\cot 3x + x(\cos x^2)(\sin x^2)^3\,dx$.

8. Assume $I_1 = \int \frac{\cos x}{\cos x + \sin x}\, dx$ and $I_2 = \int \frac{\sin x}{\cos x + \sin x}\, dx$. By considering the integrals $I_1 + I_2$ and $I_1 - I_2$, find I_1 and I_2. Similarly, find $\int \frac{\cos x}{a\cos x + b\sin x}\, dx$ where a, b are two nonzero constants and $\int \frac{\cos x + 3}{3\cos x + 4\sin x + 25}\, dx$.

9. Use a suitable substitution to find each of the following integrals:

   (a) $\int \frac{1}{1+\sqrt{t}}\, dt$; (b) $\int \sqrt{1 - x^2}\, dx$; (c) $\int \sqrt{a^2 - x^2}\, dx$;

   (d) $\int \frac{1}{\sqrt{a^2 - x^2}}\, dx$; (e) $\int \sqrt{a^2 + x^2}\, dx$; (f) $\int \sqrt{t^2 - a^2}\, dt$;

   (g) $\int \frac{1}{\sqrt{1+e^y}}\, dy$; (h) $\int \frac{1}{\sqrt{x^2+1}}\, dx$; (i) $\int \frac{1}{x(x^7+1)}\, dx$. [Hint: let $t = \frac{1}{x}$.]

10. Use integration by parts to find each of the following integrals:

    (a) $\int x \ln(1-x)\, dx$; (b) $\int \arcsin x\, dx$;

    (c) $\int \frac{\arctan e^x}{e^x}\, dx$; (d) $\int x^2 e^{-x}\, dx$;

    (e) $\int (x \sin^2 x)\, dx$; (f) $\int \sin(\ln x)\, dx$;

    (g) $\int \arctan x\, dx$; (h) $\int (\ln x)^2\, dx$;

    (i) $\int e^{\sqrt{x}}\, dx$; (j) $\int (e^{ax} \cos bx)\, dx$.

11. Use integration by parts to establish the reduction formula for:

    (a) $\int x^n \sin x\, dx = -x^n \cos x + nx^{n-1} \sin x - n(n-1) \int x^{n-2} \sin x\, dx$;

    (b) $\int (\ln x)^n\, dx = x(\ln x)^n - n \int (\ln x)^{n-1}\, dx$;

    (c) $\int x^n e^{ax}\, dx = \frac{x^n e^{ax}}{a} - \frac{n}{a} \int x^{n-1} e^{ax}\, dx,\ a \neq 0$;

    (d) $\int \sec^n x\, dx = \frac{\tan x \sec^{n-2} x}{n-1} + \frac{n-2}{n-1} \int \sec^{n-2} x\, dx$, for $n \geq 2$;

    (e) If $I_{m,n} = \int \frac{x^m}{(\ln x)^n}\, dx$, prove that $(n-1)I_{m,n} = -\frac{x^{m+1}}{(\ln x)^{n-1}} + (m+1)I_{m,n-1}$.

12. Show that, for $k > 1$,

$$\int \frac{dx}{((x-a)^2 + b^2)^k} = \frac{x-a}{2b^2(k-1)((x-a)^2 + b^2)^{k-1}} + \frac{2k-3}{2b^2(k-1)} \int \frac{dx}{((x-a)^2 + b^2)^{k-1}}.$$

13. Evaluate each of the following indefinite integrals:

    (a) $\int \frac{x^3}{x+3}\, dx$; (b) $\int \frac{x^2+1}{(x+1)^2(x-1)}\, dx$;

    (c) $\int \frac{x^2+4x+1}{x^3-x^2+x-1}\, dx$; (d) $\int \frac{2x^2+5}{(x^2+1)^2}\, dx$;

    (e) $\int \frac{2x+3}{(x-1)(x+5)}\, dx$; (f) $\int \frac{1}{x^4-1}\, dx$;

    (g) $\int \frac{dx}{\sqrt[3]{(x+1)^2(x-1)^4}}$; (h) $\int \frac{dx}{\sqrt{x}+\sqrt[4]{x}}$;

    (i) $\int \frac{1}{(1+\sqrt[3]{x})\sqrt{x}}\, dx$; (j) $\int \frac{dx}{3+\cos x}$.

14. (**Logistic modeling of population growth**) In 1838, Pierre Verhulst derived a differential equation to describe the self-limiting growth of a biological population as follows:

$$\frac{dP}{dt} = kP(M - P).$$

Notice that the rate of reproduction is proportional to both the existing population P and the amount of available resources M. Alfred J. Lotka derived this type of equation again in 1925, calling it the *law of population growth*.

(a) Find the exact solution of the logistic equation if $P(0) = P_0$.

(b) When does the population grow the fastest? Is this critical time an inflection point of $P(t)$?

(c) What happens to P when $t \to \infty$?

(d) Sketch the graph of $P(t)$, given that $P(0) = 10$, $k = 0.02$, and $M = 100$.

15. Find each of the following definite integrals:

(a) $\int_0^\pi \sin 2x \, dx$;

(b) $\int_0^2 (x-1)^{20} \, dx$;

(c) $\int_{-\pi}^\pi \frac{xe^{-x^2}}{2+x^2} + \cos^2 x \, dx$;

(d) $\int_{-1}^1 \frac{5z}{(4+z^2)^2} \, dz$;

(e) $\int_0^2 \frac{\sqrt{x}}{1+x} \, dx$;

(f) $\int_0^1 r\sqrt{1-r^2} \, dr$;

(g) $\int_{-\pi/4}^0 \tan y \sec^2 y \, dy$;

(h) $\int_1^3 \frac{e^x}{3+e^x} \, dx$;

(i) $\int_1^{e^3} \frac{dx}{x\sqrt{1+\ln x}}$;

(j) $\int_{-\pi}^\pi \frac{\cos x}{\sqrt{5+4\sin x}} \, dx$;

(k) $\int_0^1 x^3 \sqrt{x^4+9} \, dx$;

(l) $\int_0^a \sqrt{x^2+a^2} \, dx$;

(m) $\int_0^2 y \, dx$, if $y = 3\sin t$ and $x = 2\cos t$;

(n) $\int_{-\pi}^\pi (\cos nx \cos mx) \, dx$, $n, m \in \mathbf{Z}$;

(o) $\int_{-\pi}^\pi (\cos nx \sin mx) \, dx$, $n, m \in \mathbf{Z}$;

(p) $\int_{-\pi}^\pi (\sin nx \sin mx) \, dx$, $n, m \in \mathbf{Z}$.

16. Given that $a < b$ are two constants, show that $\int_a^b \frac{1}{\sqrt{(x-a)(b-x)}} \, dx = \pi$. [Hint: use the substitution $x = a\cos^2 t + b\sin^2 t$.]

17. Assume $f(x)$ is continuous everywhere. Find $\frac{dy}{dx}$ if $y = \int_0^{x^2} f(x^2 - t) \, dt$.

18. Assume $f(x) = \begin{cases} x^2, & x < 0, \\ \frac{1}{1+x}, & x \geq 0. \end{cases}$ Find $\int_0^2 f(x-1) \, dx$.

19. Assume $f(x)$ is differentiable and $f(0) = 0$. If $g(x) = \int_0^1 f(xt) \, dt$, show that $g'(0) = \frac{f'(0)}{2}$.

20. If $f(x) = \int_0^x \cos((x-t)^2) \, dt$, find $f'(x)$.

21. Using substitution, show that $\int_x^1 \frac{1}{1+t^2} \, dt = \int_1^{\frac{1}{x}} \frac{1}{1+t^2} \, dt$.

22. Show that $\int_0^1 x^m (1-x)^n \, dx = \int_0^1 x^n (1-x)^m \, dx$, where m and n are positive integers.

23. Assume $f(x)$ is continuous on $[a,b]$. Show that $\int_a^b f(x) \, dx = \int_a^b f(a+b-x) \, dx$.

24. Use the substitution $u = \pi - x$ to show that

$$\int_0^\pi xf(\sin x) \, dx = \frac{\pi}{2} \int_0^\pi f(\sin x) \, dx.$$

By the same or another method, evaluate the integral

$$\int_0^\pi \frac{x\sin x}{1+\cos^2 x} \, dx.$$

25. The **sine integral function** $\mathrm{Si}(x) = \int_0^x \frac{\sin t}{t} \, dt$ is used in optics.

(a) Find the value of $\mathrm{Si}(0)$.

(b) Find the value(s) of x such that $\text{Si}(x)$ has a local extreme value on $[-2\pi, 2\pi]$.

(c) Show that $\text{Si}(x)$ is an odd function.

26. Find the average value of the function $f(x) = \frac{\sqrt{x^2-1}}{x}$ for $1 \leqslant x \leqslant 2$.

27. Evaluate each of the following definite integrals:

(a) $\int_0^1 xe^{-x}\, dx$; (b) $\int_0^{\pi/2} x\sin 2x\, dx$; (c) $\int_0^{\pi/2} u^2 \cos 2u\, du$;

(d) $\int_{\frac{\pi}{4}}^{\frac{\pi}{3}} \frac{x}{\sin^2 x}\, dx$; (e) $\int_1^e \ln^2 x\, dx$; (f) $\int_0^\pi e^{2x} \sin 3x\, dx$;

(g) $\int_0^9 e^{\sqrt{3x+9}}\, dx$; (h) $\int_0^{\pi^2} \sin \sqrt{x}\, dx$; (i) $\int_1^4 \frac{\ln x}{\sqrt{x}}\, dx$.

28. Use the trapezoidal rule with intervals of unit length to approximate the integral $\int_1^n \ln x\, dx$, where n is a positive integer that is greater than 2. Then deduce that $n! \approx e\sqrt{n}(\frac{n}{e})^n$ (note: the famous **Stirling approximation** to $n!$ is $n! \approx 2\pi n(\frac{n}{3})^n$).

29. (**Reduction formulas**)

(a) Given that $I_n = \int_0^1 x^n e^{-x}\, dx$, show that $I_n = nI_{n-1} - e^{-1}$ for $n \geqslant 1$.

(b) If $I_n = \int_0^{\pi/3} \cos^n x\, dx$, show that $nI_n = \frac{\sqrt{3}}{2^n} + (n-1)I_{n-2}$ for $n \geqslant 2$ and find I_3.

(c) If $I_n = \int_0^{\pi/4} \tan^n x\, dx$, prove that $I_n = \frac{1}{n-1} - I_{n-2}$ for $n \geqslant 2$ and find I_4.

(d) If $I_n = \int_0^1 x^n (1-x)^{\frac{1}{2}}\, dx$, show that $(2n+3)I_n = 2nI_{n-1}$ for $n \geqslant 1$ and find I_3.

(e) If $I_{m,n} = \int_1^e x^m (\ln x)^n\, dx$, prove that $(m+1)I_{m,n} = e^{m+1} - nI_{m,n-1}$ and find $I_{1,2}$.

30. If $I_n = \int_0^{\frac{\pi}{2}} \sin^n x\, dx$, use the reduction formula derived in Example 6.3.3 to show:

(a) $I_{2n} = \frac{\pi}{2} \times \frac{1}{2} \times \frac{3}{4} \times \frac{5}{6} \times \cdots \times \frac{(2n-1)}{2n}$;

(b) $I_{2n+1} = 1 \times \frac{2}{3} \times \frac{4}{5} \times \frac{6}{7} \times \cdots \times \frac{2n}{2n+1}$.

31. Determine whether each of the following improper integrals converges and, if possible, determine the value to which it converges:

(a) $\int_1^{+\infty} \frac{1}{x^3}\, dx$; (b) $\int_1^{+\infty} \frac{1}{\sqrt{x}}\, dx$;

(c) $\int_0^{+\infty} e^{-ax}\, dx$, $a > 0$; (d) $\int_0^{+\infty} \frac{dx}{\sqrt{x}(4+x)}$;

(e) $\int_{-\infty}^{\infty} \frac{dx}{x^2+2x+2}$; (f) $\int_0^1 \frac{x}{\sqrt{1-x^2}}\, dx$;

(g) $\int_0^1 \frac{1}{(1-x)^p}\, dx$; (h) $\int_0^{+\infty} (e^{-pt} \sin at)\, dt$, $a > 0$ and $p > 0$;

(i) $\int_0^e \frac{1}{x(\ln(x))^2}\, dx$.

32. Are the following arguments true or false?

(a) Since $\frac{x}{1+x^2}$ is an odd function, $\int_{-\infty}^{\infty} \frac{x}{x^2+1}\, dx = 0$.

(b) $\int_{-\infty}^{\infty} \frac{x}{x^2+1}\, dx = \lim_{a\to\infty} \int_{-a}^{a} \frac{x}{x^2+1}\, dx = 0$, since $\frac{x}{1+x^2}$ is an odd function.

(c)

$$\int_{-\infty}^{\infty} \frac{x}{x^2+1}\, dx = \lim_{a\to\infty} \int_0^a \frac{x}{x^2+1}\, dx + \lim_{a\to-\infty} \int_a^0 \frac{x}{x^2+1}\, dx$$

$$= \lim_{a\to\infty} \frac{1}{2}\ln(1+x^2)\Big|_0^a + \lim_{a\to-\infty} \frac{1}{2}\ln(1+x^2)\Big|_a^0$$

$$= \frac{1}{2}\ln a^2 - \frac{1}{2}\ln a^2 = 0.$$

33. Suppose that $\int_{-\infty}^{\infty} f(x)\,dx$ is convergent. If a and b are two real numbers, show that

$$\int_{-\infty}^{a} f(x)\,dx + \int_{a}^{\infty} f(x)\,dx = \int_{-\infty}^{b} f(x)\,dx + \int_{b}^{\infty} f(x)\,dx.$$

34. Determine the convergence of each of the following improper integrals:

   (a) $\int_{1}^{+\infty} \frac{1}{x^4+1}\,dx$; (b) $\int_{0}^{+\infty} \frac{1}{\sqrt{x^2+2x+7}}\,dx$; (c) $\int_{-\infty}^{+\infty} \frac{1}{2\sin(x^2)+3+x^2}\,dx$.

35. Express the limit $\lim_{n\to\infty} \frac{\sqrt[n]{n!}}{n}$ as an improper integral and then evaluate the limit.

36. **The gamma function** $\Gamma(x)$ is defined by the integral

$$\Gamma(x) = \int_{0}^{+\infty} e^{-t} t^{x-1}\,dt, \quad x > 0$$

   (the notion Γ is due to *Legendre*). When $x > 0$, it is also called an *Euler integral of the second kind*. The gamma function has applications in quantum physics, astrophysics, and fluid dynamics, as well as in statistics.

   (a) Use integration by parts or any other method to show that $\Gamma(x+1) = x\Gamma(x)$.

   (b) Deduce that $\Gamma(n) = (n-1)!$, where n is a nonnegative integer.

   (c) Show that $\lim_{x\to 0^+} \Gamma(x) = \infty$.

37. (**Limit comparison test**) Let $f(x)$ be a nonnegative continuous function on $[a,\infty)$, $a > 0$.

   (a) Assume $g(x)$ is another nonnegative continuous function defined on $[a,+\infty)$ with $\lim_{x\to+\infty} \frac{f(x)}{g(x)} = L$.

   (i) If $0 < L < \infty$, show that $\int_{a}^{+\infty} g(x)\,dx$ and $\int_{a}^{+\infty} f(x)\,dx$ both diverge or both converge.

   (ii) If $L = 0$, show that, if $\int_{a}^{+\infty} g(x)\,dx$ converges, so does $\int_{a}^{+\infty} f(x)\,dx$.

   (iii) If $L = +\infty$, show that, if $\int_{a}^{+\infty} g(x)\,dx$ diverges, so does $\int_{a}^{+\infty} f(x)\,dx$.

   (b) If there is a number $p > 1$ such that the $\lim_{x\to+\infty} x^p f(x)$ exists, show that the improper integral $\int_{a}^{+\infty} f(x)\,dx$ converges.

   (c) If $\lim_{x\to+\infty} xf(x)$ exists (or is $+\infty$), show that the improper integral $\int_{a}^{+\infty} f(x)\,dx$ diverges.

38. (**Absolute convergence**) If the improper integral $\int_{a}^{+\infty} |f(x)|\,dx$ converges, we say that the improper integral $\int_{a}^{+\infty} f(x)\,dx$ converges absolutely. By considering the nonnegative function $\phi(x) = f(x) + |f(x)|$ or by any other method:

   (a) show that, if $\int_{a}^{+\infty} f(x)\,dx$ converges absolutely, the improper integral $\int_{a}^{+\infty} f(x)\,dx$ itself also converges;

   (b) determine whether or not $\int_{0}^{+\infty} e^{-ax} \cos bx\,dx$ converges absolutely, where a and b are two positive constants.

39. (**Normal probability density function**) If a random variable X satisfies a **normal distribution** with mean u and standard deviation σ, then its probability density function is

$$f(x) = \frac{1}{\sigma\sqrt{2\pi}} e^{-\frac{1}{2}(\frac{x-u}{\sigma})^2}.$$

When $u = 0$ and $\sigma = 1$, it is known as the **standard normal distribution**. Use a graphing utility to:

(a) sketch the graph of the probability density function for a standard normal distribution;

(b) evaluate $\int_{-n}^{n} f(x)\,dx$ for $n = 1, 3, 10, 100$ when $u = 0$ and $\sigma = 1$;

(c) give an argument why the improper integral $\int_{-\infty}^{\infty} f(x)\,dx$ converges to 1. [Hint: consider $e^{-x^2} < e^{-x}$ and $\int_{b}^{\infty} e^{-x}\,dx \to 0$ when $b \to \infty$.]

7 Applications of the definite integral

In this chapter, you will learn about:
- *finding areas between curves;*
- *finding the volume of a solid;*
- *finding the length of curves;*
- *applying integrals to other subject areas.*

In this chapter, we show how the definite integral can be used to compute the area between two curves, the volume of a solid of revolution, the length of a curve, a distance traveled, the center of mass, fluid pressure, and probabilities. These are just a few of the applications of the definite integral.

7.1 Areas, volumes, and arc lengths

7.1.1 The area of the region between two curves

We wish to find the area A of a region R between the graphs of the functions f and g and the two lines $x = a$ and $x = b$. We already computed the area between two curves in Example 5.2.8 in Chapter 5. Using that approach and assuming $f(x) \geqslant g(x)$ for $a \leqslant x \leqslant b$, we calculate A by calculating the area A_f under $y = f(x)$ (above the x-axis) and subtracting from this the area A_g under $y = g(x)$ (above the x-axis), as seen in Figure 7.1.1, giving the result

$$A = A_f - A_g = \int_a^b f(x)\,dx - \int_a^b g(x)\,dx. \tag{7.1}$$

We now want to derive the integral formula for the area below the curve $y = f(x)$, above the curve $y = g(x)$, and between $x = a$ and $x = b$ by using the definition of a definite integral. This approach enables us to apply the results much more flexibly and generally. We use the idea of an *element of area* to calculate the area A of R. This approach is similar to the use of rectangles to approximate the area under a curve when

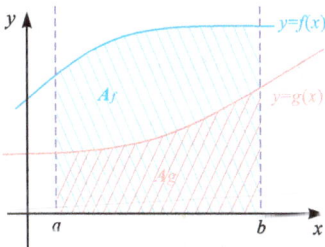

Figure 7.1.1: Area between two curves.

https://doi.org/10.1515/9783110527780-007

we interpreted the definite integral as an area. In fact, we are about to give an abbreviated version of the process used to define the definite integral given in Chapter 5. This abbreviated form could, if needed, be developed into a full derivation of the definite integral using limits of Riemann sums. Furthermore, the same approach will be used in many different applications to find a solution involving a definite integral.

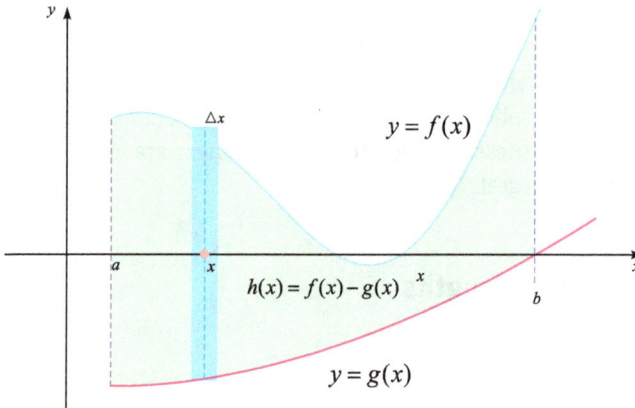

Figure 7.1.2: Vertical area element.

Assuming $f(x) \geqslant g(x)$ for $a \leqslant x \leqslant b$, a typical element of area is a thin rectangle of height $h(x) = f(x) - g(x)$ and width Δx, as shown in Figure 7.1.2. The coordinate variable x locates this rectangle. The height $h(x)$ of the rectangle varies with x. The area of this element is

$$\Delta A = h(x)\Delta x = (f(x) - g(x))\Delta x.$$

Summing the areas for all elements that fill the space from $x = a$ to $x = b$ gives the following approximate area A of the region R between $y = f(x)$ and $y = g(x)$:

$$A \approx \sum \Delta A = \sum (f(x) - g(x))\Delta x.$$

The limit as $\Delta x \to 0$ (as the width of the element goes to zero) is defined to be the actual area and this is, in fact, the limit of a Riemann sum, so it becomes the following definite integral:

$$A = \lim_{\Delta x \to 0} \sum (f(x) - g(x))\Delta x = \int_a^b (f(x) - g(x))\, dx. \tag{7.2}$$

Similarly, if the region is bounded by two curves $x = \phi(y)$ and $x = \psi(y)$ and the two lines $y = c$ and $y = d$, as shown in Figure 7.1.3, then we could choose horizontal area

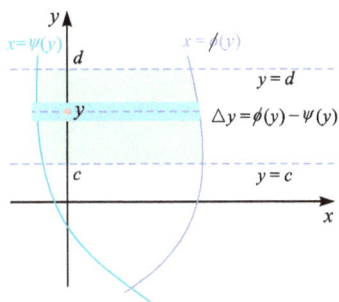

Figure 7.1.3: Horizontal area element.

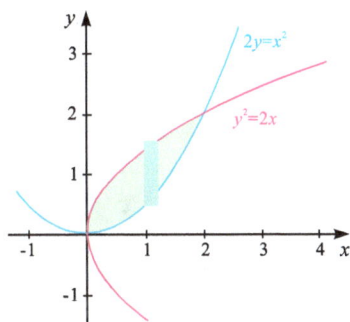

Figure 7.1.4: Vertical area element for the shaded region bounded by curves in Example 7.1.1.

elements. The area of the region is then given by

$$A = \lim_{\Delta y \to 0} \sum (\phi(y) - \psi(y))\Delta y = \int_c^d (\phi(y) - \psi(y))\,dy. \tag{7.3}$$

Example 7.1.1. Find the area of the region between the two parabolas $x^2 = 2y$ and $y^2 = 2x$, illustrated in Figure 7.1.4.

Solution. If $y^2 = 2x$, then $y = \pm\sqrt{2x}$ and the part of this curve above the x-axis is given by $y = \sqrt{2x}$. Hence the required area is below $y = \sqrt{2x}$ and above $y = \frac{x^2}{2}$, between $x = 0$ and $x = 2$. We compute the area using equation (7.2) as follows:

$$A = \int_0^2 dA = \int_0^2 \left(\sqrt{2x} - \frac{x^2}{2}\right) dx$$

$$= \left(\sqrt{2} \cdot \frac{x^{\frac{3}{2}}}{\frac{3}{2}} - \frac{1}{2} \cdot \frac{x^3}{3}\right)\Big|_0^2 = \frac{4}{3}.$$

Example 7.1.2. Calculate the area of the region between the graphs of the functions f and g for $0 \leqslant x \leqslant 30$, when

$$f(x) = \frac{6}{5}x^{1/2} - \frac{1}{25}x^{3/2} \quad \text{and} \quad g(x) = \frac{1}{30}x^{4/3} - x^{1/3}.$$

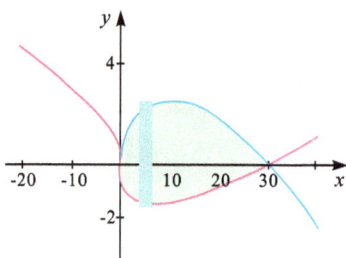

Figure 7.1.5: Vertical area element for the shaded region bounded by curves in Example 7.1.2.

Solution. As can be seen in Figure 7.1.5, the curve at the top, $y = \frac{6}{5}x^{1/2} - \frac{1}{25}x^{3/2}$, is always above the curve at the bottom, $y = \frac{1}{30}x^{4/3} - x^{1/3}$. Hence the area between the curves is

$$A = \int_a^b dA = \int_a^b (f(x) - g(x))\,dx$$

$$= \int_0^{30} \left(\frac{6}{5}x^{1/2} - \frac{1}{25}x^{3/2}\right) - \left(\frac{1}{30}x^{4/3} - x^{1/3}\right) dx$$

$$= \left(\frac{6}{5}\cdot\frac{2}{3}x^{3/2} - \frac{1}{25}\cdot\frac{2}{5}x^{5/2} - \frac{1}{30}\cdot\frac{3}{7}x^{7/3} + \frac{3}{4}x^{4/3}\right)\Big|_0^{30}$$

$$\approx 82.544.$$

Example 7.1.3. Find the area between the curves $x = y^2$ and $x = y + 2$, using horizontal area elements.

Solution. The curves intersect at the points $(1, -1)$ and $(4, 2)$. Figure 7.1.6 shows the curves and one horizontal element. Summing the areas of the horizontal elements of width $(y + 2) - y^2$ and thickness dy gives the area

$$A = \int_{-1}^2 (y + 2 - y^2)\,dy = \left(\frac{1}{2}y^2 + 2y - \frac{1}{3}y^3\right)\Big|_{-1}^2 = \frac{9}{2}.$$

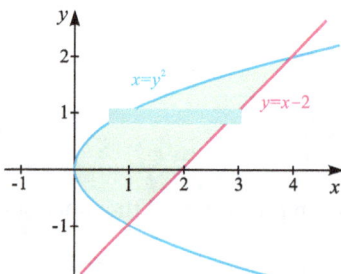

Figure 7.1.6: Horizontal area element for the shaded region bounded by curves in Example 7.1.3.

7.1.2 Volumes of solids

Volumes of revolutions

A solid of revolution is generated by revolving a two-dimensional region R about an axis, such as the y- or x-axis. Suppose a body of revolution is generated by revolving the region R about the x-axis. The area element of R is a rectangle located by the coordinate x with height $y = f(x)$ and thickness Δx. When the area element is rotated about the x-axis, it creates a cylindrical element of radius $f(x)$, thickness Δx, and volume $\Delta V = \pi f^2(x)\Delta x = \pi y^2 \Delta x$, as in Figure 7.1.7 (a). As x varies from a to b, the area elements sweep out the region R and the corresponding cylindrical elements fill the solid of revolution. The approximate volume V of the body of revolution is the sum of the volumes of the cylindrical elements

$$V \approx \sum \Delta V = \sum \pi y^2 \Delta x.$$

Taking the limit as $\Delta x \to 0$, the actual volume of the body of revolution is

$$V = \int_a^b dV = \int_a^b \pi f^2(x)\, dx = \int_a^b \pi y^2\, dx.$$

Similarly, revolving the region bounded by $x = g(y)$ between $y = c$ and $y = d$ about the y-axis (see Figure 7.1.7 (b)) gives the following volume of the revolution:

$$V = \int_c^d dV = \int_c^d \pi g^2(y)\, dy = \int_c^d \pi x^2\, dy.$$

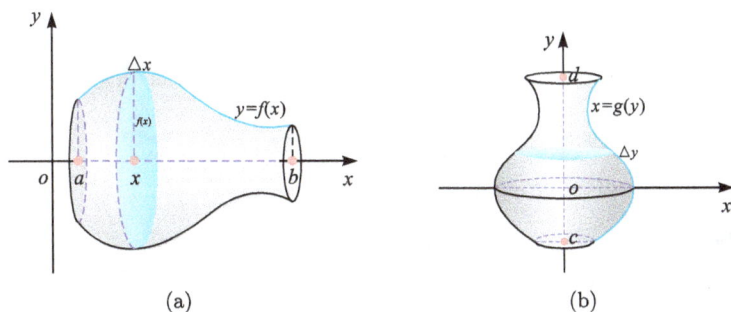

(a)   (b)

Figure 7.1.7: Revolutions about x/y-axis.

Example 7.1.4. Calculate the volume of a right circular cone of height h and base radius r.

Solution. The cone is generated by revolving the region below the line segment $y = \frac{r}{h}x$ ($0 \leqslant x \leqslant h$) and above the x-axis about the x-axis. The above formula gives the follow-

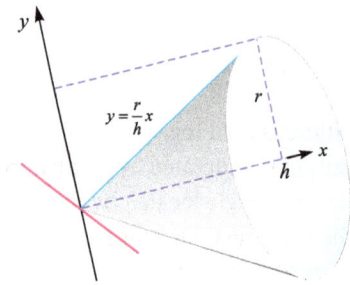

Figure 7.1.8: A circular cone is obtained by rotating a triangular region about the x-axis.

ing volume:

$$V = \int_0^h \pi y^2 \, dx = \pi \frac{r^2}{h^2} \int_0^h x^2 \, dx = \frac{\pi r^2}{h^2} \times \frac{1}{3} x^3 \Big|_0^h$$

$$= \frac{\pi r^2}{3h^2}(h^3 - 0) = \frac{1}{3}\pi r^2 h \, \text{unit}^3.$$

Figure 7.1.8 shows a cone.

Volume of a solid with known areas of cross sections

Suppose we want to find the volume of a solid that is not a volume of revolution like the ones shown in Figure 7.1.9.

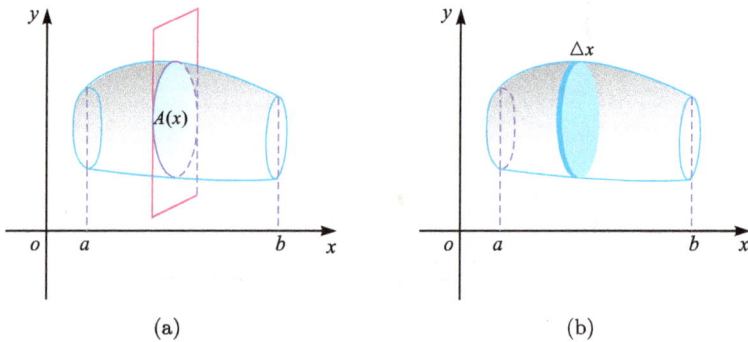

Figure 7.1.9: Cross-sections of solids.

Suppose that the cross section of the solid at each point x in the interval is a region $R(x)$ of known area $A(x)$. We partition the interval $[a, b]$ into subintervals of length Δx and create a slice of the solid between $x = a$ and $x = b$. When Δx is small, the volume element of this slice is therefore $\Delta V \approx A(x)\Delta x$. The Riemann sum for all slices between $x = a$ and $x = b$ is

$$\sum \Delta V \approx \sum A(x)\Delta x$$

and this approximates the volume of the solid. The exact volume of the solid is defined as the limit as $\|\Delta x\| \to 0$. That is, the volume of a solid of known integrable cross section area $A(x)$ from $x = a$ to $x = b$ is the integral of A from a to b.

$$V = \int_a^b A(x)\,dx.$$

Example 7.1.5. A solid has its base R in the xy-plane. The region R is bounded by the x-axis, the y-axis, and the lines $y = x$ and $x = 2$. Each cross section perpendicular to the x-axis is a half circle, as seen in Figure 7.1.10. Find the volume of the solid.

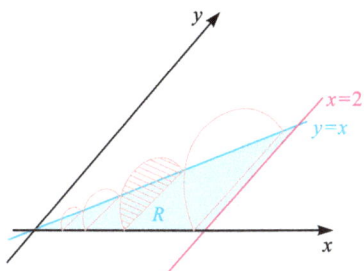

Figure 7.1.10: A solid with base R and cross sections are half circles.

Solution. The area of a half circle is $\pi r^2/2$. At each x in $(0, 2)$, the diameter r of the half circle would be y. The area of the cross section perpendicular to the x-axis is

$$A(x) = \frac{\pi(\frac{y}{2})^2}{2} = \frac{\pi y^2}{8} = \frac{\pi x^2}{8}.$$

Therefore, the volume of the solid is

$$V = \int_0^2 \frac{\pi x^2}{8}\,dx = \frac{\pi x^3}{24}\Big|_{x=0}^{x=2} = \frac{\pi}{24}(2^3 - 2^0) = \frac{\pi}{3}\,\text{unit}^3.$$

7.1.3 Arc length

Curves in Cartesian form

If the first derivative of a function $f(x)$ is continuous, then the function $f(x)$ is called *smooth* and its graph is a *smooth curve*.

For a smooth curve in the plane from $x = a$ to $x = b$ and any partition of the interval, we approximate the length element of the curve by the hypotenuse of the right triangle as in Figure 7.1.11. The length element Δs is therefore

$$\Delta s \approx \sqrt{(\Delta x)^2 + (\Delta y)^2} = \sqrt{1 + \left(\frac{\Delta y}{\Delta x}\right)^2}\,\Delta x. \qquad (7.4)$$

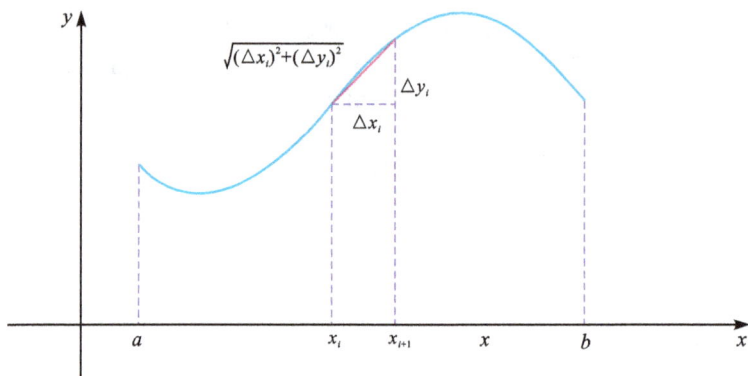

Figure 7.1.11: Arc length element.

We add all the length elements to make the following Riemann sum:

$$s = \sum \Delta s \approx \sum \sqrt{1 + \left(\frac{\Delta y}{\Delta x}\right)^2}\, \Delta x.$$

Taking the limit as the norms of the subdivisions go to zero gives the length of the curve as a definite integral. We have

$$s = \int_a^b \sqrt{1 + (f'(x))^2}\, dx = \int_a^b \sqrt{1 + \left(\frac{dy}{dx}\right)^2}\, dx. \tag{7.5}$$

In equation (7.4), we could also have written the expression for Δs as

$$\Delta s \approx \sqrt{(\Delta x)^2 + (\Delta y)^2} = \sqrt{1 + \left(\frac{\Delta x}{\Delta y}\right)^2}\, \Delta y.$$

This would have lead to the following formula for the length of the curve $x = g(y)$ between $y = c$ and $y = d$:

$$s = \int_c^d \sqrt{1 + \left(\frac{dx}{dy}\right)^2}\, dy. \tag{7.6}$$

Example 7.1.6. Calculate the length of the arc $y = \frac{1}{2}(e^x + e^{-x})$ for $0 \leqslant x \leqslant a$.

Solution. We have

$$y'(x) = \frac{1}{2}(e^x - e^{-x}).$$

Hence,

$$s = \int_0^a \sqrt{1 + \left(\frac{1}{2}(e^x - e^{-x})\right)^2}\, dx = \int_0^a \sqrt{1 + \frac{1}{4}(e^{2x} - 2 + e^{-2x})}\, dx$$

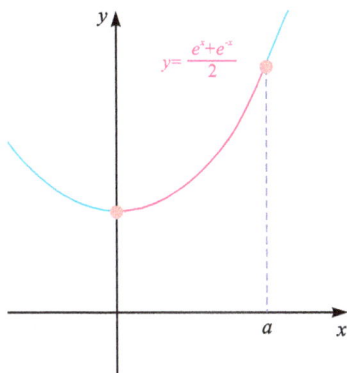

Figure 7.1.12: Graph of $y = \frac{e^x + e^{-x}}{2}$.

$$= \int_0^a \sqrt{\frac{1}{4}e^{2x} + \frac{1}{2} + \frac{1}{4}e^{-2x}}\, dx = \int_0^a \sqrt{\frac{1}{4}(e^x + e^{-x})^2}\, dx$$

$$= \frac{1}{2}\int_0^a (e^x + e^{-x})\, dx = \frac{1}{2}(e^x - e^{-x})\Big|_0^a = \frac{1}{2}(e^a - e^{-a}).$$

Figure 7.1.12 shows the arc.

Curves in parametric form

If the arc is described parametrically by

$$\begin{cases} x = f(t) \\ y = g(t), \end{cases} \qquad t_0 \leqslant t \leqslant t_1,$$

then we derive a formula for the arc length using differentials instead of Δx_i and Δy_i. The differential of the arc length is

$$ds = \sqrt{dx^2 + dy^2} = \sqrt{\left(\frac{dx}{dt}\right)^2 + \left(\frac{dy}{dt}\right)^2}\, dt = \sqrt{(f'(t))^2 + (g'(t))^2}\, dt,$$

so

$$s = \int_{t_0}^{t_1} \sqrt{(f'(t))^2 + (g'(t))^2}\, dt.$$

Example 7.1.7. Find the arc length of the graph described by

$$\begin{cases} x = a(t - \sin t) \\ y = a(1 - \cos t) \end{cases} \qquad \text{for } 0 \leqslant t \leqslant 2\pi,\ a > 0.$$

Solution. We have

$$ds = \sqrt{\left(\frac{dx}{dt}\right)^2 + \left(\frac{dy}{dt}\right)^2}\, dt = \sqrt{a^2(1 - \cos t)^2 + a^2 \sin^2 t}\, dt$$

$$= a\sqrt{2(1 - \cos t)}\, dt = 2a \sin \frac{t}{2}\, dt \quad \text{for } 0 \leqslant t \leqslant 2\pi.$$

Hence,

$$s = \int_0^{2\pi} 2a \sin \frac{t}{2}\, dt = -4a \cos \frac{t}{2}\Big|_{t=0}^{t=2\pi}$$

$$= -4a\left(\cos \frac{2\pi}{2} - \cos \frac{0}{2}\right) = -4a(-1 - 1) = 8a.$$

The graph is shown in Figure 7.1.13.

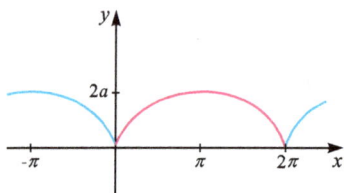

Figure 7.1.13: Graph of a cycloid.

Curves in polar form

Suppose the curve is described by $r = r(\theta)$ in *polar coordinates*. Then $x = r \cos \theta$ and $y = r \sin \theta$. Therefore, the parametric equations for the curve are

$$\begin{cases} x = r(\theta) \cos \theta \\ y = r(\theta) \sin \theta \end{cases}$$

and

$$ds = \sqrt{[x'(\theta)]^2 + [y'(\theta)]^2}\, d\theta$$

$$= \sqrt{[r' \cos \theta - r \sin \theta]^2 + [r' \sin \theta + r \cos \theta]^2}\, d\theta$$

$$= \sqrt{\begin{array}{l} r'^2 \cos^2 \theta + r^2 \sin^2 \theta - 2rr' \cos \theta \sin \theta \\ + r'^2 \sin^2 \theta + r^2 \cos^2 \theta + 2rr' \cos \theta \sin \theta \end{array}}\, d\theta.$$

This simplifies to

$$ds = \sqrt{r^2 + (r')^2}\, d\theta.$$

Example 7.1.8. Find the circumference of the circle centered at $(0,0)$ with radius R, given by the polar equation $r = R$.

Solution. The circumference c is given by

$$c = \int_0^{2\pi} \sqrt{r^2 + (r')^2}\, d\theta = \int_0^{2\pi} \sqrt{R^2 + 0^2}\, d\theta$$

$$= \int_0^{2\pi} R\, d\theta = R\theta \Big|_{\theta=0}^{\theta=2\pi} = R(2\pi - 0)$$

$$= 2\pi R.$$

Figure 7.1.14 shows the graph of a circle with radius R.

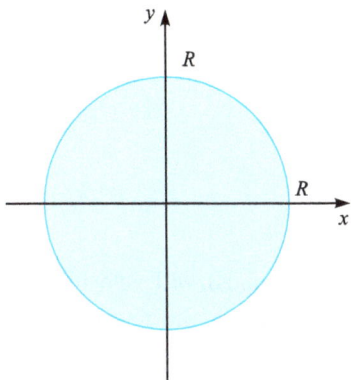

Figure 7.1.14: A circle $x^2 + y^2 = R^2$.

Example 7.1.9. Find the length of the arc described by $r = a(1 + \cos\theta)$ for $a > 0$.

Solution. By symmetry, the desired arc length is two times the portion in the first and second quadrants, so

$$s = 2\int_0^{\pi} \sqrt{r^2 + (r')^2}\, d\theta = 2\int_0^{\pi} \sqrt{a^2(1+\cos\theta)^2 + (-a\sin\theta)^2}\, d\theta$$

$$= 2\int_0^{\pi} 2a\left|\cos\frac{\theta}{2}\right| d\theta = 4a\int_0^{\pi} \cos\frac{\theta}{2}\, d\theta$$

$$= 8a\sin\frac{\theta}{2}\Big|_0^{\pi} = 8a\left(\sin\frac{\pi}{2} - \sin 0\right) = 8a.$$

Figure 7.1.15 shows the graph of $r = 1 + \cos\theta$.

NOTE. In general, it is very difficult to evaluate exactly the integral giving the length of a curve. In most cases, the integrals have to be approximated by using rules such as the trapezoidal rule or Simpson's rule. The examples given above were carefully chosen exceptions.

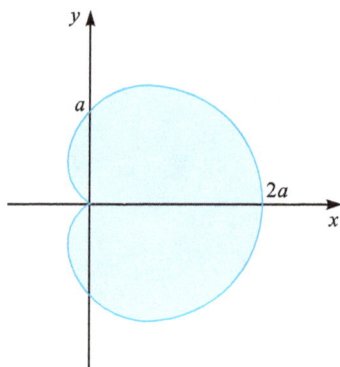

Figure 7.1.15: Graph of the cardioid $r = a(1 + \cos\theta)$.

7.2 Applications in other disciplines

7.2.1 Displacement and distance

Consider a particle moving along a straight line (say, on the x-axis), with velocity function $v(t)$, with known position at time t_0. Let $s(t)$ be the position of the object at time t (the position is its distance from a reference point, such as the origin on the x-axis). Then the net position change during time interval $[t_0, t]$ is given by $s(t) - s(t_0)$. However, we know that $\frac{ds(t)}{dt} = v(t)$, so

$$s(t) = s(t_0) + \int_{t_0}^{t} v(u)\, du$$

gives the position of the particle at any time t.

Similarly, since the velocity is an antiderivative of the acceleration function $a(t)$, that is, $\frac{dv(t)}{dt} = a(t)$, we write

$$v(t) = v(t_0) + \int_{t_0}^{t} a(u)\, du.$$

Example 7.2.1. Suppose an object is moving along the x-axis with velocity $v(t) = t^2 - 2t - 3$ for $t \geqslant 0$. When $t = 0$, its position is $x = 3$. Let $x(t)$ be the position function and $a(t)$ be the acceleration function. Find:

1. $x(5)$ and $a(5)$;
2. when the object reverses its direction;
3. the distance traveled by the object from $t = 1$ to $t = 5$.

Solution. Since $x(0) = 3$, we know

$$x(5) = x(0) + \int_0^5 (t^2 - 2t - 3)\, dt$$

$$= 3 + \left(\frac{t^3}{3} - t^2 - 3t \right) \Big|_{x=0}^{x=5}$$

$$= 3 + \frac{5^3}{3} - 5^2 - 3(5) - \left(\frac{0^3}{3} - 0^2 - 3(0) \right)$$

$$= \frac{14}{3} \text{ units.}$$

As regards the acceleration of the object at $t = 5$, we know

$$a(5) = \frac{dv}{dt} \Big|_{t=5} = (t^2 - 2t - 3)' \big|_{t=5}$$

$$= (2t - 2) \big|_{t=5} = 2(5) - 2 = 8 \text{ units.}$$

Since $v(t) = (t - 3)(t + 1)$, when $t = 3$, $v(3) = 0$. The velocity $v(t)$ changes from negative to positive at $t = 3$, which means the particle changes from moving towards the left to moving towards the right, so at $t = 3$, the object reverses its direction.

One needs to be careful when evaluating distance traveled. The integral $\int_1^5 v(t)\,dt$ gives the net change in the position of the particle instead of the distance traveled by the particle. To find the distance traveled, we use

distance traveled from $t = 1$ to $t = 5$

$$= \int_1^5 |v(t)|\,dt = \int_1^3 |v(t)|\,dt + \int_3^5 |v(t)|\,dt$$

$$= -\int_1^3 (t^2 - 2t - 3)\,dt + \int_3^5 (t^2 - 2t - 3)\,dt$$

$$= -\left(\frac{t^3}{3} - t^2 - 3t \right) \Big|_{t=1}^{t=3} + \left(\frac{t^3}{3} - t^2 - 3t \right) \Big|_{t=3}^{t=5}$$

$$= -\left(\frac{3^3}{3} - 3^2 - 3(3) \right) + \left(\frac{1^3}{3} - 1^2 - 3(1) \right)$$

$$+ \frac{5^3}{3} - 5^2 - 3(5) - \left(\frac{3^3}{3} - 3^2 - 3(3) \right)$$

$$= 16 \text{ units.}$$

7.2.2 Work done by a force

The work done by a force is defined as force times displacement, assuming that the force is constant. If the force is a function $F(x)$ that varies with x, the position of the object on which the force is acting varies. Then the work done in moving an object from $x = a$ to $x = b$ along a straight line (the x-axis) is given by

$$W - \int_a^b F(x)\,dx.$$

Example 7.2.2. Find the work done by the force $F(x) = \sin(2x)$ Newton along the x-axis from $x = \pi$ meters to $x = \frac{3}{2}\pi$ meters.

Solution.

$$W = \int_{\pi}^{\frac{3\pi}{2}} \sin(2x)\, dx = -\frac{1}{2}\cos 2x \Big|_{x=\pi}^{x=\frac{3\pi}{2}}$$

$$= -\frac{1}{2}\left(\cos\left(\frac{3\pi}{2} \times 2\right) - \cos 2\pi\right) = -\frac{1}{2}(-1-1) = 1\,\text{Joule}.$$

7.2.3 Fluid pressure

Each dam is built thicker at the bottom than at the top because the pressure of the water against it increases with depth. Physicists have found that, in any liquid, the fluid pressure p depends on the depth h as follows:

$$\text{pressure} = \rho \times g \times h = \frac{\text{Force}}{\text{Area}},$$

where ρ is the density of liquid (mass per volume) and $g \approx 9.8$ is the gravitational acceleration.

Example 7.2.3. Figure 7.2.1 shows a conical tank with height 5 meters and base radius 1 meter is filled with water ($\rho = 1\,000\,\text{kg/m}^3$). Find the total force against the wall of the tank exerted by the water.

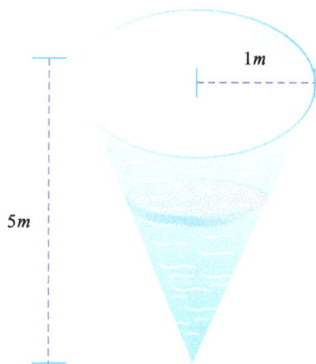

1 m

5 m

Figure 7.2.1: A conic tank.

Solution. We partition the water into thin slabs by planes parallel to the base of tank. The typical slab between y and $y + \Delta y$ has an area of approximately

$$\Delta A \approx 2\pi x \sqrt{\Delta x^2 + \Delta y^2}$$

$$= 2\pi x \sqrt{1 + \left(\frac{\Delta x}{\Delta y}\right)^2}\, \Delta y$$

$$= 2\pi x \sqrt{1 + \left(\frac{1}{5}\right)^2}\, \Delta y$$

$$= \frac{\sqrt{26}}{5} \cdot 2\pi x \Delta y.$$

The force against the wall by this slab is

$$\Delta F = \rho \times g \times (5 - y) \times \Delta A = \left(\frac{\sqrt{26}}{5}\right) \rho \times g \times (5 - y) \times 2\pi x \Delta y,$$

so the total force is

$$F = \frac{\sqrt{26}}{5} \int_0^5 \rho \times g \times (5 - y) \times 2\pi x \, dy$$

$$= \frac{\sqrt{26}}{5} 2\pi\rho g \int_0^5 (5 - y) \times \frac{y}{5} \, dy$$

$$= \left(\frac{\sqrt{26}}{5}\right) \frac{2\pi\rho g}{5} \cdot \int_0^5 5y - y^2 \, dy$$

$$= \frac{\sqrt{26}}{5} \frac{2\pi\rho g}{5} \left(\frac{5}{2}y^2 - \frac{y^3}{3}\right)\Big|_0^5$$

$$= \frac{\sqrt{26}}{5} \frac{2\pi\rho g}{5} \left(\frac{5}{2} \cdot 5^2 - \frac{5^3}{3}\right)$$

$$\approx 261\,512 \, \text{units.}$$

7.2.4 Center of mass

Suppose that a rod has three uniform components, a 5 kg weight on the left, a 10 kg weight in the middle, and a 2 kg weight on the right. Where should a fulcrum be placed so that the rod balances? This requires finding the *center of mass* of the compounded rod. We first assign a scale to the rod so that we can denote locations on the rod simply as x-coordinates. As seen in Figure 7.2.2, the weights are at $x = 2$, $x = 5$, and $x = 8$.

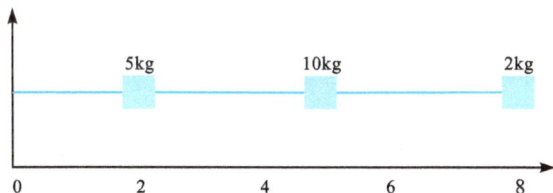

Figure 7.2.2: Center of mass: compounded rod.

If we place the fulcrum at $x = 4$, then each weight applies a force to the rod that tends to rotate it around the fulcrum. This effect is measured by a quantity called the *moment* (or *torque*), which is defined as the force times the distance. Obviously, the moments are not balanced, since the clockwise moment is $5g \times 2 = 10g$, whereas the counterclockwise moment is $10g \times 1 + 2g \times 4 = 18g$. To get the rod to balance, we need to move the fulcrum somewhere to the right. Let $\bar{x}$ be the position where the fulcrum should be. Then, equating the clockwise and counterclockwise moments, we have

$$5g(\bar{x}-2) + 10g(5-\bar{x}) = 2g(8-\bar{x}),$$
$$5\bar{x} - 10 + 50 - 10\bar{x} = 16 - 2\bar{x},$$
$$\bar{x} = \frac{34}{3}.$$

This is the position of the center of mass of the rod.

If a plane lamina has a uniform density at each point in the lamina, then the center of mass of the lamina is a purely geometric quantity. In such a case, the center of mass is called the *centroid*.

Example 7.2.4. Find the centroid of a uniform lamina which has its shape formed by $y = x^2$, the x-axis, and the line $x = 2$.

Solution. This is a two-dimensional problem. We need to find the coordinates $(\bar{x}, \bar{y})$ of the center of mass of the lamina. Luckily, we can find $\bar{x}$ and $\bar{y}$ independently.

Partition the lamina into small strips, as shown in Figure 7.2.3. For a typical strip, its mass is approximated by

$$\Delta m \approx \delta \cdot y \cdot \Delta x,$$

where δ is the uniform density function. The total moments around the y-axis exerted by these strips are given by

$$M_y = \sum (\Delta m \cdot g \cdot x) \approx \sum (\delta \cdot y \cdot \Delta x \cdot g \cdot x).$$

Taking the limit as the norm of the partition tends to 0 gives the integral

$$M_y = \int_0^2 \delta g x y \, dx = \delta g \int_0^2 x y \, dx.$$

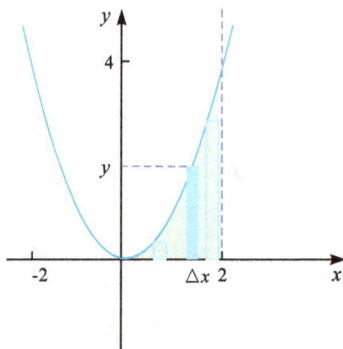

Figure 7.2.3: Center of mass: a lamina.

The total mass of the lamina is

$$m = \delta \int_0^2 y \, dx,$$

so

$$\delta g \int_0^2 xy \, dx = mg\bar{x},$$

$$\delta g \int_0^2 xy \, dx = \delta \int_0^2 y \, dx \, g\bar{x},$$

$$\bar{x} = \frac{\int_0^2 xy \, dx}{\int_0^2 y \, dx} = \frac{\int_0^2 x \cdot x^2 \, dx}{\int_0^2 x^2 \, dx} = \frac{(\frac{1}{4}x^4)|_0^2}{(\frac{1}{3}x^3)|_0^2} = \frac{3}{2}.$$

Similarly, we determine the y-coordinate of the lamina as follows:

$$\bar{y} = \frac{1}{2} \frac{\int_0^2 y^2 \, dx}{\int_0^2 y \, dx} = \frac{1}{2} \frac{\int_0^2 (x^2)^2 \, dx}{\int_0^2 x^2 \, dx}$$

$$= \frac{1}{2} \frac{(\frac{1}{5}x^5)|_0^2}{(\frac{1}{3}x^3)|_0^2} = \frac{1}{2} \times \frac{3}{5} \times 2^2 = \frac{6}{5},$$

so the centroid is at the point $(\frac{3}{2}, \frac{6}{5})$.

7.2.5 Probability

In probability theory, the probability density function $f(x)$ of a random variable X satisfies:

1. $f(x) \geq 0$;
2. $\int_{-\infty}^{\infty} f(x) \, dx = 1$;
3. $P(a < X < b) = \int_a^b f(x) \, dx$.

Figure 7.2.4 shows the graph of a probability density function $f(x)$.

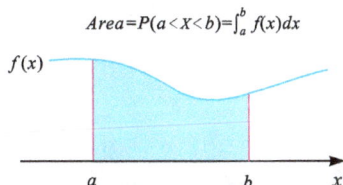

Figure 7.2.4: Area of the shaded region is $P(a < X < b)$.

Example 7.2.5. If the random variable X satisfies the negative exponential distribution with probability density function

$$f(t) = \begin{cases} 3e^{-3t}, & t \geqslant 0 \\ 0, & \text{otherwise,} \end{cases}$$

find $P(X > 5)$.

Solution. We have

$$
\begin{aligned}
P(X > 5) &= \int_5^\infty 3e^{-3t}\, dt = \lim_{b\to\infty} \int_5^b 3e^{-3t}\, dt \\
&= \lim_{b\to\infty} \left(-e^{-3t}\right)_5^b \\
&= -\lim_{b\to\infty}\left(e^{-3b} - e^{-15}\right) \\
&= -\left(0 - e^{-15}\right) \\
&= e^{-15} \\
&\approx 3.059 \times 10^{-7}.
\end{aligned}
$$

7.3 Exercises

1. Find the area of each of the shaded regions.

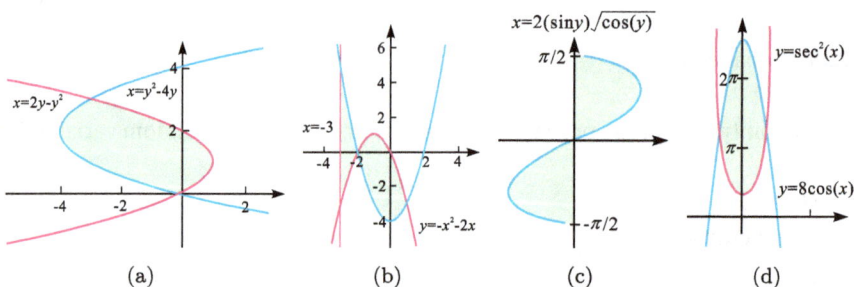

(a)   (b)   (c)   (d)

2. Sketch the following curves and then find the area of the enclosed regions:
    (a) $y = 5x - x^2$ and $y = x$;
    (b) $x = y^2 - 2$, $y = -1$, $y = 1$, and $x = e^y$;
    (c) $y = 12 - x^2$ and $y = x^2 - 6$;
    (d) $x = a\cos^3 t$ and $y = a\sin^3 t$, $0 \leqslant t \leqslant 2\pi$;
    (e) $y = \cos x$, $x = 0$, and $y = \sin x$;
    (f) $y = \cos \pi x$ and $y = 4x^2 - 1$;
    (g) $x = a(t - \sin t)$ and $y = a(1 - \cos t)$ for $0 \leqslant t \leqslant 2\pi$.
3. **(Tschirnhaus' cubic)** In geometry, the curve defined by the polar equation $r = a\sec^3(\theta/3)$ is known as Tschirnhaus' cubic. It was studied by Von Tschirnhaus,

L'Hôpital, and Catalan. It is sometimes known as L'Hôpital's cubic or the trisectrix of Catalan. Let $t = \tan \theta/3$. Then, by using the triple angle formula, one can obtain its parametric form

$$x = a(1 - 3t^2) \quad \text{and} \quad y = at(3 - t^2).$$

If you eliminate the parameter t, you find that the Cartesian equation of the curve is

$$27ay^2 = (a - x)(8a + x)^2.$$

Sketch the curve in the case where $a = 1$ and you will see that part of the curve forms a loop. Find the area enclosed by the loop.

4. (**Area of sectors**) The area of a region bounded by the polar curve $r = r(\theta)$ and two half lines $\theta = \alpha$ and $\theta = \beta$ can be shown to be $\frac{1}{2} \int_{\alpha}^{\beta} r^2(\theta) \, d\theta$. Now:
   (a) find the area of the circle $r = R$;
   (b) find the area of the region enclosed by the cardioid $r = a(1 + \cos \theta)$;
   (c) find the area of the region inside the curves $r = 1 + \sin \theta$ and $r = 3 \sin \theta$.

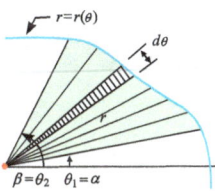

Question 4

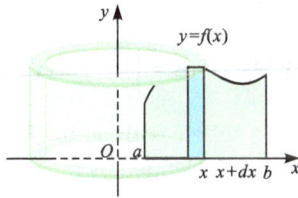

Question 11

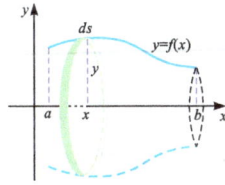

Question 12

5. Sketch the curve defined by the polar equation $r = a \sin 2\theta$, for $0 \leqslant \theta \leqslant \frac{\pi}{2}$, where a is a positive constant. Then find the area of the loop.

6. Sketch the regions bounded by the following curves, sketch the solids obtained by rotating the regions about the specific line, and sketch a typical disk or washer:
   (a) $y = 2 - \frac{1}{2}x$, $y = 0$, $x = 1$, $x = 2$; about the x-axis;
   (b) $y = \arcsin x$, $x = 1$, $y = 0$; about the x-axis;
   (c) $y = 1 + \sec x$, $y = 3$; about $y = 1$;
   (d) $x = y^2$, $x = 1 - y^2$; about $x = 3$;
   (e) $x = a(t - \sin t)$, $y = a(1 - \cos t)$, $y = 0$, for $0 \leqslant t \leqslant 2\pi$; about $y = 2a$.

7. The region bounded by $y = \sqrt{\frac{x+3}{(x+1)(x+2)^2}}$, the x-axis, and $0 \leqslant x \leqslant 1$ is rotated about the x-axis. Find the volume of the resulting solid.

8. The region bounded by $y = \frac{x}{\sqrt[4]{(1+x^2)^5}}$, the x-axis, and $0 < x < 1$ is rotated about the x-axis. Find the volume of the resulting solid.

9. A solid has a base R bounded by $y = \sqrt{4 - x^2}$ and the x-axis. Find the volume of the solid if each cross section perpendicular:

(a) to the y-axis is a square;

(b) to the x-axis is a semi-circle whose diameter lies in R.

10. The base of a solid is the region between the curve $y = 2\sqrt{\sin x}$ and the interval $[0, \pi]$ on the x-axis. The cross sections perpendicular to the x-axis are equilateral triangles with bases running from the x-axis to the curve. Find the volume of the solid.

11. (**Shell method**) In the *disk method*, we sum the volumes of infinitely many infinitesimal disks to find the total volume of a solid. The shell method considers a representative "shell" with volume $2\pi x y\, dx$ (about the y-axis) or $2\pi y x\, dy$ (about the x-axis). The total volume is given by

$$V = \int_a^b 2\pi x y\, dx \quad \text{(about the y-axis)}, \quad \text{or} \quad V = \int_c^d 2\pi y x\, dy \quad \text{(about the x-axis)}.$$

(a) Find the volume of the solid obtained by rotating the region enclosed by $y = (x-1)(x-2)^2$ and the x-axis about the y-axis.

(b) Find the volume of the solid obtained by rotating the region enclosed by $y = \sin(x^2)$ and the x-axis for $0 \leqslant x \leqslant \sqrt{\pi}$ about the y-axis.

12. (**Surface area**) It can be shown that the surface area obtained by rotating the curve $y = f(x)$, for $a \leqslant x \leqslant b$, about the x-axis is

$$\int_a^b 2\pi f(x)\, ds = \int_a^b 2\pi f(x) \sqrt{1 + [f'(x)]^2}\, dx.$$

Find the surface area of a sphere with radius R.

13. (**Surface area**) The curve C is defined by the parametric equations $x = t^3 - 3t$ and $y = 3t^2 + 1$. The arc of C, joining the point where $t = 0$ to the point where $t = \sqrt{3}$, is rotated about the y-axis. Find the surface area of the resulting solid.

14. (**Length of curves**) Find the length of each of the following curves:

(a) $y = \frac{2}{3}(x^2 - 1)^{3/2}$, $1 \leqslant x \leqslant 3$;

(b) $y = \ln \cos x$, $0 \leqslant x < \pi/3$;

(c) $x = t^3, y = \frac{3t^2}{2}, 0 \leqslant t \leqslant \sqrt{3}$;

(d) $x = t - 8\sqrt{t}, y = \frac{16}{3}\sqrt[4]{t^3}, 1 \leqslant t \leqslant 4$;

(e) $x = 8\cos t + 8t\sin t, y = 8\sin t - 8t\cos t, 0 \leqslant t \leqslant \frac{\pi}{2}$;

(f) $r = e^\theta, 0 \leqslant \theta \leqslant \frac{\pi}{4}$;

(g) $r = a(1 + \cos\theta), 0 \leqslant \theta \leqslant 2\pi$;

(h) $f'(x) = \sqrt{x^2 e^{2x} - 1}$, between $x = 0$ and $x = 3$;

(i) $y = \frac{3}{8}x^{\frac{4}{3}} - \frac{3}{4}x^{\frac{2}{3}}, 1 \leqslant x \leqslant 8$.

15. (**Average value**) The two variables x and y are related by $x^4 + y^4 = 1$ for $0 < x < 1$ and $0 < y < 1$. (a) Find $\frac{dy}{dx}$ and $\frac{d^2y}{dx^2}$. (b) Given that $y(a_1) = b_1$ and $y(a_2) = b_2$ where $0 < a_1 < a_2 < 1$, find the mean value of $\frac{d^2y}{dx^2}$ with respect to x over the interval $[a_1, a_2]$.

16. (**Volume of an infinite solid**) The cross sections of a solid horn perpendicular to the x-axis are circular disks with diameters reaching from the x-axis to the curve $y = 1/x$, for $1 \leqslant x < \infty$. Find the volume of the horn.

17. (**Particle motion**) A particle moves along the x-axis with velocity $v(t) = 2t - \sin \pi t$ for $t \geq 0$. The particle starts in the origin. Find the position of the particle when $t = 4$.

18. (**Flying distance**) A bird is flying at 10 m/sec at an altitude of 20 m. Sadly, the bird is accidentally shot by a bullet and starts to drop to the ground. The parabolic trajectory of the falling body is modeled by $h(x) = 20 - \frac{x^2}{5}$, where $h(x)$ is the height of the falling body above the ground and x is the horizontal distance it travels. Find the distance traveled by the falling body in the air.

19. (**Centroid**) Find the coordinates of the centroid of the region bounded by the x-axis, the line $x = 1$, and the curve $y = xe^{-x}$.

20. (**Water pressure**) A dam has the shape of a trapezoid. The height is 20 m and the width is 50 m at the top and 30 m at the bottom. Find the force exerted on the dam due to the hydro pressure if the water level is 4 m from the top of the dam.

21. (**Fundraising**). Contributions in response to a fundraising campaign are coming at the rate of $1\,500te^{-0.25t}$ dollars per week. How much money will be raised after 10 weeks?

22. (**Future value**) Money is transferred into an account at the rate of $r(t) = 500t \ln \sqrt{1+t}$ dollars per year. If the account pays 5% interest compounded continuously, how much will accumulate in the account over a six-year period?

23. (**Population growth**) The rate of change of the population of a certain city is modeled as $r(t) = 10\,000(t+1)e^{-0.2t}$ thousand people per year. If the current population is 1.5 million, what will the population be 10 years from now?

24. (**Probability**) If $f(x) = \begin{cases} k \sin \frac{\pi x}{10}, & \text{when } 0 \leq x \leq 10 \\ 0, & \text{otherwise} \end{cases}$ is the probability density function of a random variable X, then find:
    (a) the value of k;   (b) $P(X < 4)$;   (c) the mean of X.

8 Infinite series, sequences, and approximations

In this chapter, you will learn about:
- *infinite series;*
- *tests for convergence;*
- *alternating series;*
- *power series;*
- *Taylor series;*
- *Fourier series.*

An *infinite sequence* is a set of infinite numbers arranged in a particular order. A *series* (also called an *infinite series*) can be thought of as the sum of the numbers in a sequence. Of course, it is not possible to actually add infinitely many numbers but, nevertheless, it is possible to give a meaning to an infinite sum and, in some cases, find a value for it.

Sequences and series are very important in calculus, in other areas of study, and in their applications. For example, people have used a sequence of regular polygons with five, six, seven, … sides to approximate the circumference of a circle to any desired accuracy. Many very important functions that arise in mathematical physics, chemistry, and engineering, such as Bessel functions, are defined as infinite series. Physicists also deal with series when studying fields as diverse as optics, special relativity, and electromagnetism, where they often analyze phenomena by replacing a function with the terms from a series representation of the function. Hence, it is important to be familiar with the basic concepts of sequences, series, and especially the meaning of a convergent sequence and convergent series.

8.1 Infinite sequences

You have already encountered infinite sequences in Chapter 2. For reference, some basic definitions are repeated here.

Definition 8.1.1. An *infinite sequence*, or simply a *sequence*, is a list of infinitely many numbers a_n in the particular order given by increasing values of the index $n = 1, 2, 3, 4, \ldots$ of the following form:

$$\{a_n\} = \{a_n\}_{n=1}^{\infty} = a_1, a_2, a_3, a_4, a_5, \ldots, a_n, \ldots.$$

NOTE. Any other letters can be used in place of "a" and "n" without changing this definition.

A sequence $\{a_n\}$ has *limit a* (a number) if the numbers a_n are arbitrarily close to a for all sufficiently large values of the index n.

https://doi.org/10.1515/9783110527780-008

This is similar to the limit of a function of the form $\lim_{x \to \infty} f(x) = a$, except here we replace the variable x by integer values of n. The formal definition is repeated here.

Definition 8.1.2. A sequence $\{a_n\}$ is *convergent* with *limit* a if, for any number $\varepsilon > 0$ (no matter how small), there is an integer N (that depends on ε) such that

$$|a_n - a| < \varepsilon \quad \text{for all } n \geqslant N.$$

In this case, we write $\lim_{n \to \infty} a_n = a$. A sequence that is not convergent is called *divergent*.

A particular sequence is often specified by giving a formula for the nth term that is true for $n = 1, 2, 3 \ldots$, as in the next example.

Example 8.1.1. Show that the sequence

$$\{s_n\} = \left\{ 1 - \frac{1}{2^n} \right\} = \frac{1}{2}, \frac{3}{4}, \frac{7}{8}, \frac{15}{16}, \ldots$$

is convergent with limit $s = 1$.

Solution. Since

$$|s_n - s| = \left| \left(1 - \frac{1}{2^n} \right) - 1 \right| = \left(\frac{1}{2} \right)^n$$

for any $\varepsilon > 0$, we need to show that $(1/2)^n < \varepsilon$ for all sufficiently large n. This is probably obvious to you, but we prove it formally as follows. We find this N by taking logarithms of both sides, which preserves the inequality because the logarithmic function $\ln(x)$ is an increasing function for all $x > 0$. We have

$$\ln \left(\frac{1}{2} \right)^n < \ln \varepsilon \quad \Longleftrightarrow \quad -n \ln 2 < \ln \varepsilon \quad \Longleftrightarrow \quad n > -\frac{\ln \varepsilon}{\ln 2}.$$

Hence, $|s_n - s| = (1/2)^n < \varepsilon$ for all $n \geqslant N$ if N is chosen to be any integer larger than $-\frac{2 \ln \varepsilon}{\ln 2} \approx -2.885\,4 \ln \varepsilon$ (note that $\ln \varepsilon < 0$ if $\varepsilon < 1$).

We repeat here the theorem on bounded, monotonic sequences $\{a_n\}$, because it will be used extensively in this chapter.

Theorem 8.1.1 (Bounded monotonic sequence theorem). *If a sequence $\{a_n\}$ is bounded and monotonic, then it must have a limit L. That is, $\lim_{n \to \infty} a_n = L$.*

8.2 Infinite series

8.2.1 Definition of infinite series

Suppose $a_1, a_2, a_3, \ldots, a_n, \ldots$ is a sequence of numbers and we add the terms of the sequence, giving an expression of the following form:

$$\sum_{k=1}^{\infty} a_k = a_1 + a_2 + a_3 + \cdots + a_n + \cdots.$$

This infinite sum is called an *infinite series* (or just a *series*).

NOTE. A series can also be denoted by $\sum a_k$, or $\sum_{k=1}^{\infty} a_k$. A series can also start at other values, such as $k = 0$ or $k = 2$, or even at negative values, such as $\sum_{k=-1}^{\infty} a_{k+1}$. The subscript can be replaced by any other letter such as n or i without changing the series, so $\sum_{k=1}^{\infty} a_k$, $\sum_{n=1}^{\infty} a_n$, $\sum_{i=1}^{\infty} a_i$, and $\sum_{j=1}^{\infty} a_j$ all denote the same series. In this chapter, most often we will use k or n as the index, i.e., $\sum_{k=1}^{\infty} a_k$ and $\sum_{n=1}^{\infty} a_n$.

It would be physically impossible to add infinitely many numbers, but we can sometimes give a meaning to it, as follows. We construct the *partial sum* s_n that is the sum of the first n terms a_k, so we have

$$s_1 = a_1, \quad s_2 = a_1 + a_2, \quad s_3 = a_1 + a_2 + a_3, \quad \ldots.$$

In general,

$$s_n = a_1 + a_2 + \cdots + a_n = \sum_{k=1}^{n} a_k.$$

These partial sums form a new infinite sequence $\{s_n\}$, which may or may not have a limit.

Definition 8.2.1. Let $\sum_{k=1}^{\infty} a_k$ be an infinite series with partial sums $s_n = \sum_{k=1}^{n} a_k$. If the sequence of partial sums $\{s_n\}$ is convergent, so that $\lim_{n \to \infty} s_n = s$ exists as a real number, then we say that the series $\sum a_k$ is *convergent* and has *sum s* and we write

$$s = \sum_{k=1}^{\infty} a_k = a_1 + a_2 + a_3 + \cdots + a_n + \cdots.$$

If $\sum a_k$ is not convergent, then the series $\sum a_k$ is called *divergent*.

Thus, when we write $\sum_{k=1}^{\infty} a_k = s$, we mean that, by adding sufficiently many terms from the start of the series, we get as close as we like to the number s. In limit terms, we write

$$\lim_{n \to \infty} (a_1 + a_2 + a_3 + a_4 + \cdots + a_n) = s.$$

That is, if $\sum_{k=1}^{\infty} a_k = s$, then $s_n \approx s$ when n is large and the difference

$$r = s - s_n = a_{n+1} + a_{n+2} + a_{n+3} + \cdots,$$

called the *remainder*, approaches zero as $n \to \infty$.

Example 8.2.1. Show that the following two infinite series are divergent:
(a) $\sum_{k=1}^{\infty} k = 1 + 2 + 3 + \cdots + n + \cdots$;
(b) $\sum_{k=1}^{\infty} (-1)^{k+1} = 1 - 1 + 1 - 1 + 1 - 1 + \cdots + (-1)^{n-1} + \cdots$.

Solution. Series (a) has partial sums $s_n = n(n+1)/2$ and $\lim_{n\to\infty} s_n = +\infty$, so it diverges to $+\infty$. Series (b) has partial sums $s_n = 1$ if n is odd and $s_n = 0$ if n is even. Hence, $\lim_{n\to\infty} s_n$ does not exist, so the series diverges.

Example 8.2.2 (Telescoping series). Show that the sum of the infinite series $\sum_{k=1}^{\infty} \frac{1}{k(k+1)}$ is 1.

Solution. We use the definition of a convergent series and compute the partial sums. We note that the method of partial fractions enables us to write

$$\frac{1}{k(k+1)} = \frac{1}{k} - \frac{1}{k+1},$$

so

$$s_n = \sum_{k=1}^{n} \frac{1}{k(k+1)} = \frac{1}{1\times 2} + \frac{1}{2\times 3} + \cdots + \frac{1}{n\times(n+1)}$$
$$= \left(1 - \frac{1}{2}\right) + \left(\frac{1}{2} - \frac{1}{3}\right) + \left(\frac{1}{3} - \frac{1}{4}\right) + \cdots + \left(\frac{1}{n} - \frac{1}{n+1}\right)$$
$$= 1 - \frac{1}{n+1} \quad \text{because all other terms cancel each other.}$$

Hence, $\lim_{n\to\infty} s_n = \lim_{n\to\infty}(1 - \frac{1}{n+1}) = 1 - 0 = 1$, so the given series is convergent and $\sum_{k=1}^{\infty} \frac{1}{k(k+1)} = 1$.

Definition 8.2.2 (Geometric series). Let $a \neq 0$ and q be any fixed real numbers. The infinite series

$$a + aq + aq^2 + \cdots + aq^n + \cdots = \sum_{k=0}^{\infty} aq^k$$

is called a *geometric series* and the number q is called the *common ratio* of the series.

Example 8.2.3. Show that the geometric series

$$\sum_{k=0}^{\infty} aq^k = a + aq + aq^2 + \cdots + aq^n + \cdots \begin{cases} \text{converges to } \frac{a}{1-q}, & \text{when } |q| < 1 \\ \text{diverges,} & \text{when } |q| \geq 1. \end{cases}$$

Solution. If $q \neq 1$, we write the nth partial sum s_n and then multiply both sides of this by q, to obtain

$$s_n = a + aq + aq^2 + \cdots + aq^{n-1},$$
$$qs_n = aq + aq^2 + \cdots + aq^{n-1} + aq^n.$$

Subtracting the second equation from the first, we obtain the following equation for s_n:

$$s_n - qs_n = a - aq^n \quad \Longrightarrow \quad s_n = \frac{a(1-q^n)}{1-q}.$$

If $-1 < q < 1$, we know that $q^n \to 0$ as $n \to \infty$, so

$$\lim_{n\to\infty} s_n = \lim_{n\to\infty} \frac{a(1-q^n)}{1-q} = \frac{a}{1-q}.$$

Thus, when $|q| < 1$, the geometric series is convergent and its sum is $\frac{a}{1-q}$, so

$$\frac{a}{1-q} = a + aq + aq^2 + \cdots + aq^n + \cdots \quad \text{for } |q| < 1.$$

If $|q| > 1$ (that is, $q < -1$ or $q > 1$), then $q^n \to \pm\infty$, so $\lim_{n\to\infty} s_n$ does not exist and the sequence $\{aq^n\}$ is divergent.

If $q = -1$, then $q^n = 1$ if n is even and $q^n = -1$ if n is odd, so $\lim_{n\to\infty} s_n$ does not exist and the series diverges.

If $q = 1$, then the nth partial sum is $s_n = a + a + a + \cdots + a = na \to \infty$ (or $-\infty$). Therefore, $\lim_{n\to\infty} s_n$ does not exist and the geometric series diverges.

Therefore, the geometric series $\sum aq^k$ converges when $|q| < 1$ and diverges when $|q| \geq 1$.

Example 8.2.4. Determine whether or not the series $\sum_{k=1}^{\infty} (\frac{3}{\pi})^k$ is convergent. If it converges, find the sum of the series.

Solution. This is a geometric series with common ratio $3/\pi < 1$ and first term $3/\pi$. Therefore, it converges to

$$\frac{\frac{3}{\pi}}{1 - \frac{3}{\pi}} = \frac{3}{\pi - 3} \approx 21.188.$$

8.2.2 Properties of convergent series

Since the convergence of series is defined by a limit, we expect that the properties of convergent series are similar to those of limits.

Theorem 8.2.1. *If $\sum u_n$ and $\sum v_n$ are convergent series and $c \in \mathbb{R}$ is any constant, then $\sum c u_n$ and $\sum (u_n \pm v_n)$ are also convergent series and*

$$\sum c u_n = c \sum u_n, \quad \sum (u_n \pm v_n) = \sum u_n \pm \sum v_n.$$

Proof. These properties of convergent series follow from the corresponding limit laws. $\qquad\square$

Example 8.2.5. Find the sum of the series $\sum_{n=1}^{\infty} \left(\frac{3}{2n(n+1)} + \frac{1}{3^n} \right)$.

Solution. The series $\sum 1/3^n$ is a geometric series with $a = \frac{1}{3}$ and $q = \frac{1}{3}$, so

$$\sum_{n=1}^{\infty} \frac{1}{3^n} = \frac{\frac{1}{3}}{1 - \frac{1}{3}} = \frac{1}{2}.$$

A previous example showed that

$$\sum_{n=1}^{\infty} \frac{1}{n(n+1)} = 1,$$

so, by Theorem 8.2.1, the given series is convergent and

$$\sum_{n=1}^{\infty} \left(\frac{3}{2n(n+1)} + \frac{1}{3^n} \right) = \frac{3}{2} \sum_{n=1}^{\infty} \frac{1}{n(n+1)} + \sum_{n=1}^{\infty} \frac{1}{3^n}$$
$$= \frac{3}{2} + \frac{1}{2}$$
$$= 2.$$

NOTE. Changing a finite number of terms of an infinite series does not affect the convergence or divergence, but it does affect the sum.

This follows from the result that, for any integer N,

$$\sum_{n=1}^{\infty} a_n = \sum_{n=1}^{N-1} a_n + \sum_{n=N}^{\infty} a_n.$$

This shows that $\sum_{n=1}^{\infty} a_n$ converges if and only if $\sum_{n=N}^{\infty} a_n$ converges, because convergence only depends on the values a_n as $n \to \infty$. This means that we can add a finite number of terms to a series or delete/change a finite number of terms without alternating the series' convergence or divergence, although usually this will change the sum of a convergent series.

A test for divergence

Suppose that the infinite series $\sum_{k=1}^{\infty} a_k$ converges with sum s. Then we have

$$\lim_{n \to \infty} s_n = \lim_{n \to \infty} \sum_{k=1}^{n} a_k = s.$$

Also note that, for $n \geqslant 2$,

$$a_n = \sum_{k=1}^{n} a_k - \sum_{k=1}^{n-1} a_k = s_n - s_{n-1}.$$

Combining the above two equations, we have

$$\lim_{n \to \infty} a_n = \lim_{n \to \infty} (s_n - s_{n-1}) = \lim_{n \to \infty} s_n - \lim_{n \to \infty} s_{n-1} = s - s = 0.$$

This establishes the next theorem.

Theorem 8.2.2 (The nth term divergence test). *If the infinite series $\sum_{n=1}^{\infty} a_n$ converges, then the sequence $\{a_n\}$ of terms of the series has limit 0. Equivalently, if $\lim_{n \to \infty} a_n \neq 0$, then $\sum a_n$ is divergent.*

Example 8.2.6. Show that the following series all diverge:
(a) $\sum_{n=1}^{\infty} \frac{n^2}{3n^2+4}$; (b) $\sum_{m=1}^{\infty} (-1)^m$; (c) $\sum_{k=1}^{\infty} (\frac{k}{k+1})^k$.

Solution. These series all diverge by the nth term divergence test because:
(a) $\lim_{n \to \infty} \frac{n^2}{3n^2+4} = \lim_{n \to \infty} \frac{1}{3+4/n^2} = \frac{1}{3} \neq 0$;
(b) $\lim_{m \to \infty} (-1)^m$ does not exist;
(c) $\lim_{k \to \infty} (\frac{k}{k+1})^k = \lim_{k \to \infty} (\frac{1}{1+\frac{1}{k}})^k = \lim_{k \to \infty} \frac{1}{(1+\frac{1}{k})^k} = \frac{1}{e} \neq 0$.

The converse of the nth term divergence test is not necessarily true. That is, knowing that the limit of the terms of a series is zero does not ensure that the series converges. For instance, the harmonic series $\sum 1/n$ satisfies $\lim_{n \to \infty} 1/n = 0$, yet $\sum 1/n$ diverges, as shown in the next example.

Example 8.2.7. Show that the *harmonic series*

$$1 + \frac{1}{2} + \frac{1}{3} + \cdots + \frac{1}{n} + \cdots = \sum_{k=1}^{\infty} \frac{1}{k}$$

is divergent.

Proof. We show that the partial sums s_n become arbitrarily large and can therefore not approach a limiting value.

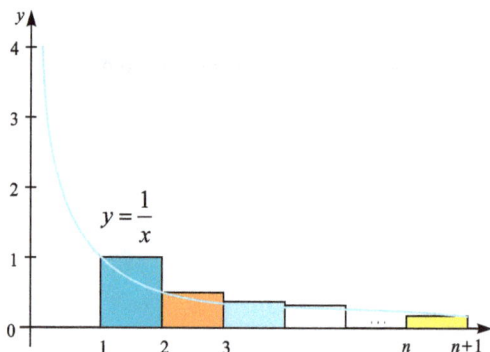

Figure 8.2.1: Graph of $y = \frac{1}{x}$ and rectangles with area $1, \frac{1}{2}, \frac{1}{3}, \ldots, \frac{1}{n}$.

As seen in Figure 8.2.1, the partial sum s_n, which is the sum of the areas of n rectangles, all with base 1, is larger than the area below the curve $y = 1/x$, above the x-axis, and between the two lines $x = 1$ and $x = n + 1$. Therefore,

$$s_n \geq \int_1^{n+1} \frac{1}{x}\, dx = \ln |x|_1^n = \ln n - \ln 1 = \ln n,$$

so $s_n \to \infty$ as $n \to \infty$. Thus $\lim_{n \to \infty} s_n$ does not exist and $\sum 1/n$ diverges. □

An alternative solution
If the harmonic series converges with sum s, then its partial sums s_n and s_{2n} both have limits s. This means

$$s_{2n} - s_n \to 0 \quad \text{as} \quad n \to 0.$$

However,

$$s_{2n} - s_n = \underbrace{\frac{1}{n+1} + \frac{1}{n+2} + \cdots + \frac{1}{2n}}_{n \text{ times}}$$

$$> \frac{1}{2n} + \frac{1}{2n} + \cdots + \frac{1}{2n} = \frac{1}{2} \nrightarrow 0.$$

This is a contradiction to the assumption that $s_{2n} - s_n \to 0$. Therefore, the harmonic series is divergent (proof by contradiction).

NOTE. The terms in the harmonic series correspond to the nodes on a vibrating string that produce multiples of the fundamental frequency, the lowest note or pitch one hears when a string is plucked. For example, 1/2 (one node divides the string into two parts) produces the harmonic that is twice the fundamental frequency, 1/3 produces a frequency that is three time the fundamental frequency, etc.

In Chapter 2, we saw the sequence form of Cauchy's theorem, which gives sufficient and necessary conditions for a convergent sequence. Now we give the series form in the following theorem.

Theorem 8.2.3. *An infinite series $\sum u_n$ converges if and only if, for any positive number ε, there exists an integer N (depending on ε) such that, whenever $n \geqslant N$, we have*

$$|s_{n+p} - s_n| = |u_{n+1} + u_{n+2} + \cdots + u_{n+p}| < \varepsilon \quad \text{for any positive integer } p.$$

This theorem states that, if a series is convergent, the "tail" of a series, the terms after u_N, must be arbitrarily small, if N is large enough.

8.3 Tests for convergence

8.3.1 Series with nonnegative terms

A series $\sum_{k=1}^{\infty} a_k$ with nonnegative terms satisfies $a_k \geqslant 0$ for all k. For each positive integer n, the partial sum s_{n+1} of a series $\sum a_k$ with nonnegative terms satisfies

$$s_{n+1} = \sum_{k=1}^{n+1} a_k = \left(\sum_{k=1}^{n} a_k\right) + a_{n+1} \geqslant \sum_{k=1}^{n} a_k = s_n,$$

so

$$s_1 \leqslant s_2 \leqslant \cdots \leqslant s_n \leqslant \cdots.$$

That is, the sequence $\{s_n\}$ of partial sums is an increasing sequence and is therefore a monotonic sequence. By the bounded monotone sequence theorem, we know that an increasing sequence $\{s_n\}$ converges if and only if the sequence $\{s_n\}$ is bounded above. If the sequence $\{s_n\}$ is not bounded above, then $\lim_{n\to\infty} s_n = +\infty$, in which case the sequence of partial sums $\{s_n\}$ and the series $\sum a_k$ both diverge to $+\infty$.

We have already seen that the partial sums of the harmonic series $\sum_{k=1}^{\infty} 1/k$ are not bounded above. Therefore, it diverges to $+\infty$.

We now give a number of tests that allow us to detect whether or not a series $\sum a_n$ with nonnegative terms is convergent. Each test is effective for a particular class of infinite series. These tests do not help us to find the sum of a series.

The p-series and integral test
Definition 8.3.1. A p-series is a series of nonnegative terms of the following form:

$$1 + \frac{1}{2^p} + \frac{1}{3^p} + \cdots + \frac{1}{n^p} + \cdots = \sum_{k=1}^{\infty} \frac{1}{k^p},$$

where p is a nonzero constant.

The harmonic series is a p-series with $p = 1$ and we already know that it diverges. Now we investigate p-series, for $p > 0$, $p \neq 1$, so that $\frac{1}{n^p} > \frac{1}{(n+1)^p}$ for all n. We compare the partial sums s_n to some areas of the plane as follows. Construct the following rectangles in a Cartesian coordinate system in $\mathbf{R}^2$: rectangle 1 with base the interval $[1, 2]$ on the x-axis and height 1, rectangle 2 with base the interval $[2, 3]$ on the x-axis and height $\frac{1}{2^p}$, rectangle 3 with base the interval $[3, 4]$ and height $\frac{1}{3^p}$, etc. Continue this process until you have n rectangles, the last with base $[n, n+1]$ on the x-axis and height $\frac{1}{n^p}$. The total area enclosed by these rectangles is the same as the nth partial sum of the following p-series:

$$1 \cdot \frac{1}{1^p} + 1 \cdot \frac{1}{2^p} + 1 \cdot \frac{1}{3^p} + \cdots + 1 \cdot \frac{1}{n^p} = s_n.$$

The graph of $y = \frac{1}{x^p}$ and the rectangles are shown in Figure 8.3.1 (a).

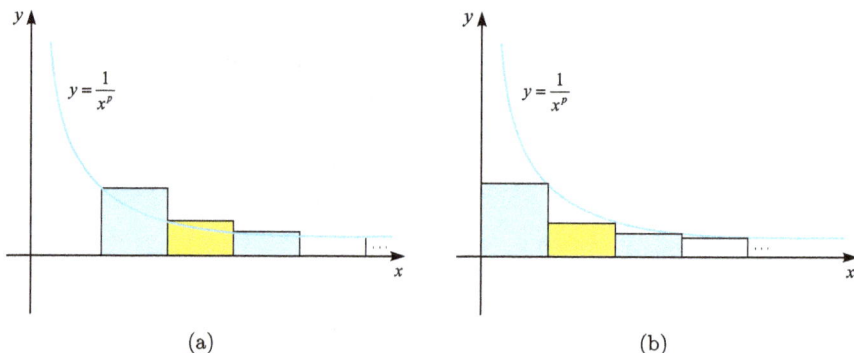

(a)   (b)

Figure 8.3.1: Graph of $y = \frac{1}{x^p}$ and rectangles with area $1, \frac{1}{2^p}, \frac{1}{3^p}, \ldots$.

The graph of $y = \frac{1}{x^p}$, for $x \geqslant 1$, is decreasing and does not go above the rectangles. Hence,

$$s_n \geqslant \int_1^{n+1} \frac{1}{x^p} \, dx = \left[\frac{1}{1-p} x^{1-p} \right]_1^{n+1} = \frac{1}{1-p} \left((n+1)^{1-p} - 1 \right).$$

Notice that, when $0 < p < 1$, the integral $\frac{1}{1-p}((n+1)^{1-p} - 1) \to \infty$ as $n \to \infty$, so $s_n \to \infty$ and the p-series diverges.

If we move all of the rectangles one unit to the left, then they are all below $y = \frac{1}{x^p}$ for $x \geqslant 1$, except for the first rectangle of area 1, as seen in Figure 8.3.1 (b). Therefore, when $p > 1$,

$$s_n \leqslant 1 + \int_1^n \frac{1}{x^p} \, dx = 1 + \left[\frac{1}{1-p} x^{1-p} \right]_1^n$$

$$= 1 + \frac{1}{1-p} (n^{1-p} - 1) = 1 + \frac{1}{1-p} \left(\frac{1}{n^{p-1}} - 1 \right)$$

$$< 1 + \frac{1}{1-p}(0-1) = 1 + \frac{1}{p-1} = \frac{p}{p-1}.$$

In this case, $\{s_n\}$ is monotonic and bounded above, so a p-series converges when $p > 1$. These results are summarized in the following theorem.

Theorem 8.3.1 (The p-series). *Let p be a real number. The infinite series*

$$\sum_{k=1}^{\infty} \frac{1}{k^p} \ \text{converges if } p > 1 \text{ and diverges if } p \leqslant 1.$$

The same geometric reasoning that we have used for p-series proves the following generalization.

Theorem 8.3.2 (Integral test). *Let f be a continuous, positive, decreasing function on $[1, \infty)$ and let $a_n = f(n)$. Then the series $\sum_{n=1}^{\infty} a_n$ is convergent if and only if the improper integral $\int_1^{\infty} f(x)\,dx$ is convergent. In other words:*
1. *If $\int_1^{\infty} f(x)\,dx$ exists as a real number, then $\sum_{n=1}^{\infty} a_n$ is convergent. In this case, if the sum $s = \sum_{n=1}^{\infty} a_n$, then $\int_1^{\infty} f(x)\,dx \leqslant s \leqslant a_1 + \int_1^{\infty} f(x)\,dx$.*
2. *If $\int_1^{\infty} f(x)\,dx$ does not exist as a real number, then $\sum_{n=1}^{\infty} a_n$ is divergent.*

The comparison test
If two infinite series have similar terms, then knowing the behavior of one of the series might be enough to determine the behavior of the other. The comparison test shows one way that this can be done.

Theorem 8.3.3 (The comparison test). *Let $\sum_{k=1}^{\infty} u_k$ and $\sum_{k=1}^{\infty} v_k$ be two series of non-negative terms such that $u_k \leqslant v_k$ for all k. Then:*
1. *if $\sum_{k=1}^{\infty} v_k$ converges, then $\sum_{k=1}^{\infty} u_k$ converges;*
2. *if $\sum_{k=1}^{\infty} u_k$ diverges, then $\sum_{k=1}^{\infty} v_k$ diverges.*

Proof. If s_{u_n} and s_{v_n} designate the nth partial sums of $\sum u_k$ and $\sum v_k$, respectively, then

$$s_{u_n} = u_1 + u_2 + \cdots + u_n \leqslant s_{v_n} = v_1 + v_2 + \cdots + v_n.$$

If $\sum v_k$ converges, this means the partial sum of $\sum u_k$ is bounded above, so $\sum u_k$ converges. If $\sum u_k$ diverges, this means the partial sum of $\sum u_k$ gets larger and larger as $k \to \infty$, so the partial sum of $\sum v_k$ is unbounded. Therefore, $\sum v_k$ must be divergent. □

NOTE. In using the comparison test, we must, of course, have a series $\sum c_k$ that is known to converge or diverge for the purpose of comparison. Often, we use either a p-series ($\sum 1/k^p$) or a geometric series $\sum a q^{k-1}$.

(*Group discussion) In Theorem 8.3.3, if we replace the condition "$u_k \leq v_k$ for all k" by "$u_k \leq cv_k$ for sufficiently large k (where c is a positive constant)," how will this affect the results of the theorem?

Example 8.3.1. Determine whether the series $\sum_{n=1}^{\infty} \frac{3}{2n^2+7n+3}$ converges or diverges.

Solution. For large n the dominant term in the denominator is $2n^2$, so we compare the given series with the series $\sum \frac{1}{n^2}$. Observe that the nth term is bounded above by

$$\frac{3}{2n^2 + 7n + 3} < \frac{3}{2n^2} = \frac{3}{2}\frac{1}{n^2}.$$

We know that $\frac{3}{2}\sum_{n=1}^{\infty} 1/n^2$ is convergent because it is a multiple of a p-series with $p = 2 > 1$. Therefore, $\sum_{n=1}^{\infty} 3/(2n^2 + 7n + 3)$ is convergent by the comparison test.

Example 8.3.2. Determine whether the infinite series $\sum_{k=1}^{\infty} \frac{1}{\sqrt{2k^2+3}}$ converges or diverges.

Solution. The dominant part of the kth term is $\frac{1}{\sqrt{2k^2}} = \frac{1}{\sqrt{2}k}$, which is a multiple of the kth term of the divergent harmonic series, suggesting divergence. To confirm this, compare it with the harmonic series as follows. For $k \geq 2$,

$$\frac{1}{\sqrt{2k^2 + 3}} \geq \frac{1}{\sqrt{2k^2 + k^2}} = \frac{1}{\sqrt{3}}\frac{1}{k}.$$

Since $\frac{1}{\sqrt{3}}\sum_{k=1}^{\infty} \frac{1}{k}$ diverges, the given series $\sum_{k=1}^{\infty} \frac{1}{\sqrt{2k^2+3}}$ also diverges by the comparison test.

NOTE. In this example, the comparison test works even though the inequality holds only for $k \geq 2$. This is because the behavior of the first few terms of a series does not matter in the comparison test (or any other test for convergence). That is, as long as the test applies to all terms after a certain point in the series, the conclusions of the test will still be true.

Theorem 8.3.4 (Limit comparison test). *Let $\sum_{k=1}^{\infty} a_k$ and $\sum_{k=1}^{\infty} b_k$ be two series of non-negative terms. Suppose that*

$$\lim_{k \to \infty} \frac{a_k}{b_k} = L, \quad \text{where } L \neq 0 \text{ and } L \neq \infty. \tag{8.1}$$

Then $\sum_{k=1}^{\infty} a_k$ and $\sum_{k=1}^{\infty} b_k$ both converge or both diverge.

Proof. Let A and B be any positive numbers such that $A < L < B$. Because a_n/b_n is close to L for large n, there is an integer N such that, for all $n > N$, we have

$$A < \frac{a_n}{b_n} < B,$$

so

$$Ab_n < a_n < Bb_n \quad \text{for all } n > N.$$

Hence, if $\sum a_n$ converges, so does $\sum Ab_n$. By the comparison test, $\sum b_n$ also converges. If $\sum a_n$ diverges, so does $\sum Bb_n$, thus $\sum b_n$ diverges. □

(*Group discussion) In the limit comparison test theorem, if $L = 0$ and $\sum b_k$ converges, does the series $\sum a_k$ have to converge? If $L = +\infty$ and $\sum b_k$ diverges, does the series $\sum a_k$ have to diverge?

Example 8.3.3. Determine whether the series $\sum_{n=1}^{\infty} \frac{1}{3^n-1}$ converges or diverges.

Solution. The nth term is close to $\frac{1}{3^n}$ for large n, so we apply the limit comparison test with

$$a_n = \frac{1}{3^n - 1} \quad \text{and} \quad b_n = \frac{1}{3^n}.$$

We obtain

$$L = \lim_{n\to\infty} \frac{a_n}{b_n} = \lim_{n\to\infty} \frac{1/(3^n - 1)}{1/3^n} = \lim_{n\to\infty} \frac{3^n}{3^n - 1} = 1 > 0.$$

Since the limit $L = 1$ and $\sum 1/3^n$ is a convergent geometric series, it follows that $\sum \frac{1}{3^n-1}$ converges by the limit comparison test.

Example 8.3.4. Determine whether the series $\sum_{n=1}^{\infty} \frac{n^2+3n}{\sqrt{8+n^5}}$ converges or diverges.

Solution. The dominant part of the numerator is n^2 and the dominant part of the denominator is $\sqrt{n^5} = n^{5/2}$. This suggests to compare $a_n = \frac{n^2+3n}{\sqrt{8+n^5}}$ with $b_n = \frac{n^2}{n^{5/2}} = \frac{1}{n^{1/2}}$, which is the nth term of the divergent p-series $\sum 1/n^{\frac{1}{2}}$. We have

$$\lim_{n\to\infty} \frac{a_n}{b_n} = \lim_{n\to\infty} \frac{n^2 + 3n}{\sqrt{8 + n^5}} \times n^{1/2}$$

$$= \lim_{n\to\infty} \frac{n^{5/2} + 3n^{3/2}}{\sqrt{8 + n^5}}$$

$$= \lim_{n\to\infty} \frac{1 + \frac{3}{n}}{\sqrt{\frac{8}{n^5} + 1}} = 1.$$

Since $\sum b_n = \sum 1/n^{1/2}$ is divergent, the given series diverges by the limit comparison test.

To effectively use the limit comparison test, we need to have a collection of series whose behavior we already know. The p-series belongs to the most useful series for this purpose.

d'Alembert's ratio test

Jean-Baptiste d'Alembert
(1717–1783) was a French mathematician, mechanician, physicist, philosopher, and music theorist. The wave equation is sometimes referred to as d'Alembert's equation. The ratio test was first published by him and is sometimes known as the d'Alembert ratio test. http://en.wikipedia.org/wiki/Jean_le_Rond_d%27Alembert

Theorem 8.3.5 (The ratio test). *Let $\sum_{k=1}^{\infty} a_k$ be a series of nonnegative terms and suppose*

$$\lim_{k \to \infty} \frac{a_{k+1}}{a_k} = \rho,$$

where $0 \leqslant \rho \leqslant \infty$. Then:
(1) *if $\rho < 1$, then $\sum_{k=1}^{\infty} a_k$ converges;*
(2) *if $\rho > 1$, then $\sum_{k=1}^{\infty} a_k$ diverges;*
(3) *if $\rho = 1$, the ratio test gives no information about the convergence or divergence of $\sum_{k=1}^{\infty} a_k$.*

Proof. (1) If $\frac{a_{n+1}}{a_n} \to \rho < 1$, then we can choose a number r such that $\rho < r < 1$. Since

$$\lim_{n \to \infty} \frac{a_{n+1}}{a_n} = \rho < r,$$

there exists an integer N such that, for all $n \geqslant N$, we have $\frac{a_{n+1}}{a_n} < r$. That is, for all $n \geqslant N$,

$$a_{n+1} < r a_n.$$

Hence, $a_{N+1} < r a_N$, $a_{N+2} < r a_{N+1} < r^2 a_N$, $a_{N+3} < r a_{N+2} < r^3 a_N$, and, in general,

$$a_{N+k} < r^k a_N.$$

However, the series $\sum_{k=0}^{\infty} a_N r^k$ (or equivalently $\sum_{k=N}^{\infty} a_N r^{k-N}$) is convergent since it is a geometric series with $0 < r < 1$. Since $a_k \leqslant a_N r^{k-N}$ for each $k > N$, by the comparison test, $\sum_{k=N}^{\infty} a_k$ is also convergent. Adding back the first $N-1$ terms to the series $\sum_{k=N}^{\infty} a_k$ gives $\sum_{k=1}^{\infty} a_k$, the original series, which must also be convergent.

(2) If $\lim_{n \to \infty} a_{n+1}/a_n \to \rho > 1$, then the ratio a_{n+1}/a_n will be greater than 1 for all sufficiently large n, say, $n \geqslant M$. This means that $a_{n+1} > a_n$ whenever $n \geqslant M$, so $\lim_{n \to \infty} a_n \neq 0$. Therefore, $\sum a_n$ diverges by the nth term divergence test.

(3) If $\lim_{n\to\infty} a_{n+1}/a_n = 1$, then we give two examples. The limit of the ratio is 1 for the convergent series $\sum 1/n^2$, because

$$\lim_{n\to\infty} \frac{a_{n+1}}{a_n} = \lim_{n\to\infty} \frac{\frac{1}{(n+1)^2}}{\frac{1}{n^2}} = 1.$$

The limit of the ratio is 1 for the divergent series $\sum 1/n$, because

$$\lim_{n\to\infty} \frac{a_{n+1}}{a_n} = \lim_{n\to\infty} \frac{\frac{1}{n+1}}{\frac{1}{n}} = 1.$$

Hence, the test fails to give any information about convergence when $\lim_{n\to\infty} a_{n+1}/a_n = 1$. $\qquad\square$

Example 8.3.5. Show that the series

$$\sum_{k=1}^{\infty} \frac{1}{k!} = 1 + \frac{1}{2!} + \frac{1}{3!} + \cdots + \frac{1}{n!} + \cdots$$

converges.

Solution. We use the ratio test with $a_n = \frac{1}{n!}$. We have

$$\lim_{n\to\infty} \frac{a_{n+1}}{a_n} = \lim_{n\to\infty} \frac{\frac{1}{(n+1)!}}{\frac{1}{n!}} = \lim_{n\to\infty} \frac{1}{n+1} = 0 < 1.$$

Therefore, by the ratio test, the given series is convergent.

Example 8.3.6. Test the convergence of the series

$$\sum_{n=1}^{\infty} \frac{n!}{n^n} = \frac{1!}{1^1} + \frac{2!}{2^2} + \frac{3!}{3^3} + \cdots + \frac{n!}{n^n} + \cdots .$$

Solution. The terms $a_n = \frac{n!}{n^n}$ are positive and we have

$$\frac{a_{n+1}}{a_n} = \frac{\frac{(n+1)!}{(n+1)^{n+1}}}{\frac{n!}{n^n}} = \frac{(n+1)!}{(n+1)^{n+1}} \times \frac{n^n}{n!} = \frac{(n+1)n!}{(n+1)^{n+1}} \times \frac{n^n}{n!}$$

$$= \left(\frac{n}{n+1}\right)^n = \left(\frac{1}{1+\frac{1}{n}}\right)^n.$$

Because

$$\lim_{n\to\infty} \frac{a_{n+1}}{a_n} = \lim_{n\to\infty} \left(\frac{1}{1+\frac{1}{n}}\right)^n = \lim_{n\to\infty} \frac{1}{(1+\frac{1}{n})^n} = \frac{1}{e} < 1,$$

the given series is convergent by the ratio test.

(*Group activity) Raabe's test: assume $\lim_{n\to\infty} n(|\frac{a_n}{a_{n+1}}| - 1) = k$. Then, if $k < 1$, $\sum a_n$ diverges and, if $k > 1$, $\sum a_n$ converges. This is an extension of the ratio test and is due to Joseph Ludwig Raabe (1801–1859, a Swiss mathematician). Why does it work? Use this test to show that $\sum \frac{1}{n^p}$ converges when $p > 1$ and diverges when $p < 1$.

The root test (Cauchy's radical test)

Theorem 8.3.6 (The root test). *Let $\sum_{k=1}^{\infty} a_k$ be a series of nonnegative terms and suppose*

$$\lim_{n\to\infty} \sqrt[n]{a_n} = \rho,$$

where $0 \leqslant \rho \leqslant \infty$. Then:
(1) *if $\rho < 1$, then $\sum_{k=1}^{\infty} a_k$ converges;*
(2) *if $\rho > 1$, then $\sum_{k=1}^{\infty} a_k$ diverges;*
(3) *if $\rho = 1$, the test fails to give any information and $\sum_{k=1}^{\infty} a_k$ may converge or diverge.*

Although the precise proof is not given here, intuitively we have $\sqrt[n]{a_n} \approx \rho$ when n is large. Therefore, $a_n \approx \rho^n$ for sufficiently large n. By the limit comparison test and the theorem for geometric series, we have conclusions (1) and (2). For (3), we also have two examples $\sum 1/n$ and $\sum 1/n^2$. Both $\lim_{n\to\infty} \sqrt[n]{1/n}$ and $\lim_{n\to\infty} \sqrt[n]{1/n^2}$ are 1, but one of them converges and the other diverges. Therefore, if ρ is 1, the root test is inconclusive.

Example 8.3.7. Test the convergence of the series $\sum_{n=1}^{\infty} \frac{2+(-1)^n}{3^n}$.

Solution. We have

$$a_n = \frac{2 + (-1)^n}{3^n} \quad \text{and} \quad \frac{1}{3^n} \leqslant \frac{2 + (-1)^n}{3^n} \leqslant \frac{3}{3^n}.$$

Hence,

$$\frac{1}{3} \leqslant \sqrt[n]{\frac{2 + (-1)^n}{3^n}} \leqslant \frac{\sqrt[n]{3}}{3},$$

so $\lim_{n\to\infty} \sqrt[n]{a_n} = \frac{1}{3}$. Thus, the given series converges by the root test.

Example 8.3.8. Show that the following series is convergent:

$$1 + \frac{1}{2^2} + \frac{1}{3^3} + \cdots + \frac{1}{n^n} + \cdots.$$

Solution. We use the root test with $a_n = \frac{1}{n^n}$, since

$$\sqrt[n]{a_n} = \sqrt[n]{\frac{1}{n^n}} = \frac{1}{n} \to 0 < 1 \quad \text{as} \quad n \to \infty.$$

Therefore, we conclude that the series $\sum \frac{1}{n^n}$ is convergent.

8.3.2 Series with negative and positive terms

Alternating series test

If the terms of a series alternate in sign and decrease to 0 in absolute value, we immediately conclude that the series converges. The full result is the following.

Theorem 8.3.7 (Alternating series test). *Let $\{u_n\}_{n=1}^{\infty}$ be a sequence of positive real numbers satisfying*

$$u_1 \geqslant u_2 \geqslant \cdots \geqslant u_n \geqslant \cdots$$

(a decreasing sequence) and $\lim_{n \to \infty} u_n = 0$. *Then the* alternating series

$$\sum_{k=1}^{\infty} (-1)^{k-1} u_k = u_1 - u_2 + u_3 - u_4 + \cdots$$

formed from these numbers is convergent.

Proof. We first consider the even partial sums:

$$s_2 = u_1 - u_2 \geqslant 0,$$
$$s_4 = s_2 + (u_3 - u_4) \geqslant s_2.$$

In general, $s_{2n} = s_{2n-2} + (u_{2n-1} - u_{2n}) \geqslant s_{2n-2}$ since $u_{2n-1} \geqslant u_{2n}$, so

$$0 \leqslant s_2 \leqslant s_4 \leqslant \cdots \leqslant s_{2n} \leqslant \cdots.$$

However, we can also write

$$s_2 = u_1 - u_2, s_4 = u_1 - (u_2 - u_3) - u_4 \quad \text{and in general}$$
$$s_{2n} = u_1 - (u_2 - u_3) - (u_4 - u_5) - \cdots - (u_{2n-2} - u_{2n-1}) - u_{2n}.$$

Every term in the parentheses is positive, so $s_{2n} \leqslant u_1$ for all n. Therefore, the sequence $\{s_{2n}\}$ of even partial sums is increasing and bounded above. It is therefore convergent by the bounded monotonic sequence theorem.

Let $\lim_{n \to \infty} s_{2n} = s$. We compute the following limit of the odd partial sums:

$$\lim_{n \to \infty} s_{2n+1} = \lim_{n \to \infty} (s_{2n} + u_{2n+1})$$
$$= \lim_{n \to \infty} s_{2n} + \lim_{n \to \infty} u_{2n+1} = s + 0 = s.$$

Since both the even and the odd partial sums converge to s, the alternating series $\sum_{k=1}^{\infty} (-1)^{k-1} u_k$ is convergent with sum s. Figure 8.3.2 illustrates this idea. □

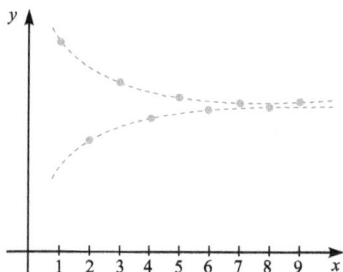

Figure 8.3.2: Graph of the sequence $\{s_n = \sum_{k=1}^{n}(-1)^{k-1}\frac{1}{k}\}$.

Example 8.3.9. Show that the following alternating harmonic series converges:

$$\sum_{k=1}^{\infty}(-1)^{k-1}\frac{1}{k} = 1 - \frac{1}{2} + \frac{1}{3} - \frac{1}{4} + \cdots.$$

Proof. If $b_k = \frac{1}{k}$, for $k = 1, 2, 3, \ldots$, then (i) $\{b_k\}$ is decreasing since $b_{k+1} = \frac{1}{k+1} < b_k = \frac{1}{k}$ and (ii) $\lim_{k \to \infty} b_k = \lim_{k \to \infty} \frac{1}{k} = 0$. Hence, the conditions of the alternating series test are satisfied, so the series $\sum_{k=1}^{\infty}(-1)^{k-1}b_k = \sum_{k=1}^{\infty}(-1)^{k-1}\frac{1}{k}$ is convergent. □

Example 8.3.10. Decide whether or not the series $\sum_{n=1}^{\infty}\frac{(-1)^n n}{4n+1}$ is convergent.

Solution. It is an alternating series but it does not satisfy the conditions of the alternating series test because, with $b_n = \frac{n}{4n+1}$,

$$\lim_{n \to \infty} b_n = \lim_{n \to \infty}\frac{n}{4n+1} = \lim_{n \to \infty}\frac{1}{4+\frac{1}{n}} = \frac{1}{4} \neq 0.$$

The nth term divergence test shows that this series diverges.

Example 8.3.11. Test the series $\sum_{n=1}^{\infty}(-1)^{n+1}\frac{n^2}{n^3+2}$ for convergence or divergence.

Solution. It is obvious that the series is alternating. If $b_n = \frac{n^2}{n^3+2}$, then

$$\lim_{n \to \infty} b_n = \lim_{n \to \infty}\frac{n^2}{n^3+2} = 0.$$

To verify that $\{b_n\} = \{\frac{n^2}{n^3+2}\}$ is decreasing, we compute the derivative of the function $f(x) = \frac{x^2}{x^3+2}$ and find

$$f'(x) = \frac{x(4-x^3)}{(x^3+2)^2} < 0 \quad \text{for all } x > \sqrt[3]{4}.$$

This means that $f(x)$ is decreasing for $x \geq 2$, so, for all $n \geq 2$,

$$f(n) \geq f(n+1) \implies b_n \geq b_{n+1}.$$

Then $\{\frac{n^2}{n^3+2}\}$ is decreasing for $n \geqslant 2$. Thus the given series is convergent by the alternating series test.

NOTE. Again, we only checked whether the conditions of the alternating series test are satisfied for all $n \geqslant 2$, since the first few terms of any series do not affect its convergence, as seen in Figure 8.3.3.

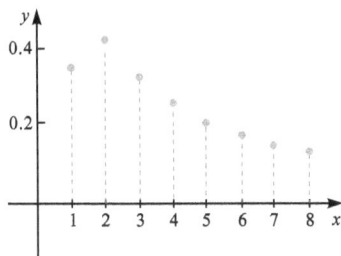

Figure 8.3.3: Graph of the sequence $\{\frac{n^2}{n^3+2}\}$.

Absolute and conditional convergence

A series $\sum a_n$ is called *absolutely convergent* if the series of absolute values $\sum |a_n|$ is convergent. A series $\sum a_n$ is called *conditionally convergent* if it is convergent but not absolutely convergent.

Example 8.3.12. The series $\sum_{n=1}^{\infty} (-1)^{n-1} \frac{1}{n^2}$ is absolutely convergent, because

$$\sum_{n=1}^{\infty} \left| \frac{(-1)^{n-1}}{n^2} \right| = \sum_{n=1}^{\infty} \frac{1}{n^2} = 1 + \frac{1}{2^2} + \frac{1}{3^2} + \cdots$$

is a convergent p-series with $p = 2$.

Example 8.3.13. The series $\sum (-1)^{n-1} \frac{1}{n}$ is conditionally convergent, since the alternating harmonic series $\sum_{n=1}^{\infty} (-1)^{n-1} \frac{1}{n}$ was shown to be convergent in a previous example, but it is not absolutely convergent because the corresponding series of absolute values is

$$\sum_{n=1}^{\infty} \left| \frac{(-1)^{n-1}}{n} \right| = \sum_{n=1}^{\infty} \frac{1}{n} = 1 + \frac{1}{2} + \frac{1}{3} + \cdots.$$

This is the harmonic series, a divergent p-series with $p - 1$.

Theorem 8.3.8 (The absolute convergence test). *Let $\sum_{k=1}^{\infty} a_k$ be an infinite series. If the series $\sum_{k=1}^{\infty} |a_k|$ converges, then the series $\sum_{k=1}^{\infty} a_k$ also converges.*

Proof. Observe the inequality

$$0 \leqslant (a_k + |a_k|) \leqslant 2|a_k|.$$

It is true because $|a_k|$ is either a_k or $-a_k$. If $\sum a_k$ is absolutely convergent, then $\sum |a_k|$ is convergent, so $\sum 2|a_k|$ is convergent. Therefore, by the comparison test, $\sum (a_k + |a_k|)$ is convergent. Then

$$\sum a_k = \sum (a_k + |a_k|) - \sum |a_k|$$

is the difference of two convergent series and it is therefore convergent. □

Example 8.3.14. Determine whether the following series is convergent or divergent:

$$\sum_{n=1}^{\infty} \frac{\sin nx}{n^2} = \frac{\sin 1}{1^2} + \frac{\sin 2}{2^2} + \cdots.$$

Solution. This series has both positive and negative terms, but it is not alternating. We apply the comparison test to the series of absolute values as follows:

$$\sum_{n=1}^{\infty} \left| \frac{\sin nx}{n^2} \right| = \sum_{n=1}^{\infty} \frac{|\sin nx|}{n^2}.$$

Since

$$\frac{|\sin nx|}{n^2} \leqslant \frac{1}{n^2}$$

and $\sum 1/n^2$ is convergent, we find that $\sum |\sin nx|/n^2$ is convergent by the comparison test. Thus the given series $\sum (\sin nx)/n^2$ is absolutely convergent and therefore convergent by the absolute convergence test.

NOTE. If we take the absolute values of the terms of any series $\sum a_n$, then we create a series $\sum |a_n|$ with nonnegative terms. Then we can use any of the tests for convergence of series with nonnegative terms in determining whether the series is absolutely convergent: the ratio test, root test, comparison test, integral test, etc.

Example 8.3.15. Test the series $\sum_{n=1}^{\infty} (-1)^n \frac{2^{n^2}}{n!}$ for absolute convergence.

Solution. We use the ratio test as follows:

$$\left| \frac{a_{n+1}}{a_n} \right| = \left| \frac{\frac{(-1)^{n+1} 2^{(n+1)^2}}{(n+1)!}}{\frac{(-1)^n 2^{n^2}}{n!}} \right| = \frac{2^{(n+1)^2 - n^2}}{(n+1)} = \frac{2^{2n+1}}{n+1} > 1, \quad \text{when } n > 1.$$

Thus $\lim_{n \to \infty} |a_{n+1}/a_n| > 1$ and therefore $\sum_{n=1}^{\infty} 2^{n^2}/n!$ diverges, so the series $\sum_{n=1}^{\infty} (-1)^n 2^{n^2}/n!$ is not absolutely convergent. Since $|a_{n+1}| > |a_n|$, it follows from the nth term divergence test that the series $\sum_{n=1}^{\infty} (-1)^n 2^{n^2}/n!$ diverges.

There are many interesting properties of series that converge absolutely. We now state without proof the following theorem.

Theorem 8.3.9. (1) *If $\sum a_n$ is absolutely convergent, then any new series $\sum b_n$ formed by rearranging the order of the terms in $\sum a_k$ is also absolutely convergent, with the same sum.*

(2) *If $\sum a_n = A$ and $\sum b_n = B$ are both absolutely convergent, then the product $\sum_{i,k}^{\infty} a_i b_k$ (of all possible product pairs a_i and b_k from the two series) is also absolutely convergent, with sum AB.*

8.4 Power series and Taylor series

8.4.1 Power series

If $u_1(x), u_2(x), \ldots, u_n(x), \ldots$ are real-valued functions defined on an interval **I**, then we can form the infinite series with these functions as

$$\sum_{k=1}^{\infty} u_k(x) = u_1(x) + u_2(x) + \cdots + u_k(x) + \cdots. \tag{8.2}$$

Such a series may start at other k-values, such as $k = 0$. Since each function value is a number, we can use all of the properties and convergence tests for infinite series of numbers developed in the previous sections. In particular, for a specific number $x_0 \in \mathbf{I}$, the series (8.2) becomes an infinite series with constant values as

$$\sum_{k=1}^{\infty} u_k(x_0) = u_1(x_0) + u_2(x_0) + \cdots + u_k(x_0) + \cdots. \tag{8.3}$$

For this particular value of $x = x_0$, this series may converge or diverge. If it converges, then x_0 is called a *convergent point*; otherwise it is called a *divergent point*. The set of all convergent points is called the *convergent set* of the series. For any point x in the convergent set, the infinite series in equation (8.3) must have a sum, which we write as $s(x)$. That is, for each x in the convergent set,

$$s(x) = \sum_{k=1}^{\infty} u_k(x) = u_1(x) + u_2(x) + \cdots + u_n(x) + \cdots.$$

In this case, $s(x)$ is a function of x defined for all x in the convergent set and it is called the *sum function*. We define the partial sum function $s_n(x)$ for $n = 1, 2, \ldots$ by

$$s_n(x) = (u_1(x) + u_2(x) + \cdots + u_n(x)) = \sum_{k=1}^{n} u_k(x).$$

From the previous theories of infinite series, $s(x)$ is the limiting value of the partial sums, for each x in the convergent set, written

$$s(x) = \lim_{n \to \infty} (u_1(x) + u_2(x) + \cdots + u_n(x)) = \lim_{n \to \infty} s_n(x).$$

The *remainder* for the partial sum $s_n(x)$ is $r_n(x) = s(x) - s_n(x)$.

NOTE. The Weierstrass function $f(x) = \sum_{n=0}^{\infty} a^n \cos(b^n \pi x)$, where $0 < a < 1$, b is an odd integer, and $ab > 1 + 3\pi/2$, is defined as a series of real-valued functions. This function is a continuous function but nowhere differentiable. The graph of Weierstrass function is illustrated in Figure 8.4.1. It was presented and proved by Karl Weierstrass on 18 July 1872.

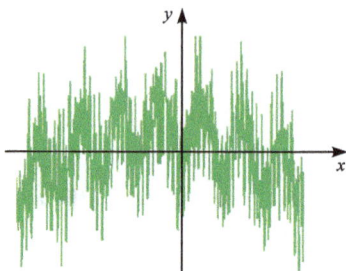

Figure 8.4.1: Illustration of a Weierstrass function.

We will restrict our attention for the next few sections to a special case where each function term is a particular type of the polynomial $u_n(x) = c_n(x-a)^n$, for some constants a and c_n. Such a series is called a *power series*. The formal definition is as follows.

Definition 8.4.1. Let a be a real number and let $\{c_k\}_{k=0}^{\infty}$ be a sequence of real numbers. Then an infinite series of the form

$$\sum_{k=0}^{\infty} c_k(x-a)^k = c_0 + c_1(x-a) + c_2(x-a)^2 + \cdots \tag{8.4}$$

is called a *power series about a*. The numbers $c_0, c_1, \ldots$ are the *coefficients* of the power series.

Notice that, when $x = a$, all of the terms are 0 for $n \geqslant 1$, so the power series always converges when $x = a$. If $a = 0$, then this power series is about the origin and it looks like the following infinite polynomial:

$$\sum_{k=0}^{\infty} = c_0 + c_1 x + c_2 x^2 + \cdots. \tag{8.5}$$

Example 8.4.1. For what values of x does the series

$$\sum_{k=1}^{\infty} (-1)^{k-1} \frac{x^k}{k} = \frac{x}{1} - \frac{x^2}{2} + \frac{x^3}{3} - \frac{x^4}{4} + \cdots$$

converge?

Solution. Let $a_n = (-1)^{n-1}\frac{x^n}{n}$ and apply the ratio test as follows:

$$\lim_{n\to\infty}\left|\frac{a_{n+1}}{a_n}\right| = \lim_{n\to\infty}\left|\frac{(-1)^n\frac{x^{n+1}}{n+1}}{(-1)^{n-1}\frac{x^n}{n}}\right| = \lim_{n\to\infty}\left|(-1)\frac{x^{n+1}}{n+1}\times\frac{n}{x^n}\right|$$

$$= \lim_{n\to\infty}\left|\frac{nx}{n+1}\right| = |x|\lim_{n\to\infty}\frac{n}{n+1} = |x|.$$

By the ratio test, the series converges absolutely if $|x| < 1$ and diverges if $|x| > 1$. The values $x = 1$ and $x = -1$ are not covered by this, so we investigate these separately. When $x = 1$,

$$\sum_{k=1}^{\infty}(-1)^{k-1}\frac{x^k}{k} = \sum_{k=1}^{\infty}(-1)^{k-1}\frac{(1)^k}{k} = \sum_{k=1}^{\infty}(-1)^{k-1}\frac{1}{k}$$

and this series converges by the alternating series test.
When $x = -1$,

$$\sum_{k=1}^{\infty}(-1)^{k-1}\frac{x^k}{k} = \sum_{k=1}^{\infty}(-1)^{k-1}\frac{(-1)^k}{k} = -\sum_{k=1}^{\infty}\frac{1}{k}$$

and this series diverges since it is a multiple of the harmonic series.

Therefore, when $x \in (-1,1]$ the series converges and elsewhere, when $x \le -1$ or $1 < x$, the series diverges. Figure 8.4.2 shows the graph of $f_1(x) = \sum_{k=1}^{10}(-1)^{k-1}\frac{x^k}{k}$, $f_2(x) = \sum_{k=1}^{50}(-1)^{k-1}\frac{x^k}{k}$, and $f_3(x) = \sum_{k=1}^{100}(-1)^{k-1}\frac{x^k}{k}$ in a $[-2,2] \times [-2,2]$ window.

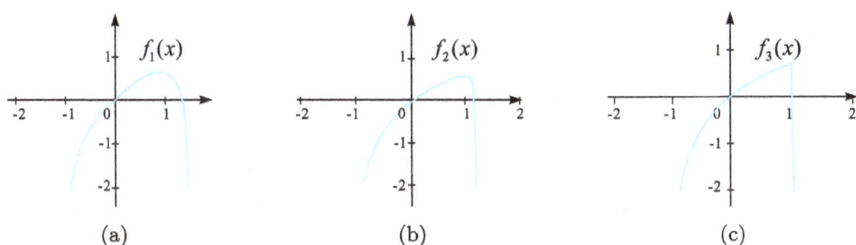

(a)   (b)   (c)

Figure 8.4.2: Graphs of partial sums ($n = 10, 50, 100$) of $\sum_{k=1}^{\infty}(-1)^{k-1}\frac{x^k}{k}$.

In this example, the set of x for which the given series converges is an interval. This is not a coincidence, since it is true for every power series, as shown in the following theorems.

Theorem 8.4.1 (Abel). *If the power series $\sum c_n x^n$ is convergent when $x = x_0 \ne 0$, then the power series $\sum c_n x^n$ is absolutely convergent at any x satisfying $|x| < |x_0|$. Similarly, if $\sum c_n x^n$ is divergent when $x = x_0$, then the power series $\sum c_n x^n$ is divergent at any x satisfying $|x| > |x_0|$.*

Niels Henrik Abel (1802–1829) was a Norwegian mathematician. His most famous single result is the first complete proof demonstrating the impossibility of solving the general quintic equation in radicals. He was also an innovator in the field of elliptic functions and discoverer of Abelian functions. Despite his achievements, Abel was largely unrecognized during his lifetime; he made his discoveries while living in poverty and died at the age of 26. http://en.wikipedia.org/wiki/Niels_Henrik_Abel

Proof. Suppose $\sum c_n x_0^n$ is convergent and $|x| < |x_0|$. Then the nth term $c_n x_0^n \to 0$ as $n \to \infty$. This means there exists a bound M such that $|c_n x_0^n| < M$ for all $n = 0, 1, 2, \ldots$. Now, for any x satisfying $|x| < |x_0|$, we have, for all n,

$$|c_n x^n| = \left| c_n x_0^n \cdot \frac{x^n}{x_0^n} \right| = |c_n x_0^n| \cdot \left| \frac{x}{x_0} \right|^n \leq M \left| \frac{x}{x_0} \right|^n. \tag{8.6}$$

Hence, if $|x| < |x_0|$, then $|x/x_0| < 1$ and the geometric series $\sum M|x/x_0|^n$ converges. Equation (8.6) and the comparison test show that $\sum |c_n x^n|$ converges. Thus, $\sum c_n x^n$ is also convergent.

For the second part of the theorem, suppose $\sum c_n x_0^n$ is divergent. If there is a value of x_1 such that $|x_1| > |x_0|$ and $\sum c_n x_1^n$ converges, then, by the first part of the theorem, it would follow that $\sum c_n x_0^n$ is convergent, which is not true. Hence, we have a proof by contradiction that $\sum c_n x^n$ is divergent for $|x| > |x_0|$. This completes the proof of the theorem. □

From this theorem, we deduce the following theorem.

Theorem 8.4.2. *If the set of values of x for which the power series $\sum c_n x^n$ converges is not $\{0\}$ or $(-\infty, +\infty)$, then there must be a positive number R such that, when $x \in (-R, R)$, the series $\sum c_n x^n$ converges and, when $x < -R$ or $x > R$, the series $\sum c_n x^n$ diverges.*

NOTE. When $x = R$ or $x = -R$, the theorem gives no information and the series $\sum c_n x^n$ may converge or diverge.

The number R is called the *radius of convergence* of the power series. By convention, the radius of convergence is $R = 0$ if $\sum c_n x^n$ is convergent only when $x = 0$ and $R = \infty$ if $\sum c_n x^n$ converges for all real numbers x. The *interval of convergence* of a power series is the interval that consists of all values of x for which the series converges, which must be $(-R, R)$, $(-R, R]$, $[-R, R)$, or $[-R, R]$. Similar results hold for the more general form of a power series $\sum c_n (x - a)^n$.

Theorem 8.4.3. *For a given power series $\sum c_n (x - a)^n$, there are only three possibilities:*
1. *the series converges only when $x = a$;*
2. *the series converges for all x;*
3. *there is a positive number R such that the series converges if $|x - a| < R$ and diverges if $|x - a| > R$.*

NOTE. For the more general series $\sum c_n(c-a)^n$, the number R from the theorem is called the radius of convergence and the series converges for $a - R < x < a + R$. Hence, the interval of convergence is $(a - R, a + R)$, $[a - R, a + R)$, $(a - R, a + R]$, or $[a - R, a + R]$.

Example 8.4.2. For what values of x is the series $\sum_{n=0}^{\infty} \frac{x^n}{n!}$ convergent?

Solution. We use the ratio test. If $a_n = \frac{x^n}{n!}$ denotes the nth term of the series, then, for $x \neq 0$, we have

$$\lim_{n \to \infty} \left| \frac{a_{n+1}}{a_n} \right| = \lim_{n \to \infty} \left| \frac{x^{n+1}}{(n+1)!} \frac{n!}{x^n} \right| = \lim_{n \to \infty} \frac{|x|}{n+1} = 0 < 1.$$

By the ratio test, the series converges absolutely for all values of x. Hence, the series $\sum_{n=0}^{\infty} \frac{x^n}{n!}$ converges for all x, the radius of converges is infinity, and the interval of convergence is $(-\infty, \infty)$.

Example 8.4.3. Find the radius of convergence and the interval of convergence of the series $\sum_{n=1}^{\infty} \frac{(x-1)^n}{n}$.

Solution. Let $a_n = \frac{(x-1)^n}{n}$. Then

$$\lim_{n \to \infty} \left| \frac{a_{n+1}}{a_n} \right| = \lim_{n \to \infty} \left| \frac{\frac{(x-1)^{n+1}}{n+1}}{\frac{(x-1)^n}{n}} \right| = \lim_{n \to \infty} \left| \frac{n}{n+1} (x-1) \right| = |x-1|.$$

By the ratio test, the given series is absolutely convergent and therefore convergent when $|x - 1| < 1$ and divergent when $|x - 1| > 1$. Now $|x - 1| < 1$ is equivalent to

$$-1 < x - 1 < 1 \quad \Leftrightarrow \quad 0 < x < 2,$$

so the series converges when $0 < x < 2$ and diverges when $x < 0$ or $x > 2$.

The ratio test gives no information when $|x - 1| = 1$, so we must consider $x = 0$ and $x = 2$ separately. When $x = 0$, the series becomes $\sum (-1)^n \frac{1}{n}$ and this is the convergent alternating harmonic series. If $x = 2$, the series is $\sum 1/n$, which is the divergent harmonic series. Thus, the given power series converges for $0 \leqslant x < 2$. Hence, the radius of convergence is 1 and the interval of convergence is $[0, 2)$.

When the German astronomer Friedrich Bessel (1784–1846) solved Kepler's equation for describing planetary motion, he introduced Bessel functions, one of which is described in the next example. A Bessel function is defined as a power series. As a matter of fact, Bessel functions have been applied in many different physical situations, including the determination of the temperature distribution in a circular plate and the shape of a vibrating drumhead.

Friedrich Wilhelm Bessel
(1784–1846) was a German astronomer and mathematician. He was the first astronomer who determined reliable values for the distance from the sun to another star by the method of parallax. http://en.wikipedia.org/wiki/Friedrich_Bessel

Example 8.4.4. Find the domain of the Bessel function of order 0 defined by

$$J_0(x) = \sum_{n=0}^{\infty} \frac{(-1)^n x^{2n}}{2^{2n}(n!)^2}.$$

Solution. Let $a_n = \frac{(-1)^n x^{2n}}{2^{2n}(n!)^2}$. Then

$$\lim_{n\to\infty}\left|\frac{a_{n+1}}{a_n}\right| = \lim_{n\to\infty}\left|\frac{(-1)^{n+1}x^{2(n+1)}}{2^{2(n+1)}((n+1)!)^2} \times \frac{(-1)^n x^{2n}}{2^{2n}(n!)^2}\right|$$

$$= \lim_{n\to\infty}\frac{x^2}{4(n+1)^2} = 0 \quad \text{for all } x.$$

Thus, by the ratio test, the given series converges for all values of x. In other words, the domain of the Bessel function J_0 can be taken to be $(-\infty, \infty)$.

Example 8.4.5. Find the radius of convergence and the interval of convergence of the series

$$\sum_{n=0}^{\infty} \frac{(-3)^n x^n}{\sqrt{n+1}}.$$

Solution. Let $a_n = \frac{(-3)^n x^n}{\sqrt{n+1}}$. Then

$$\lim_{n\to\infty}\left|\frac{a_{n+1}}{a_n}\right| = \lim_{n\to\infty}\left|\frac{(-3)^{n+1}x^{n+1}}{\sqrt{n+2}}\frac{\sqrt{n+1}}{(-3)^n x^n}\right| = \lim_{n\to\infty}\left|-3x\sqrt{\frac{n+1}{n+2}}\right|$$

$$= \lim_{n\to\infty} 3|x|\sqrt{\frac{1+(1/n)}{1+(2/n)}} = 3|x|.$$

By the ratio test, the given series converges if $3|x| < 1$ and diverges if $3|x| > 1$. Thus, the series converges if $|x| < 1/3$ and diverges if $|x| > 1/3$. This means that the radius of convergence is $R = 1/3$.

We must now test the convergence at the endpoints of the interval of convergence. If $x = -1/3$, the series becomes

$$\sum_{n=0}^{\infty} \frac{(-3)^n(-\frac{1}{3})^n}{\sqrt{n+1}} = \sum_{n=0}^{\infty} \frac{1}{\sqrt{n+1}}$$

$$= \frac{1}{\sqrt{1}} + \frac{1}{\sqrt{2}} + \frac{1}{\sqrt{3}} + \cdots.$$

This is a p-series with $p = 1/2 < 1$ and it diverges. If $x = 1/3$, the series is

$$\sum_{n=0}^{\infty} \frac{(-3)(\frac{1}{3})^n}{\sqrt{n+1}} = \sum_{n=0}^{\infty} \frac{(-1)^n}{\sqrt{n+1}}.$$

This series converges by the alternating series test. Therefore, the given power series converges when $-1/3 < x \leqslant 1/3$ and the interval of convergence is $(-1/3, 1/3]$.

Example 8.4.6. Find the radius of convergence of the series $\sum_{n=0}^{\infty} \frac{(2n)!}{(n!)^2} x^{2n}$.

Solution. Let $a_n = \frac{(2n)!}{(n!)^2} x^{2n}$. Using the ratio test, we have

$$\lim_{n\to\infty} \frac{|a_{n+1}|}{|a_n|} = \lim_{n\to\infty} \left| \frac{\frac{(2(n+1))!}{((n+1)!)^2} x^{2(n+1)}}{\frac{(2n)!}{(n!)^2} x^{2n}} \right| = \lim_{n\to\infty} \left| \frac{(2n+2)(2n+1)}{(n+1)^2} x^2 \right|$$

$$= 2x^2 \lim_{n\to\infty} \left| \frac{(2 + \frac{1}{n})}{(1 + \frac{1}{n})} \right| = 4x^2.$$

The series is absolutely convergent, so it is convergent, when $4x^2 < 1$ or $x \in (-1/2, 1/2)$. It is divergent when $4x^2 > 1$. Therefore, the radius of convergence is $1/2$.

8.4.2 Working with power series

The following properties are important properties of power series and are given without proof.

Theorem 8.4.4. *Suppose that $\sum_{k=0}^{\infty} a_k(x - c)^k = s(x)$ for $x \in (c - R, c + R)$, where R is the radius of convergence. Then $s(x)$ is continuous for $x \in (c - R, c + R)$.*

Theorem 8.4.5 (Term-by-term integration and differentiation of power series). *Suppose that $\sum_{k=0}^{\infty} a_k x^k = s(x)$ for $x \in (-R, R)$, where R is the radius of convergence. Then the series can be integrated or differentiated term by term for any $x \in (-R, R)$ as follows:*

$$\int_0^x s(t)\, dt = \int_0^x \sum_{k=0}^{\infty} a_k t^k\, dt = \sum_{k=0}^{\infty} \int_0^x a_k t^k\, dt = \sum_{k=0}^{\infty} a_k \frac{x^{k+1}}{k+1},$$

$$\frac{ds(x)}{dx} = \frac{d}{dx}\left(\sum_{k=0}^{\infty} a_k x^k\right) = \sum_{k=0}^{\infty} \frac{d}{dx}(a_k x^k) = \sum_{k=1}^{\infty} k a_k x^{k-1}.$$

These two series have the same radius of convergence, R, as the original series $\sum_{k=0}^{\infty} a_k x^k$.
The same applies to a series in the form $\sum_{k=0}^{\infty} a_k(x - c)^k$, for $x \in (c - R, c + R)$, where R is the radius of convergence.

Example 8.4.7. Find the sum of the series $\sum_{n=1}^{\infty} nx^n$ and use that sum to find the sum of the series $\sum_{n=1}^{\infty} \frac{n}{2^n}$.

Solution. Since

$$\lim_{n\to\infty} \left| \frac{(n+1)x^{n+1}}{nx^n} \right| = \lim_{n\to\infty} \left| \frac{n+1}{n} x \right| = |x|,$$

when $|x| < 1$, this series converges. When $x = 1$ or $x = -1$, this series becomes $\sum n$ or $\sum n(-1)^n$. Both of them diverge by the nth term divergence test. Therefore, the radius of convergence of this series is 1 and the interval of convergence is $(-1, 1)$. Let

$$s(x) = \sum_{n=1}^{\infty} nx^n \quad \text{for } x \in (-1, 1).$$

Dividing both sides by x and then integrating term by term creates a geometric series for which we compute the following sum:

$$\frac{s(x)}{x} = \sum_{n=1}^{\infty} nx^{n-1},$$

$$\int \frac{s(x)}{x} dx = \sum_{n=1}^{+\infty} \int nx^{n-1} dx = \sum_{n=1}^{\infty} x^n + C = \frac{x}{1-x} + C.$$

Differentiating this result ($\int \frac{s(x)}{x} dx = \frac{x}{1-x} + C$) creates a formula for $s(x)$ as follows:

$$\frac{s(x)}{x} = \left(\frac{x}{1-x} \right)' = \frac{1}{(1-x)^2},$$

$$s(x) = \frac{x}{(1-x)^2} \quad \text{for } x \in (-1, 1).$$

Using $x = \frac{1}{2}$ in this equation gives $\sum_{n=1}^{\infty} \frac{n}{2^n} = 2$.

It is worth mentioning that term-by-term integration and term-by-term differentiation work for power series because a power series is uniformly convergent on its interval of convergence $(-R, R)$. This is described, for the reader's reference, in the following definitions and theorems, given without proof.

Definition 8.4.2. Suppose $\sum_{k=0}^{\infty} u_k(x)$ is a series with function terms $u_k(x)$, all defined on some interval **I** and convergent for $x \in$ **I**. Let $s(x) = \sum u_k(x)$ be the sum function and let $s_n(x) = \sum_{k=0}^{n-1} u_k(x)$ be the partial sum function for each n. Suppose further that, for any positive number ε, there exists a natural number N (which depends on ε but not on x) such that, for every $x \in$ **I**, whenever $n \geq N$, we have

$$|s(x) - s_n(x)| = |u_n(x) + u_{n+1}(x) + u_{n+2}(x) + \cdots| < \varepsilon.$$

Then we say that $\sum u_k(x)$ *converges uniformly* to the function $s(x)$ on the interval **I**.

Theorem 8.4.6. *Suppose that the series $\sum_{k=0}^{\infty} u_k(x)$, defined on $\mathbf{I}$, satisfies the following two conditions:*

(1) *for each $k = 1, 2, 3, \ldots$, there is a constant $a_k > 0$ for which $|u_k(x)| \leqslant a_k$ for all $x \in \mathbf{I}$ (a_k is an upper bound for $|u_k(x)|$ on $\mathbf{I}$);*

(2) *the series $\sum_{k=0}^{\infty} a_k$, with nonnegative terms, converges.*

Then the series $\sum u_k(x)$ uniformly converges on the interval $\mathbf{I}$.

The next two theorems show the connection between power series and uniform convergence.

Theorem 8.4.7. *If all $u_k(x)$ are continuous on some interval $\mathbf{I}$ and $\sum_{k=0}^{\infty} u_k(x)$ converges uniformly to $s(x)$ on $\mathbf{I}$, then $s(x)$ is continuous on $\mathbf{I}$ and the series $\sum u_k(x)$ can be integrated term by term. If further all $u_k(x)$ are differentiable on $\mathbf{I}$ and the series of differentiated functions $\sum u_k'(x)$ is also uniformly convergent, then $\sum u_k(x)$ can be differentiated term by term, so that $s'(x) = \sum u_k'(x)$.*

Theorem 8.4.8. *The power series $\sum a_n x^n$ with radius of convergence $R > 0$ is uniformly convergent on any closed interval $[a, b] \subset (-R, R)$ and can be differentiated or integrated term by term.*

Example 8.4.8. The series

$$\frac{\sin x}{1^2} + \frac{\sin(2^2 x)}{2^2} + \frac{\sin(3^2 x)}{3^2} + \cdots + \frac{\sin(n^2 x)}{n^2} + \cdots$$

is uniformly convergent on any closed interval, since we have

$$\left| \frac{\sin(n^2 x)}{n^2} \right| \leqslant \frac{1}{n^2}$$

and $\sum 1/n^2$ converges. However, if the series is differentiated term by term, the differentiated series becomes

$$\cos x + \cos 2^2 x + \cdots + \cos n^2 x + \cdots,$$

which does not converge, since the nth term does not tend to 0.

8.4.3 Taylor series

Recall the one variable Taylor theorem.

Theorem 8.4.9. *If f is differentiable up to order $n + 1$ in an open interval $\mathbf{I}$ containing a, then, for each x in $\mathbf{I}$, there exists a number c (which may depend on both x and a)*

between x and a such that

$$f(x) = P_n(x) + R_n(x),\qquad(8.7)$$

where $P_n(x)$ is a polynomial approximation called the Taylor polynomial of degree n to *$f(x)$ and*

$$P_n(x) = f(a) + f'(a)(x - a) + \frac{f''(a)}{2!}(x - a)^2 + \cdots + \frac{f^{(n)}(a)}{n!}(x - a)^n.$$

The remainder R_n *is given by*

$$R_n(x) = \frac{f^{(n+1)}(c)}{(n + 1)!}(x - a)^{n+1}\quad\text{(called the Lagrange remainder).}$$

This is a generalization of the mean value theorem. The higher order Taylor polynomials usually provide increasingly better polynomial approximations to $f(x)$ for x-values in a neighborhood of a.

Based on this theorem, we define the Taylor series, which is a power series.

Definition 8.4.3. Suppose that f has derivatives of all orders on some open interval **I** containing a. The *Taylor series of f at the point a* is the power series

$$\sum_{k=0}^{\infty} \frac{f^{(k)}(a)}{k!}(x - a)^k = f(a) + f'(a)(x - a) + \frac{f''(a)}{2!}(x - a)^2$$

$$+ \cdots + \frac{f^{(k)}(a)}{k!}(x - a)^k + \cdots.$$

For the special case $a = 0$, the Taylor series becomes

$$\sum_{n=0}^{\infty} \frac{f^{(n)}(0)}{n!}x^n = f(0) + \frac{f'(0)}{1!}x + \frac{f''(0)}{2!}x^2 + \cdots + \frac{f^{(n)}(0)}{n!}x^n + \cdots.\qquad(8.8)$$

This case arises frequently enough that it is given a special name, the *Maclaurin series for f*.

Given a Taylor series of a function f, we will usually be interested in answering two questions: "For what values of x does the Taylor series converge?" and "If it converges, does it converge to the function $f(x)$ on the interval of convergence?" Let us first look into an example.

Example 8.4.9. Find the Maclaurin series for $\sin x$.

Solution. The function $\sin x$ has derivatives of all orders and for all x-values. The derivatives are

$$f(x) = \sin x, \quad f'(x) = \cos x, \quad f''(x) = -\sin x,$$
$$f'''(x) = -\cos x, \quad f^{(4)}(x) = \sin x, \quad \dots .$$

The fourth derivative is $\sin x$ again, so by induction, we give the following formula for the nth derivative:

$$f^{(n)}(x) = \sin\left(x + \frac{n\pi}{2}\right) \quad \text{for } n = 0, 1, 2, 3, \dots .$$

Therefore,

$$f^{(n)}(0) = \sin\left(\frac{n\pi}{2}\right) = \begin{cases} 0, & \text{when } n = 2k \\ (-1)^k, & \text{when } n = 2k + 1, \end{cases}$$

where $f^{(2k)}(0) = 0$ and $f^{(2k+1)}(0) = (-1)^k$.

This means the Maclaurin series has only odd-powered terms, and, for all x, the Maclaurin series of $\sin x$ is

$$x - \frac{x^3}{3!} + \frac{x^5}{5!} - \cdots + (-1)^k \frac{x^{2k+1}}{(2k+1)!} + \cdots .$$

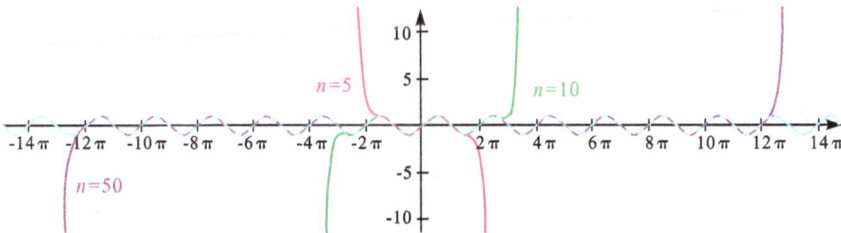

Figure 8.4.3: Graphs of partial sums ($n = 5, 10, 50$) of $\sum_{k=0}^{\infty} (-1)^k \frac{x^{2k+1}}{(2k+1)!}$.

Now let us investigate the graphs of some Maclaurin polynomials of degree 5, 10, and 50 and the graph of $\sin x$. It is easy to see that the larger degree of the polynomial, the better approximation of the graph to $\sin x$, as shown in Figure 8.4.3. We expect, as $n \to \infty$, the MacLaurin series to tend to $\sin x$. In fact, we have, by the Taylor theorem,

$$|R_n(x)| = \left| \sin\left(c + \frac{(n+1)\pi}{2}\right) \frac{x^{n+1}}{(n+1)!} \right|$$
$$\leqslant \frac{|x|^{n+1}}{(n+1)!} \to 0 \quad \text{as} \quad n \to \infty, \text{ for any } x.$$

Hence, for every value of x, we have

$$\lim_{n \to \infty} |\sin x - P_n(x)| = \lim_{n \to \infty} |R_n(x)| = 0,$$

so

$$\sin x = x - \frac{x^3}{3!} + \frac{x^5}{5!} - \cdots + (-1)^k \frac{x^{2k+1}}{(2k+1)!} + \cdots = \sum_{k=0}^{\infty} (-1)^k \frac{x^{2k+1}}{(2k+1)!}.$$

This means that the Maclaurin series for $\sin x$ actually converges to $\sin x$ for all x!

In general, if $\lim_{n\to\infty} R_n(x) = 0$ for all $x \in \mathbf{I}$, then the Taylor series converges to $f(x)$ on $\mathbf{I}$. This is summarized in the following theorem.

Theorem 8.4.10. *Let a be a real number in the domain of a function f and suppose that f has derivatives of all orders at a. The Taylor series $\sum_{k=0}^{\infty} \frac{f^{(k)}(a)}{k!}(x-a)^k$ of f converges to $f(x)$ in an interval $\mathbf{I}$ if and only if $\lim_{n\to\infty} R_n(x) = 0$ for all $x \in \mathbf{I}$, where $R_n(x)$ is the Lagrange remainder*

$$R_n(x) = \frac{f^{(n+1)}(c)}{(n+1)!}(x-a)^{n+1}$$

and c is between x and a.

If $|f^{(n+1)}(x)|$ is bounded by M for a set of x-values satisfying $|x-a| \leqslant d$, then the remainder $R_n(x)$ of the Taylor series satisfies *Taylor's inequality* and we have

$$|R_n(x)| = \frac{|f^{(n+1)}(c)|}{(n+1)!}|x-a|^{n+1} \leqslant \frac{M}{(n+1)!} d^{n+1} \quad \text{for } |x-a| \leqslant d.$$

This inequality is often used to estimate errors in approximations.

Example 8.4.10. Find the Maclaurin series of the function $f(x) = e^x$, find its radius of convergence R, and show that it converges to e^x for all $x \in (-R, R)$. Use this to find an infinite series for the number e.

Solution. The function $f(x) = e^x$ has derivatives of all orders throughout the interval $(-\infty, +\infty)$ and the nth derivative is $f^{(n)}(x) = e^x$ for $n = 0, 1, 2, \ldots$. Hence, $f^{(n)}(0) = e^0 = 1$ for all n and, therefore, the Maclaurin series for f is

$$\sum_{n=0}^{\infty} \frac{f^{(n)}(0)}{n!} x^n = f(0) + \frac{f'(0)}{1!}x + \frac{f''(0)}{2!}x^2 + \cdots$$

$$= \sum_{n=0}^{\infty} \frac{x^n}{n!} = 1 + \frac{x}{1!} + \frac{x^2}{2!} + \cdots.$$

You can check for yourself that the radius of convergence is $R = \infty$, using the ratio test, but this will also follow from the following analysis. In order to show that $e^x = \sum_{n=0}^{\infty} \frac{x^n}{n!}$ for all x, we show that the Lagrange remainder $R_n(x) \to 0$, for all x,

$$R_n(x) = \frac{e^c}{(n+1)!} x^{n+1} \quad \text{for some } c \text{ between } 0 \text{ and } x.$$

Since e^x is an increasing function,

$$|R_n(x)| = \left| \frac{e^c}{(n+1)!} x^{n+1} \right| \leq e^{|x|} \frac{|x|^{n+1}}{(n+1)!}.$$

However,

$$\lim_{n \to \infty} e^{|x|} \frac{|x|^{n+1}}{(n+1)!} = e^{|x|} \lim_{n \to \infty} \frac{|x|^{n+1}}{(n+1)!} = 0,$$

so $\lim_{n \to \infty} R_n(x) = 0$ for any value $x \in (-\infty, +\infty)$. We conclude that the Maclaurin series of e^x converges to e^x on $(-\infty, +\infty)$. That is,

$$e^x = \sum_{n=1}^{\infty} \frac{x^n}{n!} = 1 + \frac{x}{1!} + \frac{x^2}{2!} + \cdots \qquad \text{for all } x \in (-\infty, +\infty).$$

It follows that the radius of convergence is $R = +\infty$. When $x = 1$, the Maclaurin series becomes

$$e = 1 + \frac{1}{1!} + \frac{1}{2!} + \frac{1}{3!} + \cdots .$$

Figure 8.4.4 shows the graphs of the second, third, and fourth partial sums (Taylor polynomials) of the Maclaurin series of e^x. The highest degree Taylor polynomial gives the best approximation.

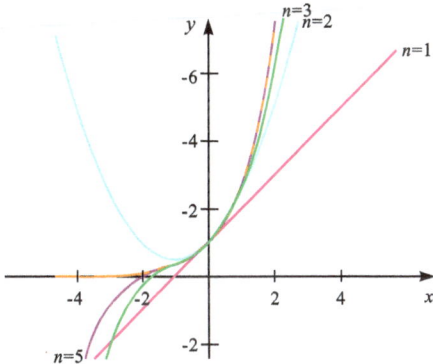

Figure 8.4.4: Graphs of e^x and its Taylor polynomials with $n = 1, 2, 3, 5$.

Example 8.4.11. Find the Maclaurin series for $(1 + x)^m$ where m is constant (this is called *binomial expansion/series*).

Solution. Since

$$f(x) = (1 + x)^m \quad \Longrightarrow \quad f(0) = 1,$$
$$f'(x) = m(1 + x)^{m-1} \quad \Longrightarrow \quad f'(0) = m,$$

$$f''(x) = m(m-1)(1+x)^{m-2} \quad \Longrightarrow \quad f''(0) = m(m-1),$$

$$\vdots$$

$$f^{(n)}(x) = m(m-1)(m-2)\cdots(m-n+1)(1+x)^{m-n},$$
$$f^{(n)}(0) = m(m-1)(m-2)\cdots(m-n+1).$$

We obtain the Maclaurin series of $(1+x)^m$ as follows:

$$1 + mx + \frac{m(m-1)}{2!}x^2 + \frac{m(m-1)(m-2)}{3!}x^3 + \cdots$$
$$+ \frac{m(m-1)(m-2)\cdots(m-n+1)}{n!}x^n + \cdots. \tag{8.9}$$

If $u_n(x) = \frac{m(m-1)(m-2)\cdots(m-n+1)}{n!}x^n$ for each n, then, applying the ratio test for absolute convergence, we see

$$\left|\frac{u_{n+1}}{u_n}\right| = \left|\frac{m-n}{n+1}x\right| \to |x| \quad \text{as} \quad n \to \infty.$$

Therefore, when $|x| < 1$, the series converges absolutely and the radius of convergence is $R = 1$. However, this does not prove that the series converges to $(1+x)^m$. The series does in fact converge to $(1+x)^m$, although the proof is not given here. Thus,

$$(1+x)^m = 1 + mx + \frac{m(m-1)}{2!}x^2 + \cdots$$
$$+ \frac{m(m-1)(m-2)\cdots(m-n+1)}{n!}x^n + \cdots \quad \text{for } |x| < 1.$$

Some special cases of the binomial expansion include:
$m = -1$:

$$\frac{1}{1+x} = 1 - x + x^2 - x^3 + x^4 - \cdots \quad \text{for } |x| < 1;$$

$m = 1/2$:

$$\sqrt{1+x} = 1 + \frac{x}{2} + \frac{\frac{1}{2}(\frac{1}{2}-1)}{2!}x^2 + \frac{\frac{1}{2}(\frac{1}{2}-1)(\frac{1}{2}-2)}{3!} + \cdots \quad \text{for } |x| < 1.$$

Example 8.4.12. Use the fact that $\frac{1}{1-x} = 1 + x + x^2 + \cdots + x^n + \cdots$ with interval of convergence $(-1, 1)$ to show that

$$\ln 2 = 1 - \frac{1}{2} + \frac{1}{3} - \frac{1}{4} + \cdots \quad \text{and} \quad \frac{\pi}{4} = 1 - \frac{1}{3} + \frac{1}{5} - \frac{1}{7} + \cdots.$$

Proof. Integrating the series term by term, we have

$$-\ln(1-x) = x + \frac{x^2}{2} + \frac{x^3}{3} + \frac{x^4}{4} + \cdots.$$

This is equivalent to

$$\ln(1-x) = -x - \frac{x^2}{2} - \frac{x^3}{3} - \frac{x^4}{4} - \cdots, \quad -1 \leqslant x < 1.$$

The interval of convergence of the integrated series $\sum \frac{x^n}{n}$ is $[-1,1)$, because when $x = -1$ the alternating series $\sum(-1)^n/n$ converges, but when $x = 1$ the series $\sum 1/n$ diverges. Thus we substitute $x = -1$ into the equation and obtain

$$\ln 2 = 1 - \frac{1}{2} + \frac{1}{3} - \frac{1}{4} + \cdots.$$

To obtain the second series, we replace x with $-x^2$ in the original series and we obtain

$$\frac{1}{1+x^2} = 1 - x^2 + x^4 - x^6 + \cdots \quad \text{for } -1 < x < 1.$$

We integrate term by term to obtain

$$\arctan x = x - \frac{x^3}{3} + \frac{x^5}{3} - \frac{x^7}{7} + \cdots \quad \text{for } -1 \leqslant x \leqslant 1.$$

This is convergent when $x = 1$ by the alternating series test, so substituting $x = 1$ we find

$$\frac{\pi}{4} = 1 - \frac{1}{3} + \frac{1}{5} - \frac{1}{7} + \cdots. \qquad \square$$

NOTE. For Taylor series, much of our effort has gone into calculating the derivatives $f^{(k)}(c)$ for $k = 0, 1, 2, \dots$. In fact, many power series/Taylor series can be found by making a substitution in a known series, or term-by-term differentiation/integration of a known series, as the previous examples illustrated. The radius of convergence of a power series before and after term-by-term differentiation/integration remains the same, but the interval of convergence may differ because convergence/divergence can change at the endpoints of the interval. This is also shown in the previous examples.

Example 8.4.13. Find the Maclaurin series for $\cos x$.

Solution. There is no need to compute $\cos^{(n)} x$ this time. From Example 8.4.9, we use the term-by-term differentiation of the Maclaurin series of $\sin x$ and obtain

$$\cos x = 1 - \frac{x^2}{2!} + \frac{x^4}{4!} - \cdots + (-1)^k \frac{x^{2k}}{(2k)!} + \cdots \quad \text{for all } x.$$

We now list some of the most useful Maclaurin series, which we have derived one way or another:

1. $\frac{1}{1-x} = 1 + x + x^2 + x^3 + \cdots = \sum_{k=0}^{\infty} x^k$, for $-1 < x < 1$;
2. $\frac{1}{1+x} = 1 - x + x^2 - \cdots + (-1)^n x^n + \cdots = \sum_{k=0}^{\infty} (-1)^k x^k$, for $|x| < 1$;

3. $\sin x = x - \frac{x^3}{3!} + \frac{x^5}{5!} - \frac{x^7}{7!} + \cdots = \sum_{k=0}^{\infty}(-1)^k \frac{x^{2k+1}}{(2k+1)!}$, for $-\infty < x < \infty$;

4. $\cos x = 1 - \frac{x^2}{2!} + \frac{x^4}{4!} - \frac{x^6}{6!} + \cdots = \sum_{k=0}^{\infty}(-1)^k \frac{x^{2k}}{(2k)!}$, for $-\infty < x < \infty$;

5. $e^x = 1 + x + \frac{x^2}{2!} + \frac{x^3}{3!} + \cdots = \sum_{k=0}^{\infty} \frac{x^k}{k!}$, for $-\infty < x < \infty$;

6. $\sqrt{1+x} = 1 + \frac{1}{2}x - \frac{1}{2^2}\frac{x^2}{2!} + \cdots = 1 + \frac{x}{2} + \sum_{k=2}^{\infty}(-1)^{k-1}\frac{1\cdot3\cdot5\cdots(2k-3)}{2^k k!}x^k$, for $-1 < x \leqslant 1$;

7. $\ln(1+x) = x - \frac{x^2}{2} + \frac{x^3}{3} - \cdots + (-1)^{k-1}\frac{x^k}{k} + \cdots = \sum_{k=1}^{\infty}(-1)^{k-1}\frac{x^k}{k}$, for $-1 < x \leqslant 1$;

8. $\arctan x = x - \frac{x^3}{3} + \frac{x^5}{5} - \cdots + (-1)^k\frac{x^{2k+1}}{2k+1} + \cdots = \sum_{k=0}^{\infty}(-1)^k\frac{x^{2k+1}}{2k+1}$, for $|x| \leqslant 1$.

We now give two more examples of deriving series from a known series by substitution.

Example 8.4.14. Find the Taylor series about $a = \frac{\pi}{2}$ for $\sin x$.

Solution. We first try the Maclaurin series of $\sin x$, replacing x with $x - \frac{\pi}{2}$, to obtain

$$\sin x = \sum_{k=0}^{\infty}(-1)^k \frac{x^{2k+1}}{(2k+1)!} \quad \text{for } x \in R \quad \Longrightarrow$$

$$\sin\left(x - \frac{\pi}{2}\right) = \sum_{k=0}^{\infty}(-1)^k \frac{(x - \frac{\pi}{2})^{2k+1}}{(2k+1)!}.$$

However, this is not the Taylor series of $\sin x$ about $x = \frac{\pi}{2}$. The trigonometric identity $\sin x = \cos(x - \frac{\pi}{2})$ suggests using the same method with the Maclaurin series for $\cos x$. Substituting $x - \frac{\pi}{2}$ in place of x in this series leads to the following required Taylor series for $\sin x$ about $x = \frac{\pi}{2}$:

$$\cos x = \sum_{k=0}^{\infty}(-1)^k \frac{x^{2k}}{(2k)!}, \quad -\infty < x < \infty \quad \Longrightarrow$$

$$\sin x = \cos\left(x - \frac{\pi}{2}\right) = \sum_{k=0}^{\infty}(-1)^k \frac{(x - \frac{\pi}{2})^{2k}}{(2k)!} \quad \text{for all } x \in R.$$

Example 8.4.15. Find the Taylor series about $c = -2$ for the function

$$f(x) = \frac{2x+1}{x^2 - 5x + 6}, \quad x \neq 2, 3.$$

Also find the radius of convergence and the interval of convergence.

Solution. Use partial fractions to simplify and rearrange the function to a function of $x + 2$ as follows:

$$\frac{2x+1}{x^2 - 5x + 6} = \frac{7}{x - 3} - \frac{5}{x - 2}$$

$$= \frac{7}{x + 2 - 5} - \frac{5}{x + 2 - 4}$$

$$= -\frac{7}{5}\left(\frac{1}{1 - \frac{x+2}{5}}\right) + \frac{5}{4}\left(\frac{1}{1 - \frac{x+2}{4}}\right).$$

We find expansions of these functions by using

$$\frac{1}{1-x} = \sum_{n=0}^{\infty} x^n, \quad -1 < x < 1.$$

Substituting gives

$$-\frac{7}{5}\left(\frac{1}{1-\frac{x+2}{5}}\right) = -\frac{7}{5}\sum_{n=0}^{\infty}\left(\frac{x+2}{5}\right)^n = -\sum_{n=0}^{\infty}\frac{7(x+2)^n}{5^{n+1}},$$

$$\frac{5}{4}\left(\frac{1}{1-\frac{x+2}{4}}\right) = \frac{5}{4}\sum_{n=0}^{\infty}\left(\frac{x+2}{4}\right)^n = \sum_{n=0}^{\infty}\frac{5(x+2)^n}{4^{n+1}}.$$

The first is valid if $|(x+2)/5| < 1 \Longrightarrow -7 < x < 3$ and the second if $|(x+2)/4| < 1 \Longrightarrow -6 < x < 2$. Hence, for $-6 < x < 2$, we have the Taylor expansion

$$\frac{2x+1}{x^2-5x+6} = -\sum_{n=0}^{\infty}\frac{7(x+2)^n}{5^{n+1}} + \sum_{n=0}^{\infty}\frac{5(x+2)^n}{4^{n+1}}$$

$$= \sum_{n=0}^{\infty}\left[-\frac{7(x+2)^n}{5^{n+1}} + \frac{5(x+2)^n}{4^{n+1}}\right]$$

$$= \sum_{n=0}^{\infty}\left[-\frac{7}{5^{n+1}} + \frac{5}{4^{n+1}}\right](x+2)^n.$$

The interval of convergence is $(-6, 2)$ and the radius of convergence is 4.

8.4.4 Applications of power series

One of the most common applications of power series is approximation. All computers and calculators give values to functions like e^x, $\sin x$, $\ln(x)$, and $\arctan x$ by computing the value of an approximation to the function, with accuracy equal to the precision required. Taylor series give the best general purpose approximation that can be achieved with a polynomial approximation.

Example 8.4.16. Calculate the number e with an error of less than 10^{-6}.

Solution. Set $x = 1$ in the Maclaurin series for $e^x = 1 + x + \frac{x^2}{2!} + \frac{x^3}{3!} + \cdots$, giving

$$e^1 = e = 1 + 1 + \frac{1}{2!} + \frac{1}{3!} + \cdots.$$

If, after n terms of this series, the remainder is less than 10^{-6}, then the partial sum with n terms will give the approximation with the required accuracy. The remainder term from equation (8.7) gives

$$R_n(1) = \frac{e^c}{(n+1)!}1^{n+1} = e^c\frac{1}{(n+1)!}, \quad \text{where } 0 < c < 1.$$

Hence, $1 \leqslant e^c \leqslant e$, so

$$\frac{1}{(n+1)!} \leqslant R_n(1) \leqslant \frac{e}{(n+1)!} < \frac{3}{(n+1)!}$$

and we will have the required accuracy if $\frac{3}{(n+1)!} < 10^{-6}$. By trial-and-error we find

$$n = 8 : \frac{3}{(n+1)!} \approx 8.26 \times 10^{-6} \quad \text{(fails)}$$

$$n = 9 : \frac{3}{(n+1)!} \approx 8.27 \times 10^{-7} \quad \text{(works)}.$$

Therefore, we choose $n = 9$ and this gives

$$e \approx 1 + 1 + \frac{1}{2!} + \frac{1}{3!} + \frac{1}{4!} + \frac{1}{5!} + \frac{1}{6!} + \frac{1}{7!} + \frac{1}{8!} + \frac{1}{9!}$$

$$e \approx 2.718\,281\,5, \quad \text{accurate to } 10^{-6} = 0.000\,001.$$

Example 8.4.17. Approximate the integral $\int_0^1 \frac{\sin x}{x}\, dx$ by using a Taylor polynomial (about 0) of degree 5.

Solution. Using the Maclaurin series of $\sin x$, we obtain

$$\sin x = x - \frac{x^3}{3!} + \frac{x^5}{5!} - \frac{x^7}{7!} + \cdots, \quad \text{so}$$

$$\frac{\sin x}{x} = 1 - \frac{x^2}{3!} + \frac{x^4}{5!} - \frac{x^6}{7!} + \cdots.$$

Therefore, the Taylor polynomial of degree 5 is $1 - \frac{x^2}{3!} + \frac{x^4}{5!}$ (the degree five term is zero). Hence,

$$\int_0^1 \frac{\sin x}{x}\, dx = \int_0^1 \left(1 - \frac{x^2}{3!} + \frac{x^4}{5!}\right) dx = x - \frac{x^3}{3 \cdot 3!} + \frac{x^5}{5 \cdot 5!} \Big|_{x=0}^{x=1}$$

$$= 1 - \frac{1^3}{3 \cdot 3!} + \frac{1^5}{5 \cdot 5!} \approx 0.946\,11.$$

We now informally derive the famous Euler formula for the complex number function e^{ix}, which is useful in many areas. The formula says

$$e^{ix} = \cos x + i \sin x, \quad \text{where } i^2 = -1.$$

If we replace x with ix in the Maclaurin series of e^x, then we are able to deduce the formula as follows:

$$e^{ix} = 1 + ix + \frac{(ix)^2}{2!} + \frac{(ix)^3}{3!} + \frac{(ix)^4}{4!} + \frac{(ix)^5}{5!} \cdots$$

$$= 1 + ix - \frac{x^2}{2!} - i\frac{x^3}{3!} + \frac{x^4}{4!} + \frac{ix^5}{5!} - \cdots$$

$$= \left(1 - \frac{x^2}{2!} + \frac{x^4}{4!} - \cdots\right) + i\left(x - \frac{x^3}{3!} + \frac{x^5}{5!} - \cdots\right)$$

$$= \cos x + i \sin x.$$

Using $x = \pi$ in Euler's formula gives the following famous identity, that connects e, i, π, 1, and 0, the five most important constants in mathematics:

$$e^{i\pi} + 1 = 0.$$

8.5 Fourier series

Joseph Fourier (1768–1830) was a French mathematician and physicist. He is best known for initiating the investigation of Fourier series and their applications to problems of heat transfer and vibrations. The Fourier transform and Fourier's law are also named in his honor. Fourier is also generally credited with the discovery of the greenhouse effect. http://en.wikipedia.org/wiki/Joseph_Fourier

While studying the problem of heat conduction in a long thin insulated rod, the French mathematician Jean-Baptiste Joseph Fourier needed to express a function $f(x)$ as a linear combination of trigonometric functions of the form $\sin kx$ and $\cos kx$ for $k = 1, 2, 3, \ldots$. He developed remarkable ideas behind the following results.

If a function $f(x)$ is defined for $x \in (-\pi, \pi)$ (except, possibly, at a finite set of x-values), $f(x)$ is the sum of a series of the following form:

$$f(x) = \frac{a_0}{2} + \sum_{k=1}^{\infty} (a_k \cos kx + b_k \sin kx). \tag{8.10}$$

This series is called a *Fourier series of f*. This type of series has a tremendous range of scientific and engineering applications in the study of heat conduction, wave phenomena, and concentrations of chemicals and pollutants, just to mention a few.

A Fourier series of $f(x)$ can be defined for any other interval of the form $(-l, l)$, in which case the series would be composed of the functions $\cos\frac{k\pi x}{l}$ and $\sin\frac{k\pi x}{l}$ with period $2l$. Then we have

$$f(x) = \frac{a_0}{2} + \sum_{k=1}^{\infty} \left(a_k \cos\frac{k\pi x}{l} + b_k \sin\frac{k\pi x}{l}\right).$$

Notice that a Fourier series on $(-l, l)$ only involves the functions $\cos\frac{n\pi x}{l}$ and $\sin\frac{n\pi x}{l}$, both of which are periodic with period $2l$. Then the sum of the series must be periodic with period $2l$. Hence, the Fourier series not only represents a function f over the interval $-l < x < l$, but it also provides a *periodic extension* of f, with period $2l$, over the entire real-number line.

8.5.1 Fourier series expansion with period 2π

For a function $f(x)$ we first of all need to find the *Fourier coefficients* $a_0, a_1, b_1, a_2, b_2, \ldots$ such that equation (8.10) is true. To determine these coefficients, one can view a Fourier series as a member of the linear space with a basis of infinite dimensions as follows:

$$\text{basis} = \{1, \cos x, \sin x, \cos 2x, \sin 2x, \cos 3x, \sin 3x, \cos 4x, \sin 4x, \ldots\}.$$

Let us first explore the inner products of a pair of elements in the basis. The following results show that these elements are orthogonal under the inner product $\int_{-\pi}^{\pi} f(x)g(x)\,dx$ of two functions $f(x)$ and $g(x)$.

Theorem 8.5.1. *If n and k are any positive integers, then:*
1. $\int_{-\pi}^{\pi} \cos kx\,dx = 0;$
2. $\int_{-\pi}^{\pi} \sin kx\,dx = 0;$
3. $\int_{-\pi}^{\pi} \cos nx \sin kx\,dx = 0;$
4. $\int_{-\pi}^{\pi} \sin nx \sin kx\,dx = \begin{cases} \pi, & \text{if } k = n \\ 0, & \text{if } k \neq n; \end{cases}$
5. $\int_{-\pi}^{\pi} \cos nx \cos kx\,dx = \begin{cases} \pi, & \text{if } k = n \\ 0, & \text{if } k \neq n. \end{cases}$

Proof. The proof of this theorem is not hard. We only give the proof of 4.

$$\int_{-\pi}^{\pi} \sin nx \sin kx\,dx = -\frac{1}{2}\int_{-\pi}^{\pi}(\cos(n+k)x - \cos(n-k)x)\,dx$$
$$= \frac{1}{2}\int_{-\pi}^{\pi}\cos(n-k)x\,dx$$
$$= \begin{cases} 0, & \text{if } n \neq k \\ \pi, & \text{if } n = k. \end{cases}$$

This is true because, if $n = k$, $\int_{-\pi}^{\pi}\cos(n-k)x\,dx = \int_{-\pi}^{\pi} 1\,dx = 2\pi$ and, if $n \neq k$, $\int_{-\pi}^{\pi}\cos(n-k)x\,dx = -\frac{1}{n-k}\sin(n-k)x|_{-\pi}^{\pi} = 0$. $\square$

Some examples of these inner products include

$$\int_{-\pi}^{\pi} \sin 2x \sin nx\,dx = \begin{cases} 0, & \text{if } n \neq 2 \\ \pi, & \text{if } n = 2 \end{cases}$$

and

$$\int_{-\pi}^{\pi} \cos 3x \cos nx\,dx = \begin{cases} 0, & \text{if } n \neq 3 \\ \pi, & \text{if } n = 3. \end{cases}$$

It can be proved that the series of equation (8.10) can be integrated term by term and we assume this is true. To obtain a_0, we integrate both sides from $x = -\pi$ to $x = \pi$, so we have

$$\int_{-\pi}^{\pi} f(x)\,dx = \int_{-\pi}^{\pi} \frac{a_0}{2}\,dx + \sum_{k=1}^{\infty} \int_{-\pi}^{\pi} (a_k \cos kx + b_k \sin kx)\,dx$$
$$= a_0 \pi.$$

Solving this for a_0 yields

$$a_0 = \frac{1}{\pi} \int_{-\pi}^{\pi} f(x)\,dx. \tag{8.11}$$

To solve for a_n, when $n \neq 0$, we multiply both sides of equation (8.10) by $\cos(nx)$ and integrate the result from $-\pi$ to π to obtain

$$\int_{-\pi}^{\pi} f(x) \cos nx\,dx$$
$$= \frac{a_0}{2} \int_{-\pi}^{\pi} \cos nx\,dx + \sum_{k=1}^{\infty} \int_{-\pi}^{\pi} (a_k \cos kx \cos nx + b_k \sin kx \cos nx)\,dx$$
$$= 0 + \sum_{k=1}^{\infty} \int_{-\pi}^{\pi} (a_k \cos kx \cos nx)\,dx + \sum_{k=1}^{\infty} \int_{-\pi}^{\pi} (b_k \sin kx \cos nx)\,dx$$
$$= \int_{-\pi}^{\pi} a_n \cos nx \cos nx\,dx = a_n \pi,$$

since all terms of the summations are zero except the single integral shown for which $k = n$ (by Theorem 8.5.1). Solving this for a_n, we have

$$a_n = \frac{1}{\pi} \int_{-\pi}^{\pi} f(x) \cos nx\,dx. \tag{8.12}$$

Similarly, when we multiply equation (8.10) by $\sin nx$ and integrate the result from $-\pi$ to π, we obtain

$$b_n = \frac{1}{\pi} \int_{-\pi}^{\pi} f(x) \sin nx\,dx. \tag{8.13}$$

Example 8.5.1. Find the Fourier series expansion of the periodic function $f(x)$, with period 2π, defined on $(-\pi, \pi]$ by

$$f(x) = \begin{cases} 0, & \text{if } -\pi < x \leq 0 \\ x, & \text{if } 0 < x \leq \pi. \end{cases}$$

Solution. We compute the Fourier coefficients by equations (8.12) and (8.13). We have

$$a_0 = \frac{1}{\pi} \int_{-\pi}^{\pi} f(x)\,dx = \frac{1}{\pi} \left(\int_{-\pi}^{0} 0\,dx + \int_{0}^{\pi} x\,dx \right) = \frac{\pi}{2},$$

$$a_n = \frac{1}{\pi} \int_{-\pi}^{\pi} f(x) \cos nx \, dx = \frac{1}{\pi} \int_0^{\pi} x \cos nx \, dx \quad \text{(use integration by parts)}$$

$$= \frac{1}{n\pi} \left([x \sin nx]_0^{\pi} - \int_0^{\pi} \sin nx \, dx \right) = \frac{1}{n\pi} \left[\frac{1}{n} \cos nx \right]_0^{\pi}$$

$$= \frac{1}{n^2\pi} [(-1)^n - 1] = \begin{cases} \frac{-2}{n^2\pi}, & \text{when } n = 1, 3, 5, \ldots \\ 0, & \text{when } n = 2, 4, 6, \ldots, \end{cases}$$

$$b_n = \frac{1}{\pi} \int_{-\pi}^{\pi} f(x) \sin nx \, dx = \frac{1}{\pi} \int_0^{\pi} x \sin nx \, dx$$

$$= -\frac{1}{n\pi} \left([x \cos nx]_0^{\pi} - \int_0^{\pi} \cos nx \, dx \right)$$

$$= -\frac{1}{n\pi} \left(\pi \cos n\pi - \left[\frac{1}{n} \sin nx \right]_0^{\pi} \right) = \frac{-1}{n} (-1)^n.$$

Hence, the Fourier series is

$$\frac{\pi}{4} + \left(-\frac{2 \cos x}{\pi} + \sin x \right) + \left(0 - \frac{\sin 2x}{2} \right) + \left(-\frac{2 \cos 3x}{3^2\pi} + \frac{\sin 3x}{3} \right) + \cdots.$$

As with Taylor series, we now have two more questions to solve: "Is the Fourier series convergent?" and "If the Fourier series converges, does it converge to $f(x)$?". We graph some functions of the first 5, 10, and 20 terms of the series and the graph of $f(x)$ in Figure 8.5.1.

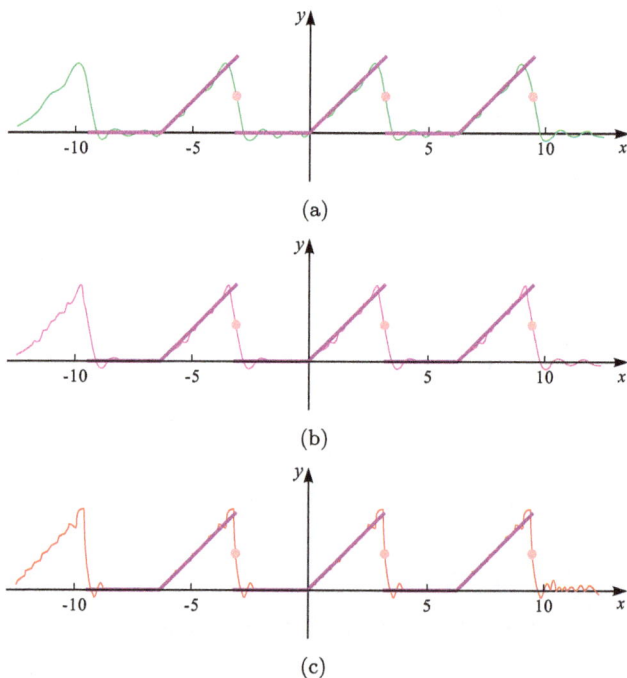

(a)

(b)

(c)

Figure 8.5.1: Graphs of $f(x)$ in Example 8.5.1 and some partial sums of its Fourier series.

If we notice that each approximating function passes through the three points $(-\pi, \pi/2)$, $(\pi, \pi/2)$, and $(3\pi, \pi/2)$, we may understand the following theorem, given without proof.

Theorem 8.5.2 (Convergence of Fourier series). *If the function f is defined for $x \in (-l, l)$ and both f and its derivative f' are continuous or piecewise continuous on the interval $(-l, l)$, then f has a Fourier series defined on $(-l, l)$ and $f(x)$ is the sum of its Fourier series at all points of continuity. At a point c where a jump discontinuity occurs in f, the Fourier series converges to the average*

$$\frac{f(c^+) + f(c^-)}{2},$$

where $f(c^+)$ stands for the right limit and $f(c^-)$ is the left limit of f at the point c.

Hence, by the convergence theorem, in Example 8.5.1, at the endpoints $x = (2k + 1)\pi$, $k = 0, \pm 1, \pm 2, \dots$, the Fourier series of f converges to the value

$$\frac{f(x^+) + f(x^-)}{2} = \frac{f(\pi^+) + f(\pi^-)}{2} = \frac{\pi}{2}.$$

Example 8.5.2. Find the Fourier series expansion on $(-\pi, \pi]$ of the function

$$f(x) = \begin{cases} -x, & \text{if } -\pi < x < 0 \\ x, & \text{if } 0 \leqslant x \leqslant \pi. \end{cases}$$

Solution. We compute the Fourier coefficients by using formulas (8.11), (8.12), and (8.13) and the fact that $f(x)$ is an even function (recall that $\int_{-\pi}^{\pi} g(x)\, dx = 2\int_0^{\pi} g(x)\, dx$ when g is an even function and $\int_{-\pi}^{\pi} g(x)\, dx = 0$ when g is an odd function). We have

$$a_0 = \frac{1}{\pi} \int_{-\pi}^{\pi} f(x)\, dx = \frac{2}{\pi} \int_0^{\pi} x\, dx = \pi,$$

$$b_n = \frac{1}{\pi} \int_{-\pi}^{\pi} f(x) \sin nx\, dx = 0 \quad (\text{since } f(x) \sin nx \text{ is an odd function}),$$

$$a_n = \frac{1}{\pi} \int_{-\pi}^{\pi} f(x) \cos nx\, dx = \frac{2}{\pi} \int_0^{\pi} x \cos nx\, dx \quad (\text{integration by parts})$$

$$= \frac{2}{n\pi} \left[[x \sin nx]_0^{\pi} - \int_0^{\pi} \sin nx\, dx \right]$$

$$= \frac{2}{n\pi} \left[\frac{1}{n} \cos nx \right]_0^{\pi} = \frac{2}{n^2 \pi} (\cos n\pi - 1)$$

$$= \frac{2}{n^2 \pi} [(-1)^n - 1].$$

If n is even, then $a_n = 0$ and if n is odd, then $a_n = -\frac{4}{n^2 \pi}$, so in the Fourier series expansion of $f(x)$ there are only terms with coefficients of the form a_{2k+1}. We have

$$\frac{a_0}{2} + \sum_{n=1}^{\infty} (a_n \cos nx + b_n \sin nx)$$

$$= \frac{\pi}{2} + \sum_{k=0}^{\infty} a_{2k+1} \cos(2k+1)x = \frac{\pi}{2} + \sum_{k=0}^{\infty} \frac{-4}{(2k+1)^2 \pi} \cos(2k+1)x$$

$$= \frac{\pi}{2} - \frac{4}{\pi} \left(\frac{\cos x}{1^2} + \frac{\cos 3x}{3^3} + \frac{\cos 5x}{5^2} + \cdots + \frac{\cos(2n+1)}{(2n+1)^2} + \cdots \right).$$

The function $f(x)$ is continuous everywhere on $-\pi < x \leqslant \pi$ (including $f(\pi^+) = f(\pi^-)$). By the convergence theorem, we have

$$f(x) = \frac{\pi}{2} - \frac{4}{\pi} \left(\frac{\cos x}{1^2} + \frac{\cos 3x}{3^2} + \frac{\cos 5x}{5^2} + \cdots \right) \quad \text{for all } x \in [-\pi, \pi]$$

and the series converges to the periodic extension of $f(x)$ on the entire x-axis.

We now use this Fourier series to find the sums of some infinite series of constants. For example, in the Fourier series from the previous example, using $x = 0$, we obtain

$$f(0) = 0 = \frac{\pi}{2} - \frac{4}{\pi} \left(\frac{\cos 0}{1^2} + \frac{\cos 0}{3^2} + \frac{\cos 0}{5^2} + \cdots \right),$$

so

$$\frac{\pi^2}{8} = 1 + \frac{1}{3^2} + \frac{1}{5^2} + \frac{1}{7^2} + \cdots.$$

If we let s be the sum of the following series with squares of all possible integers in the denominators:

$$s = 1 + \frac{1}{2^2} + \frac{1}{3^2} + \frac{1}{4^2} + \cdots,$$

then

$$s - \frac{\pi^2}{8} = \left(1 + \frac{1}{2^2} + \frac{1}{3^2} + \frac{1}{4^2} + \cdots \right) - \left(1 + \frac{1}{3^2} + \frac{1}{5^2} + \frac{1}{7^2} + \cdots \right)$$

$$= \frac{1}{2^2} + \frac{1}{4^2} + \frac{1}{6^2} + \cdots$$

$$= \frac{1}{4} \left(1 + \frac{1}{2^2} + \frac{1}{3^2} + \frac{1}{4^2} + \cdots \right) = \frac{1}{4} s.$$

Therefore, we have

$$s - \frac{\pi^2}{8} = \frac{s}{4} \implies s = \frac{\pi^2}{6},$$

so

$$\frac{\pi^2}{6} = s = 1 + \frac{1}{2^2} + \frac{1}{3^2} + \frac{1}{4^2} + \cdots.$$

8.5.2 Fourier cosine and sine series with period 2π

You may have noticed that, if $f(x)$ is an even function with period 2π, the Fourier series of f only has cosine terms and a constant term and is of the form

$$f(x) = \frac{a_0}{2} + \sum_{n=1}^{\infty} a_n \cos nx.$$

This is called a *Fourier cosine series*. This happens because $f(x)$ is an even function, so $f(x) \sin nx$ is an odd function. Therefore, all $b_n = \frac{1}{\pi} \int_{-\pi}^{\pi} f(x) \sin nx \, dx = 0$ for $n = 1, 2, 3, \ldots$. Likewise, if $f(x)$ is an odd function with period 2π, then its Fourier series expansion is of the form

$$f(x) = \sum_{n=1}^{\infty} b_n \sin nx.$$

This is called a *Fourier sine series*.

Sometimes, we need to find a Fourier cosine series or sine series of a function $f(x)$ defined on the nonsymmetric interval $0 < x < \pi$. If this is the case, we simply extend the definition of f to the whole interval $(-\pi, \pi)$ in such a way that it is the required even or odd function. If we need an *even extension* of f over $(-\pi, \pi)$, then we define $f(x) = f(-x)$ when $x \in (-\pi, 0]$. If we need an *odd extension* of f over $(-\pi, \pi)$, then we define $f(x) = -f(-x)$ when $x \in (-\pi, 0]$.

Example 8.5.3. Find the Fourier sine series and the Fourier cosine series for the function $f(x)$ defined on $[0, \pi]$ by

$$f(x) = \begin{cases} 0, & \text{when } 0 < x \leq \frac{\pi}{2} \\ 1, & \text{when } \frac{\pi}{2} < x \leq \pi. \end{cases}$$

Solution. For the Fourier sine series, we select the odd extension of the function over $(-\pi, \pi)$ with $f(x) = 0$ for $-\frac{\pi}{2} \leq x \leq 0$ and $f(x) = -1$ for $-\pi < x < -\frac{\pi}{2}$, as shown in Figure 8.5.2. We compute the Fourier sine series (and cosine series) from the values $x \in (0, \pi]$ as follows.

Since f is now an odd function on $(-\pi, \pi)$, we only need to compute b_n for $n = 1, 2, 3, \ldots$. We have

$$b_n = \frac{1}{\pi} \int_{-\pi}^{\pi} f(x) \sin nx \, dx = \frac{2}{\pi} \int_0^{\pi} f(x) \sin nx \, dx = \frac{2}{\pi} \int_{\frac{\pi}{2}}^{\pi} \sin nx \, dx$$

$$= \frac{2}{n\pi} [-\cos nx]_{\frac{\pi}{2}}^{\pi} = \frac{2}{n\pi} \left(1 + \cos \frac{n\pi}{2}\right)$$

$$= \begin{cases} \frac{2}{n\pi}, & \text{when } n \text{ is odd} \\ 0, & \text{when } n = 2, 6, 10, \ldots \\ \frac{4}{n\pi}, & \text{when } n = 4, 8, 12, \ldots, \end{cases}$$

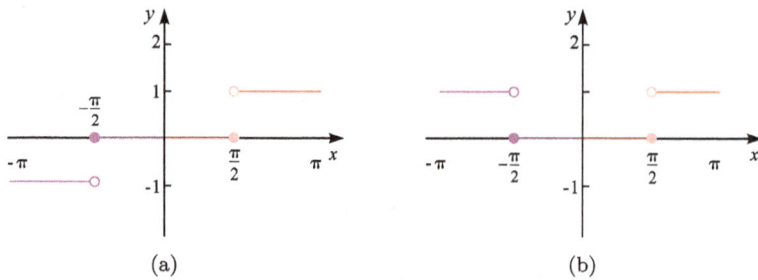

Figure 8.5.2: Odd/even expansion.

so the Fourier sine series of $f(x)$ is

$$\frac{2}{\pi}\sin x + \frac{2}{3\pi}\sin 3x + \frac{4}{4\pi}\sin 4x + \frac{2}{5\pi}\sin 5x + \frac{2}{7\pi}\sin 7x + \cdots$$

$$= \begin{cases} f(x), & \text{for } x \in (0,\pi) \text{ and } x \neq \frac{\pi}{2} \\ \frac{1}{2}, & \text{for } x = \frac{\pi}{2}. \end{cases}$$

For the Fourier cosine series, we select the even extension of the function $f(x)$ over $(-\pi,\pi)$ with $f(x) = 0$ for $-\frac{\pi}{2} \leqslant x \leqslant 0$ and $f(x) = 1$ for $-\pi < x < -\frac{\pi}{2}$. We need only to calculate a_n, for $n = 0, 1, 2, \ldots$. We have

$$a_0 = \frac{1}{\pi}\int_{-\pi}^{\pi} f(x)\,dx = \frac{2}{\pi}\int_0^{\pi} f(x)\,dx = \frac{2}{\pi}\int_{\frac{\pi}{2}}^{\pi} 1\,dx = 1,$$

$$a_n = \frac{1}{\pi}\int_{-\pi}^{\pi} f(x)\cos nx\,dx = \frac{2}{\pi}\int_0^{\pi} f(x)\cos nx\,dx = \frac{2}{\pi}\int_{\frac{\pi}{2}}^{\pi} \cos nx\,dx$$

$$= \frac{2}{\pi}\left[\frac{1}{n}\sin nx\right]_{\frac{\pi}{2}}^{\pi} = -\frac{2}{n\pi}\sin\frac{n\pi}{2}$$

$$= \begin{cases} \frac{-2}{n\pi}(-1)^{\frac{n-1}{2}}, & \text{when } n \text{ is odd} \\ 0, & \text{when } n \text{ is even.} \end{cases}$$

Hence, the desired Fourier cosine series is

$$\frac{1}{2} - \frac{2}{\pi}\cos x + \frac{2}{3\pi}\cos 3x - \frac{2}{5\pi}\cos 5x + \cdots.$$

This Fourier cosine series converges to $f(x)$ on $(0,\pi)$ when $x \neq \frac{\pi}{2}$. When $x = \frac{\pi}{2}$, this series converges to $\frac{1}{2}$.

8.5.3 The Fourier series expansion with period $2l$

In practice, very often we will be interested in a Fourier expansion of a function with a period that is not 2π. If $f(x)$ is a function defined on the interval $(-l, l)$, then the linear transformation $z = \frac{\pi x}{l}$ gives the new function $F(z) = f(\frac{lz}{\pi}) = f(x)$, defined on the interval $z \in (-\pi, \pi)$, so the previous methods can be applied to F. The Fourier series expansion for $F(z)$ is

$$F(z) = \frac{a_0}{2} + \sum_{n=1}^{\infty}(a_n \cos nz + b_n \sin nz),$$

where

$$a_n = \frac{1}{\pi}\int_{-\pi}^{\pi} F(z)\cos nz\, dz \quad \text{for } n = 0,1,2,\dots \quad \text{and}$$

$$b_n = \frac{1}{\pi}\int_{-\pi}^{\pi} F(z)\sin nz\, dz \quad \text{for } n = 1,2,3,\dots.$$

Now, substituting back $z = \frac{\pi x}{l}$ and using $f(x) = F(z)$ gives

$$a_n = \frac{1}{\pi}\int_{-\pi}^{\pi} F(z)\cos nz\, dz$$

$$= \frac{1}{\pi}\int_{-l}^{l} f(x)\cos\left(n\frac{\pi x}{l}\right)d\left(\frac{\pi x}{l}\right)$$

$$= \frac{1}{l}\int_{-l}^{l} f(x)\cos\left(\frac{n\pi x}{l}\right)dx \quad \text{for } n = 0,1,2,\dots,$$

$$b_n = \frac{1}{\pi}\int_{-\pi}^{\pi} F(z)\sin nz\, dz$$

$$= \frac{1}{\pi}\int_{-l}^{l} f(x)\sin\left(n\frac{\pi x}{l}\right)d\left(\frac{\pi x}{l}\right)$$

$$= \frac{1}{l}\int_{-l}^{l} f(x)\sin\left(\frac{n\pi x}{l}\right)dx \quad \text{for } n = 1,2,3,\dots.$$

Hence, we have proved the following result.

Theorem 8.5.3. *The Fourier series of a function $f(x)$ defined on the interval $-l < x < l$ is*

$$\frac{a_0}{2} + \sum_{n=1}^{\infty}\left(a_n \cos\left(\frac{n\pi x}{l}\right) + b_n \sin\left(\frac{n\pi x}{l}\right)\right), \tag{8.14}$$

where

$$a_n = \frac{1}{l}\int_{-l}^{l} f(x)\cos\left(\frac{n\pi x}{l}\right)dx \quad \text{for } n = 0,1,2,\dots \quad \text{and} \tag{8.15}$$

$$b_n = \frac{1}{l}\int_{l}^{l} f(x)\sin\left(\frac{n\pi x}{l}\right)dx \quad \text{for } n = 1,2,3,\dots.$$

This expansion is also true for the periodic extension of $f(x)$ over the entire real-number line. At any point x of continuity of f, the series converges to $f(x)$. That is, when x is a *point of continuity,*

$$f(x) = \frac{a_0}{2} + \sum_{n=1}^{\infty}\left(a_n \cos\frac{n\pi x}{l} + b_n \sin\frac{n\pi x}{l}\right).$$

At a point c of jump discontinuity of the periodic extension of f, the series converges to

$$\frac{1}{2}(f(c^+) + f(c^-)).$$

Example 8.5.4. Find the Fourier series with period 4 of $f(x)$, defined on the interval $(-2, 2]$ by

$$f(x) = \begin{cases} 0, & \text{when } -2 \leqslant x < 0 \\ 1, & \text{when } 0 \leqslant x < 2. \end{cases}$$

Solution. Using formula (8.15) with $l = 2$, we have

$$a_0 = \frac{1}{l}\int_{-l}^{l} f(x)\,dx = \frac{1}{2}\int_{-2}^{2} f(x)\,dx = \frac{1}{2}\left[\int_{-2}^{0} 0\,dx + \int_{0}^{2} 1\,dx\right] = 1,$$

$$a_n = \frac{1}{l}\int_{-l}^{l} f(x)\cos\frac{n\pi x}{l}\,dx$$

$$= \frac{1}{2}\left[\int_{-2}^{0} 0 \times \cos\frac{n\pi x}{2}\,dx + \int_{0}^{2} 1 \times \cos\frac{n\pi x}{2}\,dx\right]$$

$$= \frac{1}{2}\left[\frac{2}{n\pi}\sin\frac{n\pi x}{2}\right]_{0}^{2} = 0,$$

$$b_n = \frac{1}{l}\int_{-l}^{l} f(x)\sin\frac{n\pi x}{l}\,dx$$

$$= \frac{1}{2}\left[\int_{-2}^{0} 0 \times \sin\frac{n\pi x}{2}\,dx + \int_{0}^{2} 1 \times \sin\frac{n\pi x}{2}\,dx\right]$$

$$= -\left[\frac{1}{n\pi}\cos\frac{n\pi x}{2}\right]_{0}^{2} = \frac{1}{n\pi}(1 - (-1)^n)$$

$$= \begin{cases} 0, & \text{when } n = 2, 4, 6, \ldots \\ \frac{2}{n\pi}, & \text{when } n = 1, 3, 5, \ldots. \end{cases}$$

Therefore, the Fourier series of $f(x)$ is

$$\frac{1}{2} + \frac{2}{\pi}\sin\frac{\pi x}{2} + \frac{2}{3\pi}\sin\frac{3\pi x}{2} + \frac{2}{5\pi}\sin\frac{5\pi x}{2} + \cdots.$$

When $x \neq 2k$ for $k = 0, 1, 2, \ldots$, this Fourier series converges to $f(x)$. When $x = 2k$ for $k = 0, 1, 2, \ldots$, the Fourier series converges to $\frac{1}{2}$. Figure 8.5.3 shows the function $f(x)$ and several partial sums of the Fourier series of $f(x)$. There are three partial sums, with the first 5, 10, and 20 terms.

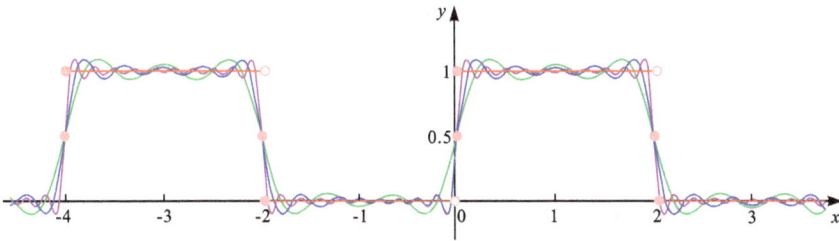

Figure 8.5.3: Graphs of $f(x)$ and some partial sums of its Fourier series.

8.5.4 Fourier series with complex terms

In electrical fields, engineers often use the series

$$\sum_{n=-\infty}^{\infty} c_n e^{i\frac{n\pi x}{l}}, \quad \text{where } c_n = \frac{1}{2l}\int_{-l}^{l} f(x)e^{-i\frac{n\pi x}{l}}\, dx, \text{ for } n \in \mathbf{Z}, \tag{8.16}$$

to represent a periodic function $f(x)$ with period $2l$. We show how this is related to the Fourier series defined in the previous subsection.

Recall Euler's theorem

$$e^{ix} = \cos x + i \sin x \quad \text{and} \quad e^{-ix} = \cos x - i \sin x.$$

Let a_0, a_n, b_n be the coefficients of the Fourier series of $f(x)$ in equation (8.15). The series (8.16) can be rewritten as

$$\sum_{n=-\infty}^{\infty} c_n e^{i\frac{n\pi x}{l}} = c_0 + \sum_{n=1}^{\infty}\left(c_n e^{i\frac{n\pi x}{l}} + c_{-n}e^{-i\frac{n\pi x}{l}}\right), \tag{8.17}$$

$$\text{where} \quad c_0 = \frac{1}{2l}\int_{-l}^{l} f(x)\, dx = \frac{a_0}{2} \tag{8.18}$$

and the general term $\left(c_n e^{i\frac{n\pi x}{l}} + c_{-n}e^{-i\frac{n\pi x}{l}}\right)$ is equal to

$$= \left[\frac{1}{2l}\int_{-l}^{l} f(x)e^{-i\frac{n\pi x}{l}}\, dx\right]e^{i\frac{n\pi x}{l}} + \left[\frac{1}{2l}\int_{-l}^{l} f(x)e^{i\frac{n\pi x}{l}}\, dx\right]e^{-i\frac{n\pi x}{l}}$$

$$= \frac{1}{2l}\left[\int_{-l}^{l} f(x)\left(\cos\frac{n\pi x}{l} - i\sin\frac{n\pi x}{l}\right) dx\right]e^{i\frac{n\pi x}{l}}$$

$$+ \frac{1}{2l}\left[\int_{-l}^{l} f(x)\left(\cos\frac{n\pi x}{l} + i\sin\frac{n\pi x}{l}\right) dx\right]e^{-i\frac{n\pi x}{l}}$$

$$= \frac{1}{2}(a_n - ib_n)e^{i\frac{n\pi x}{l}} + \frac{1}{2}(a_n + ib_n)e^{-i\frac{n\pi x}{l}}$$

$$= \frac{1}{2}(a_n - ib_n)\left(\cos\frac{n\pi x}{l} + i\sin\frac{n\pi x}{l}\right)$$

$$+ \frac{1}{2}(a_n + ib_n)\left(\cos\frac{n\pi x}{l} - i\sin\frac{n\pi x}{l}\right)$$

$$= a_n \cos\frac{n\pi x}{l} + b_n \sin\frac{n\pi x}{l}.$$

Therefore, (8.16) becomes

$$\sum_{n=-\infty}^{\infty} c_n e^{i\frac{n\pi x}{l}} = c_0 + \sum_{n=1}^{\infty}\left(c_n e^{i\frac{n\pi x}{l}} + c_{-n} e^{-i\frac{n\pi x}{l}}\right)$$

$$= \frac{a_0}{2} + \sum_{n=1}^{\infty}\left(a_n \cos\frac{n\pi x}{l} + b_n \sin\frac{n\pi x}{l}\right).$$

This means that the Fourier series in complex form (8.16) is the same as the real Fourier series defined in (8.14).

8.6 Exercises

1. Determine whether each of the following series is convergent or divergent and, if it is convergent, find its sum:
    (a) $\sum_{n=1}^{\infty}(\sqrt{n+1} - \sqrt{n})$; (b) $\sum_{n=1}^{\infty}\sin(\frac{n\pi}{3})$; (c) $\sum_{n=1}^{\infty}(-1)^n$;
    (d) $\sum_{n=1}^{\infty}(\frac{1}{2^n} + \frac{2}{3^n})$; (e) $\sum_{n=1}^{\infty}\frac{n^2}{n^2+1}$; (f) $\sum_{n=1}^{\infty}\frac{1}{(2n-1)(2n+1)}$;
    (g) $\sum_{n=1}^{\infty}\sqrt{\frac{n+1}{n}}$; (h) $\sum_{k=0}^{\infty}\frac{1}{(\ln 3)^k}$; (i) $\sum_{i=2}^{\infty}(\frac{2}{\pi})^{i-1}$;
    (j) $\sum_{j=3}^{\infty} e^j \pi^{-j}$.

2. For each of the following series, determine the values of x for which the series converges. When the series converges, it defines a function of x. What is the function?
    (a) $\sum_{n=1}^{\infty}\frac{x^n}{2^n}$; (b) $\sum_{k=1}^{\infty}\frac{(2x+1)^k}{5^k}$; (c) $\sum_{i=0}^{\infty} 3^{i+2}x^i$; (d) $\sum_{n=1}^{\infty}(\ln x)^n$.

3. Test the following series for convergence (there may be more than one correct method):
    (a) $\sum_{n=2}^{\infty}\frac{1}{n(\ln n)^2}$; (b) $\sum_{n=1}^{\infty}\frac{\ln n}{n^2}$;
    (c) $\sum_{n=3}^{\infty}\frac{1}{n(\ln n)(\ln(\ln n))}$; (d) $\sum_{k=1}^{\infty}\frac{k^2-k}{k^4-2k+3}$;
    (e) $\sum_{n=1}^{\infty}\frac{\arctan n}{\sqrt{n^3+n+1}}$; (f) $\sum_{k=1}^{\infty}\frac{k\ln k}{\sqrt{(k+2)^5}}$;
    (g) $\sum_{n=3}^{\infty}\frac{3^n}{2^n+5^n}$; (h) $\sum_{n=1}^{\infty}\frac{4n^3-3}{n(n^2+1)(\sqrt{n}+5)}$;
    (i) $\sum_{i=1}^{\infty}\sqrt{\frac{i^2+i-1}{4i^5+3i+5}}$; (j) $\sum_{n=1}^{\infty}\frac{e^n}{1+e^{2n}}$;
    (k) $\sum_{n=1}^{\infty}\frac{1+n}{1+n^2}$; (l) $\sum_{n=1}^{\infty}\frac{\ln n}{n}$;
    (m) $\sum_{n=2}^{\infty}\frac{1}{(\ln n)^{\ln n}}$; (n) $\sum_{n=1}^{\infty}\frac{1}{1+a^n}$, $(a > 0)$;
    (o) $\sum_{n=1}^{\infty} n\sin\frac{1}{n}$; (p) $\sum_{i=1}^{\infty}\sin(\frac{1}{i^2})$;
    (q) $\sum_{n=1}^{\infty}(\sqrt[n]{2} - 1)$; (r) $\sum_{n=1}^{\infty}\frac{4^n}{n!}$;
    (s) $\sum_{n=1}^{\infty}\frac{n3^n}{n+4^n}$; (t) $\sum_{j=1}^{\infty}\frac{j^5}{5^j}$;
    (u) $\sum_{n=1}^{\infty}\frac{2^n n!}{n^n}$; (v) $\sum_{n=1}^{\infty}\frac{n!2^n}{(2n)!}$;
    (w) $\sum_{n=1}^{\infty}(1 - \frac{1}{n})^{n^2}$; (x) $\sum_{n=1}^{\infty}(\frac{n}{3n-1})^{2n}$;

(y) $\sum_{n=3}^{\infty} \frac{n}{(\ln n)^n}$; (z) $\sum_{n=1}^{\infty} \frac{\ln(n+2)}{(a+\frac{1}{n})^n}$, $(a > 0)$;

(aa) $\sum_{n=1}^{\infty} \frac{1}{n^2 - \ln n}$.

4. Determine whether each of the following series converges absolutely, converges conditionally, or diverges:

(a) $\sum_{n=1}^{\infty} (-1)^{n+1} \frac{2+n}{n^2}$; (b) $\sum_{n=1}^{\infty} (-1)^{n+1} (0.99)^n$; (c) $\sum_{n=2}^{\infty} \frac{(-1)^{n+1}}{n \ln n}$;

(d) $\sum_{n=1}^{\infty} (-1)^n \frac{\sin \frac{n\pi}{2}}{n}$; (e) $\sum_{n=1}^{\infty} (-1)^{n+1} \frac{n}{4^{n-1}}$; (f) $\sum_{n=1}^{\infty} \frac{(-1)^n}{\ln(n+1)}$;

(g) $\sum_{n=1}^{\infty} \frac{\cos n\pi}{n^2}$; (h) $\sum_{n=1}^{\infty} \frac{(-1)^{n+1}}{n^2 - \ln n}$; (i) $\sum_{k=1}^{\infty} \frac{(-1)^{k-1}}{\sqrt{k}}$;

(j) $\sum_{k=1}^{\infty} \frac{(-1)^{k-1} \ln k}{k}$.

5. Give an example of an alternating series $\sum (-1)^{n-1} a_n$ satisfying $a_n > 0$ and $\lim_{n \to \infty} a_n = 0$, while $\sum (-1)^{n-1} a_n$ diverges.

6. Determine the convergence of each of the following series at $x = 1$ and $x = 3$:

(a) $\sum_{n=2}^{\infty} \frac{(x-2)^n}{n \ln n}$; (b) $\sum_{n=1}^{\infty} \frac{(x-2)^n \sin n}{2^n}$; (c) $\sum_{k=1}^{\infty} \frac{(x-2)^k}{k3^k + 2^k}$.

7. Show that $\lim_{n \to \infty} \frac{b^{3n}}{n! a^n} = 0$, where a, b are two nonzero constants. [Hint: consider the convergence of the series $\sum \frac{b^{3n}}{n! a^n}$.]

8. Assume both $\sum u_n^2$ and $\sum v_n^2$ are convergent. Prove that $\sum u_n v_n$, $\sum (u_n + v_n)^2$, $\sum (u_n - v_n)^2$, and $\sum \frac{u_n}{n}$ are all convergent.

9. If the sequence $\{x_n\}$ with positive terms is decreasing and $\sum (-1)^n x_n$ diverges, does the series $\sum \frac{1}{(1+x_n)^n}$ converge? Explain.

10. Let $f(x) = \begin{cases} \frac{\sin x}{x}, & \text{if } x \neq 0 \\ 1, & \text{if } x = 0. \end{cases}$

(a) Is $f(x)$ differentiable at $x = 0$? Justify your answer.

(b) Is there a number $c \in (\frac{\pi}{2}, \pi)$ such that $f'(c) = -\frac{4}{\pi^2}$? Explain.

(c) Is there a number $d \in (0, \frac{3\pi}{2})$ such that $f(d) = 0$? Explain.

(d) Prove that the function $y = \frac{\sin x}{x}$ decreases on $(0, 1)$.

(e) Assume $x_1 = 1$ and $x_{n+1} = \sin x_n$.

(i) Does $\lim_{n \to \infty} x_n$ exist? Explain (you may use the fact that $x > \sin x$ for all $x > 0$).

(ii) By considering the partial sum, show that the series $\sum_{n=1}^{\infty} (x_{n+1} - x_n)$ converges. Does it converge absolutely or conditionally?

(iii) Does the series $\sum_{n=1}^{\infty} (-1)^{n-1} \sin x_n$ converge? Explain.

(iv) Does the series $\sum_{n=1}^{\infty} (-1)^{n-1} \frac{x_{n+1}}{x_n}$ converge? Explain.

(v) Show that the series $\sum_{n=1}^{\infty} |\frac{x_n x_{n+2}}{x_{n+1}^2} - 1|$ converges.

11. By considering

$$\frac{1}{3} + \frac{1}{4} \geqslant \frac{1}{4} + \frac{1}{4} = \frac{1}{2}, \quad \frac{1}{5} + \frac{1}{6} + \frac{1}{7} + \frac{1}{8} \geqslant 4 \times \frac{1}{8} = \frac{1}{2}, \quad \ldots,$$

show that the partial sum $s_{2^n} \geqslant 1 + n/2$ and deduce that the harmonic series $\sum_{n=1}^{\infty} 1/n$ diverges (the original proof was due to the French philosopher **Nicolas Oresme** (1323–1382)).

12. (**The integral test remainder estimates**) Assume f is a positive, decreasing, and continuous function and $a_n = f(n)$. Show that the error in the nth partial sum s_n

of $\sum a_n$ is bounded by the improper integral

$$\left| s_n - \sum_{n=1}^{\infty} a_n \right| \le \int_n^{\infty} f(x)\, dx.$$

13. (**Alternating series remainder estimate**) If the alternating series $\sum_{k=1}^{\infty} (-1)^{k-1} a_k$
(a) satisfies $a_k \ge 0$ and $a_k \to 0$ as $k \to \infty$ and (b) $\{a_k\}$ decreases for all $n \ge N$, then
the error in the nth partial sum of $\sum (-1)^{k-1} a_k$ is bounded by a_{n+1}. We have

$$\left| s_n - \sum_{k=1}^{\infty} (-1)^{k-1} a_k \right| \le a_{n+1}.$$

Using the result above, how many terms of the series $\sum_{k=1}^{\infty} (-1)^{k-1}/k$ should we use
to approximate the true sum with error less than $1/100\,000$?

14. (**Zeno's paradox**) Zeno's paradox is about a race between Achilles and a tortoise.
The tortoise begins with a head start of 100 meters and Achilles seeks to overtake
it. After a certain elapsed time from the start, Achilles reaches point A, where the
tortoise started, but the tortoise has moved ahead to point B. After a certain further
interval of time, Achilles reaches point B, but the tortoise has moved ahead to a
point C, etc. Zeno then concluded that Achilles can never pass the tortoise. Why
is this argument wrong?

15. (**Euler's constant** $y \approx 0.577\,215\,664\dots$) The Euler constant y is defined as

$$y = \lim_{n \to +\infty} \left(1 + \frac{1}{2} + \frac{1}{3} + \cdots + \frac{1}{n} - \ln n \right).$$

(a) Show that $\ln(n+1) - \ln n \ge \frac{1}{n+1}$.
(b) By considering $\ln(n+1) \le 1 + \frac{1}{2} + \frac{1}{3} + \cdots + \frac{1}{n} \le 1 + \ln n$, or any other method,
show that the sequence $\{y_n\}$

$$y_n = 1 + \frac{1}{2} + \frac{1}{3} + \cdots + \frac{1}{n} - \ln n$$

is bounded and monotonic and therefore $\lim_{n \to \infty} y_n$ exists.
(c) Let $s_n = \sum_{k=1}^{n} (-1)^{k-1}/k$ and show that

$$s_{2n} = \left(1 + \frac{1}{2} + \frac{1}{3} + \cdots + \frac{1}{2n} \right) - \left(1 + \frac{1}{2} + \frac{1}{3} + \cdots + \frac{1}{n} \right)$$

and $s_n \to \ln 2$ as $n \to \infty$.

16. (**Riemann's rearrangement theorem**) If $\sum a_n$ is a conditionally convergent series
and m is any real number, then there is a rearrangement of $\sum a_n$ which converges
to m.

(a) Rearrange the series $1 - \frac{1}{2} + \frac{1}{3} - \frac{1}{4} + \frac{1}{5} - \cdots + (-1)^n \frac{1}{n} + \cdots$ so that it converges to 2.
(b) Prove this theorem.

17. Find the radius of convergence and the interval of convergence of the following series:

(a) $\sum_{n=1}^{\infty} x^n$;

(b) $\sum_{n=1}^{\infty} \frac{nx^n}{n+1}$;

(c) $\sum_{n=0}^{\infty} \frac{(x-5)^n}{10^n}$;

(d) $\sum_{n=1}^{\infty} \frac{n}{3^n}(x-3)^{2n}$;

(e) $\sum_{n=1}^{\infty} \frac{n(x-1)^n}{4^n(n^2+1)}$;

(f) $\sum_{n=1}^{\infty} \frac{2n-1}{2^n}x^{2n-2}$;

(g) $\sum_{n=1}^{\infty} \frac{1}{2^n n}(x-1)^n$;

(h) $\sum_{n=1}^{\infty} \frac{n(2x+3)^n}{n+1}$;

(i) $\sum_{n=1}^{\infty} \frac{3^n(x-2)^{2n+1}}{(n+2)^2}$;

(j) $\sum_{n=1}^{\infty} \frac{n!x^n}{n^n 4^n}$;

(k) $\sum_{n=0}^{\infty} \frac{2^n x^n}{\ln(n+2)}$;

(l) $\sum_{n=1}^{\infty} \frac{1}{a^n+b^n}x^n$, $(a > b > 0)$;

(m) $\sum_{n=1}^{\infty} (\frac{x+1}{3n})^n$.

18. (**Weierstrass's nowhere differentiable continuous function**) In Weierstrass's original paper, the function was defined to be

$$f(x) = \sum_{n=0}^{\infty} a^n \cos(b^n \pi x)$$

$$= \cos(\pi x) + a\cos(b\pi x) + a^2\cos(b^2\pi x) + a^3\cos(b^3\pi x) + \cdots,$$

where $0 < a < 1$, b is a positive odd integer, and $ab > 1 + 3\pi/2$. The proof that this function is continuous but nowhere differentiable was given by Weierstrass on 18 July 1872 (extended reading: http://en.wikipedia.org/wiki/Weierstrass_function). Use a graphing utility to graph the sum with its first three terms for the case $a = 2/3$ and $b = 9$. Describe the graph that you see.

19. The function $s(x)$ is defined by a series as

$$s(x) = \sum_{k=1}^{\infty} kx^{k-1} = 1 + 2x + 3x^2 + 4x^3 + \cdots + nx^{n-1} + .$$

(a) Find the domain of $s(x)$.

(b) Rewrite $s(x)$ in terms of the basic functions listed in Chapter 1.

(c) Prove the **Nicole Oresme theorem**

$$1 + \frac{1}{2} \times 2 + \frac{1}{4} \times 3 + \cdots + \frac{1}{2^{n-1}} \times n + \cdots = 4.$$

20. If a function $g(x)$ is defined by

$$g(x) = \sum_{k=2}^{\infty} \frac{(-1)^k x^k}{k(k-1)} = \frac{x^2}{1 \cdot 2} - \frac{x^3}{2 \cdot 3} + \frac{x^4}{3 \cdot 4} - \cdots + \frac{(-1)^n x^n}{n(n-1)} + \cdots,$$

write $g(x)$ in terms of the basic functions listed in Chapter 1. [Hint: use term-by-term differentiation twice.] Find the value of

$$\frac{1}{1 \cdot 2 \cdot 2^2} - \frac{1}{2 \cdot 3 \cdot 2^3} + \frac{1}{3 \cdot 4 \cdot 2^4} - \cdots + \frac{(-1)^n}{n(n-1)2^n} + \cdots.$$

21. Find the value of the following series:

(a) $\sum_{n=1}^{\infty} \frac{2^{2n+1}}{5^n}$;   (b) $\sum_{n=1}^{\infty} \frac{1}{n3^n}$;   (c) $\sum_{n=1}^{\infty} \frac{n^2}{5^n}$.

22. (**Geometric distribution**) Assume a certain product is "bad" with a probability p and "good" with a probability $q = 1-p$. Each product is independent. An inspector is checking on the product line. The number X of products up to and including the first bad product that is identified is a random variable that has a geometric distribution. Show that:
    (a) $P(X = n) = q^{n-1}p$ and $P(X \leqslant n) = 1 - q^n$;
    (b) $E(X) = \frac{1}{p}$ (the expected value of a discrete random variable X is defined as $E(X) = \sum_{n=1}^{\infty} nP(X = n)$).

23. If $f(x) = \sum_{n=1}^{\infty} \frac{x^n n!}{n^n}$, then:
    (a) find the interval of convergence;
    (b) find the radius of convergence;
    (c) use the first three terms of this series to approximate $f(-\frac{1}{2})$ and estimate the error.

24. (**Long division**) Use long division to find the first three terms of the MacLaurin series for the function $y = \frac{\ln(1+x)}{1+2x}$.

25. Find the following Taylor series for $f(x)$ centered at the given value of a, assuming that f has a power series expansion (you do not need to show that $R_n(x) \to 0$):
    (a) $f(x) = \frac{1}{x^2+3x+2}$, $a = -4$;   (b) $f(x) = \cos x$, $a = -\frac{\pi}{3}$;
    (c) $\sum_{n=1}^{\infty} \frac{(-1)^{n-1}}{2^{2n-1}} \frac{1}{(2n-1)!} x^{2n-1}$, $a = \frac{\pi}{2}$.

26. Use a Maclaurin series derived in this chapter to obtain the Maclaurin series for the following functions:
    (a) $f(x) = (1-x)\ln(1+x)$;   (b) $f(x) = \arcsin x$;
    (c) $g(x) = \frac{1+x}{(1-x)^2}$;   (d) $k(x) = xe^{-2x}$.

27. Find the sum, in terms of e, of
$$\sum_{k=1}^{\infty} \frac{k^2}{k!} = \frac{1^2}{1!} + \frac{2^2}{2!} + \frac{3^2}{3!} + \cdots + \frac{n^n}{n!} + \cdots.$$

28. How many terms of the Maclaurin series for $\ln(1+x)$ are required in order to estimate the value of $\ln 3$ to within 0.0001? Show your work.

29. Use the first five terms of the Maclaurin series for the given function $f(x)$ to estimate the value of the following integrals:
    (a) $\int_0^1 \frac{1}{1+x^4} dx$;   (b) $\int_0^{0.5} e^{x^2} dx$.

30. Assume $f(x) = \begin{cases} e^{-\frac{1}{x^2}}, & x \neq 0 \\ 0, & x = 0. \end{cases}$ Show that $f'(0) = f''(0) = f'''(0)$. Does the Taylor series for $f(x)$ converge to $f(x)$? Explain.

31. (**Bailey–Borwein–Plouffe formula**) If m is an integer that is less than 8, show that
$$\int_0^{1/\sqrt{2}} \frac{x^{m-1}}{1-x^8} dx = 2^{-\frac{m}{2}} \sum_{n=1}^{\infty} \frac{1}{16^n(8n+m)}.$$

Then prove the following Bailey–Borwein–Plouffe formula for π:

$$\pi = \sum_{n=0}^{\infty} \frac{1}{16^n}\left(\frac{4}{8n+1} - \frac{2}{8n+4} - \frac{1}{8n+5} - \frac{1}{8n+6}\right).$$

Compared with $\pi = 4\sum_{n=1}^{\infty}(-1)^{n-1}/(2n-1)$, which one converges "faster"?

32. (**Harmonic numbers**) The harmonic numbers are $H_n = 1 + 1/2 + 1/3 + \cdots + 1/n$ for $n = 1,2,3,\ldots$. Prove that

$$\sum_{n=1}^{\infty} H_n x^n = \frac{\ln(1-x)}{x-1} \quad \text{for } |x| < 1.$$

33. (**Irrational numbers e**) Write the Maclaurin series for e^x. Then:
   (a) state the series that represents e;
   (b) find $r_n = n!e - [n!e]$ where $[x]$ is the greatest integer function;
   (c) deduce that $0 < r_n < \frac{1}{n}$ and then show that e is irrational.

34. Find the Fourier series expansion for the functions over each of the following intervals:

   (a) $f(x) = \begin{cases} x, & -\pi < x < 0 \\ 2x, & 0 \leqslant x < \pi; \end{cases}$ (b) $f(x) = \begin{cases} e^x, & -\pi < x < 0 \\ 1, & 0 \leqslant x \leqslant \pi; \end{cases}$

   (c) $f(x) = \frac{x}{\pi}, -\pi \leqslant x < \pi;$ (d) $f(x) = \begin{cases} 2x+1, & -2 \leqslant x < 0 \\ 1, & 0 \leqslant x < 2. \end{cases}$

35. (**Integration/differentiation term by term**) For a piecewise continuous function, it can be shown that its Fourier series can be integrated or differentiated term by term. Assume that a piecewise continuous function defined on $[-\pi, \pi]$ has a Fourier series expansion with coefficients a_0, a_n and b_n.
   (a) Show that

$$\int_{-\pi}^{x} f(s)\,ds = \frac{1}{2}a_0(x+\pi) + \sum_{n=1}^{\infty}\frac{1}{n}(a_n(\sin nx) - b_n(\cos(nx) - \cos(n\pi))).$$

   (b) Find the Fourier series expansion for the periodic function $f(x)$ with period 2π and

$$f(x) = \frac{x(2\pi - x)}{4}, \quad x \in [0, 2\pi].$$

   (c) (**Riemann hypothesis**) The Riemann zeta function is

$$\zeta(s) = \sum_{n=1}^{\infty}\frac{1}{n^s} = 1 + \frac{1}{2^s} + \frac{1}{3^s} + \cdots + \frac{1}{n^s} + \cdots.$$

   It is known that $\zeta(1) = \infty$. For any even integer s, the value of $\zeta(s)$ is also known. For example, $\zeta(2) = \frac{\pi^2}{6}$, $\zeta(4) = \frac{\pi^4}{90}$, and $\zeta(6) = \frac{\pi^6}{945}$. However, the value of zeta for odd integers is not known. The roots of $\zeta(s)$ are all complex numbers of the form $a + bi$. The Riemann hypothesis states that all the complex roots of $\zeta(s)$ have a real part of $a = \frac{1}{2}$, that is, all the complex roots of $\zeta(s)$ lie on the line $x = \frac{1}{2}$ in the complex plane. It stands today as one of the most important unsolved problems of mathematics.

(i) Use the result found in (b) to show that $\sum_{n=1}^{\infty} \frac{1}{n^2} = \frac{\pi^2}{6}$.

(ii) Use term-by-term integration on the series found in (b) and show that
$\sum_{n=1}^{\infty} \frac{1}{n^4} = \frac{\pi^4}{90}$.

36. Find the Fourier sine and cosine expansion for the following functions:

(a) $f(x) = \begin{cases} x, & 0 \leqslant x < \frac{\pi}{2} \\ 1, & \frac{\pi}{2} \leqslant x < \pi; \end{cases}$ (b) $f(x) = \sin x$, $(0 < x \leqslant \pi)$;

(c) $f(x) = x^2$, $(0 \leqslant x \leqslant 2)$.

Index

ϵ-δ definition of limit 46

absolute convergence test 373
absolute maximum 100
absolute minimum 100
absolute value function 29
absolutely convergent 373
additive property 259
alternating series test 371
antiderivative 196, 285
arc length 239, 339
area problem 1
asymptotic 86
asymptotic function 86
average rate of change 41
average velocity 4

binomial expansion 387
Bolzano–Weierstrass theorem 78
Bolzano's theorem 102
bounded 73
bounded above 9, 73
bounded above on the interval 32
bounded below 10, 73
bounded below on the interval 32
bounded monotonic sequence theorem 356
bounded on the interval 32
boundedness theorem 101

candidate theorem 184
Cauchy's radical test 370
Cauchy's theorem 78, 84
center of the neighborhood 13
chain rule 147, 148
closed interval 12
closed interval test 185
common ratio 358
comparison theorem 324
complement 7
complex numbers 8
composite function 22
composition of functions 22
concave down 220
concave up 220
concavity 220, 221
conditionally convergent 373
constant function 18

constant multiple rule 51, 95
continuity 91
continuous 91
continuous function 91
convergence of Fourier series 397
convergent 319, 322, 356, 357
convergent point 375
convergent set 375
converges 319, 322
conversion formulas 164
cosecant function 21
cosine function 18
cotangent function 20
critical number 184
cross sections 338
curvature 238, 239

d'Alembert's ratio test 368
decimal search 234
decreasing 197
definite integral 252
definite integral of a function 250
deleted δ-neighborhood 13
dependent variable 14
derivative 121, 123
difference rule 50, 95
differentiability 131
differentiable 131
differential 167
differential approximation 173
differential calculus 4
differential dx 170
differential dy 170
differential equation 290
differentiation 123, 381
Dirac δ function 49
direct substitution rule 94
Dirichlet function 38
discontinuity 91
discontinuous functions 91
divergent 319, 322, 356, 357
diverges 319, 322
domain 14

element of area 333
elementary function 31, 98

elements 6
empty set 7
even function 34
exponential function 19
extended mean value theorem 201
extreme value theorem 99, 101

Fermat's theorem 182
first derivative test 196, 199
folium of Descartes 155
Fourier coefficients 394
Fourier series 393
function 14
fundamental elementary functions 31
fundamental theorem of calculus 4
fundamental theorem of calculus, Part I 268
fundamental theorem of calculus, Part II 272

generalized function 49
geometric series 358
global extrema 99
global maximum 100
global minimum 100
greatest integer function 31
greatest lower bound 10

half-open interval 12
harmonic series 361
Heaviside cover-up method 306
horizontal asymptote 64
horizontal line test 24

implicit differentiation 155
improper integral of the first kind 318
improper integral of the second kind 322
improper integrals 318
increasing 197
indefinite integral 286
independent variable 14
indeterminate forms 11, 203
infimum 10
infinite discontinuity 93
infinite interval 12
infinite sequence 355
infinite series 357
infinitesimal 86
infinitesimal function 86
inflection point 222
instantaneous rate of change 42

instantaneous velocity 4
integers 8
integrable 253
integral calculus 4
integral test 365
integration by parts 300
integration by substitution 293
intermediate value theorem 99, 103
intersect 7
interval of convergence 378
intervals 12
inverse cosine function 27
inverse cotangent function 27
inverse functions 23, 24
inverse secant function 27
inverse sine function 27
inverse tangent function 27
irrational numbers 8
irreducible factor 306

jump discontinuity 93

Lagrange remainder 215, 384
least upper bound 10
least upper bound property 9, 10
left δ-neighborhood 13
left derivative 128
left Riemann sum 255
left-hand derivative 128
left-hand limit 55
L'Hôpital's rule 204
limit 42
limit comparison test 366
limit laws 50
limit of a function 42
limit rules 50
limits at infinity 63
limits of sequences 69
linear approximation 169
linear function 21
linearity property 259
linearization 167, 169
local maximum 181
local minimum 181
logarithmic differentiation 159
lower bound 10

Maclaurin expansion 217
Maclaurin series 384

Maclaurin's formula 217
mean value theorem 192, 260
members 6
midpoint Riemann sum 255
monotone 73
monotone decreasing 33, 73
monotone increasing 33, 73
monotonic (or monotone) 197

natural numbers 8
neighborhood 13
Newton's method 236
nth order derivative 152
nth term divergence test 361
number line 9

odd function 34
one-sided derivatives 128
one-sided limit 55
one-to-one function 23
open interval 12
oscillating discontinuity 93
osculating circle 242

p-series 363
parameter 160
parameterization 160
parametric equations 160
partial fractions 304, 305
peak term 78
Peano remainder 215
period 35
periodic extension 393
periodic function 35
point of inflection 222
point-slope form 21
points at infinity 11
polar curves 163
polynomial function 22
power function 18
power rule 51
power series 376
product 7
product rule 50, 95
proper subset 7

quadratic approximation 210
quotient rule 50, 95

radius of convergence 378
radius of curvature 242

radius of the neighborhood 13
range 14
rational function 22
rational numbers 8
rationalized 312
real numbers 8
real plane 7
reflection laws 37
relative maximum 181
relative minimum 181
remainder 215
removable discontinuity 93
Riemann sum 253
right δ-neighborhood 13
right derivative 128
right Riemann sum 255
right-hand derivative 128
right-hand limit 55
Rolle's theorem 189

sandwich theorem 79
secant function 20
secant line 122
second derivative 152
second derivative test 225
second order derivative 152
separable differential equation 292
separation of variables 292
sequence 68
sequence form of Cauchy's theorem 85
series with nonnegative terms 363
set 6
Simpson's rule 277
sine function 18
slant asymptote 65
slope of the tangent line 122
slope-intercept form 21
smooth curve 339
squeeze theorem 78, 79
stationary point 183
strictly decreasing 34, 197
strictly increasing 34, 197
subsequence 77
sum rule 50, 95
supremum 10

tabular method 304
tangent function 20
tangent line approximation 167, 169
tangent problem 3

Taylor polynomial 215
Taylor polynomial of degree n 384
Taylor series 384
Taylor's inequality 386
Taylor's theorem 214
term-by-term differentiation 382
term-by-term integration 381, 382
the comparison test 365
the ratio test 368
the root test 370
transformations 36
trapezium rule 276
trapezoidal rule 276

unbounded above 11
unbounded above on the interval 32

unbounded below 11
unbounded below on the interval 32
uniform continuity 107
uniformly continuous 107
union 7
upper bound 9

velocity graph 1
Venn diagram 7
vertical and horizontal shifts laws 36
vertical and horizontal stretch/shrink law 36
vertical asymptote 61
vertical line test 17
volumes of revolutions 337

Weierstrass function 376